INDOOR AIR

Quality and Control

Anthony L. Hines
Tushar K. Ghosh
Sudarshan K. Loyalka
Richard C. Warder, Jr.
University of Missouri-Columbia

PTR Prentice Hall
Englewood Cliffs, New Jersey 07632

Library of Congress Cataloging-in-Publication Data
Indoor air : quality and control / Anthony L. Hines . . . [et al.].
p. cm.
Includes bibliographical references and index.
ISBN 0-13-463977-4
1. Indoor air pollution. 2. Air quality management. I. Hines, Anthony L.
TD883.1.I476 1993
628.5'3—dc20 92-29799
CIP

Editorial/production supervision
and interior design: ***Ann Sullivan and Lisa Iarkowski***
Buyer: ***Mary Elizabeth McCartney***
Acquisitions Editor: ***Michael Hays***
Editorial assistant: ***Kim Intindola***
Cover design: ***Ben Santora***

A Simon & Schuster Company
Englewood Cliffs, New Jersey 07632

The publisher offers discounts on this book when ordered in bulk quantities.
For more information, contact:
Corporate Sales Department
PTR Prentice Hall
113 Sylvan Avenue
Englewood Cliffs, NJ 07632
Phone: 201-592-2863
FAX: 201-592-2249

Printed in the United States of America
10 9 8 7 6 5 4 3 2 1

ISBN 0-13-463977-4

Prentice-Hall International (UK) Limited, *London*
Prentice-Hall of Australia Pty. Limited, *Sydney*
Prentice-Hall Canada Inc., *Toronto*
Prentice-Hall Hispanoamericana, S.A., *Mexico*
Prentice-Hall of India Private Limited, *New Delhi*
Prentice-Hall of Japan, Inc., *Tokyo*
Simon & Schuster Asia Pte. Ltd., *Singapore*
Editora Prentice-Hall do Brasil, Ltd., *Rio de Janeiro*

Contents

Preface

A large fraction of our time is spent in either our homes, the workplace, or other buildings where the air may be more polluted than the air found outdoors. Indoor air pollutants are suspected of being responsible for a host of ailments, thousands of deaths, and productivity losses that have been estimated to approach $100 billion annually. While indoor air pollution has existed since man first began living in caves and building fires, it has increased in recent years and has become a greater problem because of the added emphasis on energy conservation in the design and construction of new buildings and the retrofitting of existing structures to make them airtight. Since the cost of energy usage will continue to be a major factor in the design of buildings, engineers must find ways to reduce indoor air pollution without compromising energy efficiency.

We have written this book for those who wish to gain a better understanding of the indoor air pollution problem as well as for the practicing engineer who needs detailed information about indoor pollutants and mitigation methods. Our intent is to cover the important topics dealing with indoor pollution in sufficient depth so that the reader acquires a basic understanding of the nature of the problems and becomes better equipped to solve them. Included in the book is an assessment of risks from indoor air pollution, along with a discussion of the sources of the various pollutants, their associated health hazards and standards, and the sampling of indoor air pollutants. Topics such as volatile organics, inorganics, metals, bioaerosols, radon, and respirable particulates (including tobacco smoke and asbestos) are addressed. Control strategies specific to each class of pollutants, including existing and developing mitigation methods, are discussed.

Adsorption/absorption processes are particularly attractive since they can be used in conjunction with other air treatment techniques, such as dehumidification, for reducing the loads on air-conditioning systems. The final two chapters of the book are devoted exclusively to the design of adsorption/absorption processes and systems for indoor air pollution control. We have included numerous examples that should aid practitioners in the design of control systems applicable to specific pollutants.

We have sought to both consider the current literature and to include results from our own investigations. Our interest in the subject areas has spanned the last two decades, and we have conducted research on most of the topics discussed in the book. These interests have recently become more strongly focused on pollutant control strategies as a result of support from the Gas Research Institute, Chicago, IL, and the American Society of Heating, Refrigerating and Air-Conditioning Engineers, Inc., Atlanta, GA. The book will have served its intended purpose if it helps even a single individual to understand and to solve some of the problems associated with indoor air pollution.

We are indebted to our many past and present graduate students who have contributed to our efforts in this area. We are particularly indebted to T-W Chung, who worked on several of the topics discussed in this book, to Perapong Tekasakul, who prepared many of the illustrations, and to Davor Novosel of the Gas Research Institute for his enthusiastic support. The staff of the College of Engineering Library, in particular Judy Pallardy and Carol Romano, provided valuable assistance with the reference materials. The expert typing of Mrs. Sally Schwartz is gratefully acknowledged. Finally, for each of us, the patience of our families and the encouragement of Jo Ann, Mahua, Nirja, and Dianne has been indispensable.

Anthony L. Hines*
Tushar K. Ghosh†
Sudarshan K. Loyalka†
Richard C. Warder, Jr.†

*Honda of America Mfg., Inc.
†University of Missouri-Columbia

1

The Nature of the Problem

The presence of smog in many cities and billowing industrial smokestacks have led the general public to believe that indoor air is cleaner than outdoor air. Consequently, previous studies have focused on the development of techniques and strategies for removing air contaminants from sources such as industrial effluents and automobile exhausts. The increase in the cost of energy in the last two decades, however, has led to the improved construction and retrofitting of homes and commercial buildings for enhanced energy conservation. A reduction in the infiltration of fresh air is cost effective and is widely practiced among the various energy-saving schemes that are presently being used. As a result, a large portion of the U.S. population currently is living in tightly sealed structures in which 80% to 90% of the air is recirculated to reduce energy consumption, without full realization of the air quality problem that arises from the pollutants generated and retained indoors. Studies have shown, however, that the concentration of toxic organic chemicals indoors can be several times greater than the concentration in outdoor air. Table 1.1 shows average concentrations of selected pollutants in indoor and outdoor air.

Pollutants are introduced indoors in several ways: (1) through normal biological processes—people and pets generate carbon dioxide, moisture, odors, and microbes; (2) by combustion appliances such as wood stoves, gas stoves, furnaces, fire places, and gas heaters; (3) from the use of consumer products such as spray cans, air fresheners, spray cleaners, construction materials, furnishings, and insulation; (4) from cigarette smoke; (5) from the soil under and around buildings; and (6) from appliances such as humidifiers, air conditioners, and nebulizers.

A number of studies in the past have noted a direct relationship between the level of

TABLE 1.1 Indoor and Outdoor Concentrations of Various Pollutants

Pollutants	Indoor Concentration	Outdoor Concentration	Pollutants	Indoor Concentration	Outdoor Concentration
Volatile Organic Compounds (ppb)			***Inorganic Gaseous Compounds***		
2-Butoxyethanol	0.21	BDL	Carbon dioxide	1000 ppm	400 ppm
1,2-Dichlorobenzene	3.99	0.89	Carbon monoxide	1.73 ppm	3.3 ppm
1,4-Dioxane	1.03	0.11	Oxides of nitrogen	12.9 ppb	11.3 ppb
1,4-Dichlorobenzene	3.99	1.00	Nitric oxide	5.3 ppb	2.8 ppb
1,1,2,2-Tetrachloroethane	0.02	0.10	Sulfur dioxide	8 ppm	—
α-Pinene	0.55	0.48	Ozone	10–100 ppm	120 ppb
Acetone	7.96	6.93			
Benzaldehyde	1.58	0.54	***Heavy Metals*** (μg/m^3)		
Benzene	5.16	2.80			
Carbon tetrachloride	0.40	0.17	Lead	0–10	0.0001–5
Chloroform	0.83	0.63	Mercury	0.005–6.0	0.0032
Cyclohexane	1.38	NA	Total sulfate	5	—
Dichlorobenzene	0.09	0.01			
m, p-Xylene	8.67	3.44	***Particulates*** (g/m^3)[a]		
Dimethylbenzene	2.84	1.49			
Styrene	1.41	1.35	Total respirable particulates	23.9	33.1
Ethylbenzene	2.89	1.61			
Formaldehyde	49.40	8.29	Total suspended particulates	31.7	36.9
Methyl ethyl ketone	9.24	NA			
Octane	0.88	0.40			
Tetrachloroethene	3.06	0.85	***Microorganisms***[a] (CFU/m^3)		
Toluene	7.39	7.78			
Trichlorobenzene	0.07	0.02			
Trichloroethane	48.90	0.91	Bacillus	1053	514
Trichloroethene	1.35	0.50	Micrococcus	51	13
1,2,3-Trimethylbenzene	0.57	0.52	Staphylococcus	106	19
1,2,4-Trimethylbenzene	0.91	1.81	Penicillium	839	1152
1,3,5-Trimethylbenzene	—	0.39	Aspergillus	441	272
Undecane	0.75	0.75	Other species	135	101

BDL, below detection level; NA, not available; [a] summer month measurement; CFU, colony forming units.
Sources: Shah, 1988; Singh et al., 1981; Hawthorne et al., 1984; NAS, 1981.

indoor air pollution and health problems such as headaches, nausea, respiratory infections, allergies, humidifier fever, Legionnaire's disease, and lung cancer. Excellent reviews of the subject have been provided by Samet et al. (1987, 1988). While particulates such as bioaerosols can have an immediate impact on public health, the presence of other indoor contaminants can have longer-term effects. A number of deaths have been attributed to the exposure to these pollutants. Considering that most people spend from 80% to 90% of their time indoors, either at home, in the workplace, or in shops and supermarkets, the health problems that result from exposure to indoor pollution are likely to increase. An estimated 3% of the U. S. population (7.5 million people) suffer from asthma, and on the average 500,000 cases per year result in emergency hospitalization (Platts-Mills and Chapman, 1987). In addition, Legionella species are believed to be responsible for an estimated 5000 to 7000 deaths per year. A good review of the health risks and the attributable costs associated with exposure to microbial pollutants is provided by Burge and Hodgson (1988). Lost productivity and medical expenses due to respiratory infections may cost as much as $100 billion per year (Anonymous, 1984).

The indoor air quality problem is rapidly becoming a major concern of the general public. Phrases such as sick building syndrome (SBS) and tight building syndrome (TBS) have been used to draw the attention of the public and government agencies to these problems. The severity of the problem was highlighted in an interview of an Environmental Protection Agency (EPA) administrator, who characterized the indoor air pollution problem at the EPA's Washington headquarters building as follows: "The building struck me the first few weeks as a kind of Woody Allen joke, but it's long ceased to be funny" (*Air Pollution Report*, December 4, 1989). Also, many lawsuits have been filed against building owners and manufacturers of products, such as carpets, by people seeking compensation for illnesses that have been attributed to exposure to various indoor air pollutants. These point to an increased awareness of indoor air pollution, and unless remedial action is taken to improve indoor air quality, the number of lawsuits, as well as medical costs, are likely to increase; and, more importantly, additional lives will be lost.

Figure 1.1 shows the types of pollutants (contaminants) that are present in indoor air. These include volatile organic compounds (VOCs), radon, bioaerosols, respirable particulates, and other substances such as the carbon and nitrous oxides that originate mainly from combustion products. While the existence and the associated harmful effects of some pollutants such as formaldehyde and carbon monoxide have been recognized for a long period of time, pollutants such as radon have only now begun to receive attention.

Organic chemicals constitute a major portion of all indoor air pollutants, with more than 700 compounds having been identified. Although not all these are present at the same time in one building, as many as 50 to 60 chemicals can be expected to be found indoors at any given time. These chemicals are frequently referred to collectively as volatile organic compounds (VOCs), although many are not strictly classified by this phrase. A list of compounds that have been identified indoors is given in Table 1.2. Organic pollutants originate from paints, aerosol sprays, solvents, herbicides, pesticides, building materials, tobacco smoke, combustion products, and many other sources. The indoor sources may be continuous or intermittent, lasting from a few seconds to several hours. Building materials, such as carpets and paints, emit these pollutants continuously. Although pollutants are emitted for

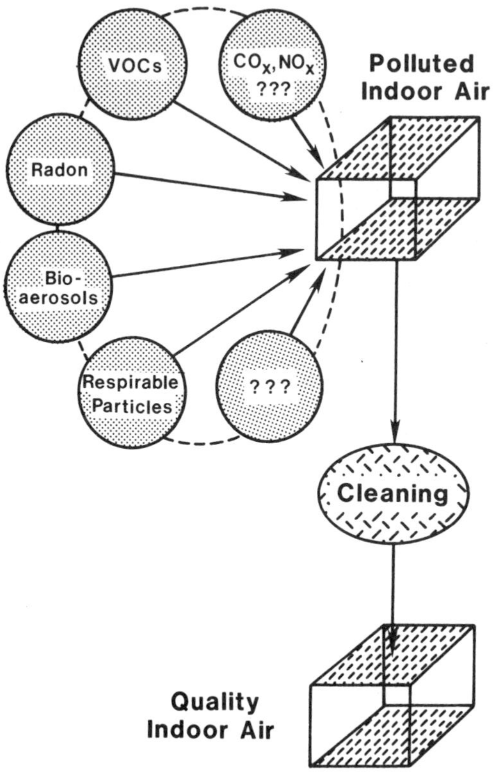

Figure 1.1 *Indoor air pollutants.*

only short periods of time during cooking or cleaning activities, they remain indoors if proper precautions are not taken. Numerous health risks have been attributed to organic pollutants, but considerable debate exists about the conclusions drawn from many of the studies reported in the literature. The health risks posed by these pollutants depend on the concentration of the pollutant, on the exposure time of the individuals, and on their state of health at the time of exposure and later. The concentration level at which these pollutants are present indoors does not pose immediate serious health problems, but continuous exposure, even at low levels, may cause serious damage, particularly to children. The cognizant agencies concerned with the indoor air pollution problem have not agreed on a single guideline regarding exposure and subsequent health risks. In view of the absence of reliable data for many compounds, the lack of a comprehensive exposure policy is understandable. The morbidity and mortality from exposure to most of the chemicals found indoors are not well documented. Even doctors typically relate common symptoms like headaches and the irritation of eyes, nose, and mucous membranes to stress or other diseases. A general observation is that the symptoms may be more frequently related to indoor pollutants.

The term *respirable particulates* is frequently used to describe particles in the size range from about 0.0005 to 10 μm that remain suspended in air for long periods of time. These include mist, smoke, dust, fibers, and other solid particles. Although certain respira-

TABLE 1.2 Volatile Organic Compounds Detected in Indoor Air

***n*-Alkanes**

n-Propane
n-Butane
n-Pentane
n-Hexane
n-Heptane
n-Octane
n-Nonane
n-Decane
n-Undecane
n-Dodecane
n-Tridecane
n-Tetradecane
n-Pentadecane
n-Hexadecane

Branched Alkanes

2-Methylpropane
2-Methylbutane
C_4H_{10}
C_5H_{12}
2,2-Dimethylbutane
2,3-Dimethylbutane
C_6H_{14}
C_7H_{16}
2-Methylpentane
3-Methylpentane
2,3-Dimethylpentane
2,4-Dimethylpentane
2-Methylhexane
3-Methylhexane
2,4-Dimethylhexane
C_8H_{18}
2-Methylheptane
2,2,4-Trimethyloctane
3-Methylheptane
4-Methylheptane
2,3,3-Trimethylpentane
C_9H_{20}
$C_{10}H_{22}$
Methylethylpentane
2,4-Dimethylheptane
2,6-Dimethylheptane
2,2,5,5-Tetramethylhexane
2,5,5-Trimethylheptane
2,3,6-Trimethylheptane
3-Methyloctane
2,5-Dimethyloctane
2-Methylnonane
2,6-Dimethyloctane
2,3,7-Trimethyloctane
4-Methyl-5-propylnonane
5-Methyldecane
4-Methyldecane
5-Ethyl-2-methylheptane
2-Methyldecane
2,2,3,4-Tetramethylpentane
3-Methyldecane
3,7-Dimethyl-undecane
$C_{12}H_{26}$
4-Methylundecane
2-Methylundecane
3-Methylundecane
$C_{13}H_{28}$
$C_{14}H_{30}$
$C_{15}H_{32}$

Alkenes and Cycloalkanes

1-Butene
1-Pentene
Cyclomethylbutadiene
Methylethylcyclopropane
Methylcyclobutane
Ethylcyclobutane
Cyclopentadiene
C_6H_{12}
Cyclopentene
2-Methyl-3-butadiene
Cyclohexane
1-Hexene
3-Methyl-1-pentene
Cyclopentane
Methylcyclopentane
5-Methyl-1-hexene
1-Heptene
Methylcyclohexane
C_7H_{14}
C_8H_{16}
2,4-Dimethyl-1,3-pentadiene
1,4-Dimethylcyclohexane
Methylethylcyclopentane
1-Octene
Ethylcyclohexane
1,1,3-Trimethylcyclohexane
C_9H_{18}
1,3,5,7-Cyclooctatetraene
5,5-Dimethyl-1-hexene
3,5,5-Trimethyl-1-hexene
1-Nonene
1-Ethyl-4-Methylcyclohexane
cis-1-Ethyl-3-methylcyclohexane
n-Propylcyclohexane
n-Butylcyclohexane
iso-Propylcyclohexane
Alkylcyclopentane
iso-Butylcyclohexane
Bicyclodecane
$C_{10}H_{20}$
1-Decene
$C_{11}H_{22}$
1-Undecene

Halogen Derivatives

Dichlorodifluoromethane
Trichlorofluoromethane
Dibromochloromethane
Trichlorofluoroethane
1,2-Dichloroethane
1,1-Dichloroethane
Chloromethane
Methylene chloride
Chloroform
Carbon tetrachloride
Freon 113
1,1,1-Trichloroethane
Trichloroethylene
1,1,2-Trichloroethane
Tetrachloroethylene
1,1,2-Trifluoro-1,2,2-trichloroethane
1,2-Dichloro-1,1,2,2-tetrafluoroethane
1,2-Dichloropropane
Chlorobenzene
1,2-Dichlorobenzene
1,4-Dichlorobenzene
Trichlorobenzene

Alcohols

Methanol
Ethanol
1-Propanol
2-Propanol
2-Methyl-1-propanol

(continued)

TABLE 1.2 Continued

Alcohols (cont.)

2-Methyl-2-propanol
1-Butanol
2-Butanol
2-Methyl-1-butanol
2-Methoxyethanol
1-Methoxy-2-propanol
2-Ethoxyethanol
1-Pentanol
2-Methyl-1-pentanol
2-Ethylcyclobutanol
1-Hexanol
2-Ethyl-1-butanol
2-Buthoxyethanol
2-*bis*(2-ethoxy)-ethanol
2-Ethyl-4-methyl-1-pentanol
4-Methyl-2-propyl-1-pentanol
2-Ethyl-1-hexanol
Octanol
1,8-Cineol
Terpeneol
Cresol
Phenol

Ethers

Vinylethylether
Methoxyvinylethylether

Aldehydes

Formaldehyde
Acetaldehyde
Propanal
Crotonaldehyde
Butanal
Pentanal
Hexanal
Heptanal
Octanal
Benzaldehyde
Phenyl acetaldehyde
α-Methylbenzene acetaldehyde
Nonanal
Ethylbenzaldehyde
2,5-Dimethylbenzaldehyde
n-Decanal

Ketones

Acetone
2-Propanone
2-Butanone
Cyclohexanone
3,3-Dimethylcyclobutanone
Methylethylketene
Methylbutylketone
4-Methyl-2-pentanone
3-Methyl-2-butanone
3-Heptanone
2-Heptanone
1-Phenylethanone
2,6-Di-*t*-butylbenzoquinone
Methyl phenyl ketone

Esters

Methylacetate
Ethylacetate
n-Propylacetate
n-Butylacetate
iso-Butylacetate
tert-Bytulacetate
2-Ethoxyethanolacetate
2-Ethoxyethylacetate
2-Methoxyethylacetate
2,2,4-Trimethylpenta-1,3-diol-isobutyrate
2,2,4-Trimethylpenta-1,3-diol-di-isobutyrate
2,2,4-Trimethylpenta-1,3-diol-1-isobutyrate
n-Butyl-*n*-butanoate
β-Butylmethacrylate
Benzylacetate
Phthalate
Diethylphthalate

Nitrogen Compounds

Nitromethane
Acetonitrile
Acrylonitrile
Pyrrole
Pyridine
4-Methylpentylnitrile
1-Butyl-1,2,4-triazole
Cyanobenzene
Benzyl cyanide
Benzothiazole
Indole

Aromatics

Benzene
Toluene
Ethylbenzene
m-Xylene
o-Xylene
p-Xylene
1-Methyl-2-Ethylbenzene
iso-Propylbenzene
n-Propylbenzene
1-Ethyl-3-methylbenzene
1-Ethyl-4-methylbenzene
1,3,5-Trimethylbenzene
Styrene
Phenylacetylene
1-Ethyl-2-methylbenzene
1,2,3-Trimethylbenzene
1,2,4-Trimethylbenzene
C_3-Alkylbenzene
C_9H_{12}, Alkylbenzene
Trimethylbenzene
2-Methyl styrene
1,1-Dimethylethylbenzene
1-Methyl-4-(1-methylethyl)- benzene
1-Methylethenyl benzene
2-Propenylbenzene
$C_{10}H_{14}$, Alkylbenzene
C_4-Alkylbenzene
1,3-Diethylbenzene
1,4-Diethylbenzene
Dimethylstyrene
1-Ethenyl-3-ethylbenzene
1-Ethenyl-4-ethylbenzene
1-Methyl-2-propylbenzene
1-Methyl-3-propylbenzene
1-Methyl-4-propylbenzene
Diethylbenzene
iso-Butylbenzene
sec-Butylbenzene
1-Methylpropylbenzene
Diethylstyrene
1-Propyl-4-isopropylbenzene
2-Ethyl-1,3-dimethylbenzene
2-Ethyl-1,4-dimethylbenzene
3-Ethyl-1,2-dimethylbenzene
4-Ethyl-1,2-dimethylbenzene
1-Ethyl-2,4-dimethylbenzene
1-Ethyl-3,5-dimethylbenzene
1,2,3,5-Tetramethylbenzene
1,2,3,4-Tetramethylbenzene
2,3-Dihydro-2-methyl-14-indene
n-Pentylbenzene

(continued)

TABLE 1.2 Continued

Aromatics (cont.)	*Terpenes*	*Miscellaneous*
C_5-Alkylbenzene	α-Pinene	Acetic acid
Indan	β-Pinene	2-Epoxy-4-methylpentane
Methylindan	Δ^3-Carene	Hexamethylcyclotrisiloxane
Dimethylindan	Limonene	Octamethyl-cyclo-tetrasiloxane
Dimethylhydroindene	4 (10)-Thujene	Siloxane
Naphthalene	Myrene	Carbonyl sulfide
Tetrathin	Ocimene	Carbon disulfide
Decahydronaphthalene	Camphene	Tetramethylsilane
Tetrahydronaphthalene	Terpene	Diphenyl ether
1-Methylnaphthalene	***Hetrocyclic Compounds***	Sulfur dioxide
2-Methylnaphthalene	Furan	Camphor
$C_{13}H_{20}$ Alkylbenzene	Methyl furan	
$C_{14}H_{22}$ Alkylbenzene	Tetrahydrofuran	
Biphenyl	2-Methyl-1,3-dioxane	
Methyldiphenyl	*p*-Dioxane	
Methylfluorene		

Source: Berglund et al., 1986.

ble particulates, such as those cited, are known to cause serious health problems by themselves, they may become even more dangerous when combined with other indoor pollutants, which can attach to a particulate and become lodged in the upper respiratory tract or lungs. In addition to indoor sources, respirable particulates may find their way indoors through the circulation of unfiltered outdoor air. Certainly, risks can be reduced by eliminating potential sources of these pollutants. For example, there is a trend toward "smoking free" public buildings and transportation systems. In addition, there are a number of mitigation techniques available for reducing or completely removing respirable particulates in the home and workplace.

Chronic obstructive lung diseases, chronic bronchitis, and emphysema rank second only to coronary artery disease as a cause of Social Security-compensated disability. The health risks associated with exposure to certain respirable particulates, such as tobacco smoke and asbestos fibers, have been discussed extensively throughout both the scientific and popular literature. Numerous studies have shown that tobacco smoke can significantly increase the levels of particulate matter and polycyclic aromatic hydrocarbons in indoor air. Yet there are differing conclusions regarding the effects of tobacco smoke on lung cancer, pulmonary function, and respiratory health. The carcinogenic potency of asbestos and man-made mineral fibers is believed to be especially significant. These cancers include the lung cancer mesothelioma, a cancer of the membrane lining the chest and abdomen, and cancers of the gastrointestinal tract. In addition, exposure to asbestos dust increases the risk of asbestosis, a fibrotic disease of the lung whereby imbedded dust fibers are surrounded by scar tissue. A recent $4 million literature review commissioned by the EPA has corroborated the position that well-maintained asbestos in public buildings poses little health risk to office workers (HEI-AR, 1991). However, there is considerable disagreement over the severity of these risks, and this situation will undoubtedly continue for some time.

One general class of airborne particulates that is a part of our ecological system is collectively referred to as organisms, microorganisms, or bioaerosols. These include viruses, bacteria, fungi, algae, mites, pollen, and numerous others that are less well known. These microorganisms require a reservoir (storage site), an amplification mechanism, and a means of dispersal. Airborne microorganisms are introduced into the body by inhalation and/or through open wounds. Fortunately, exposure to many of the bioaerosols has been studied extensively; the health risks are known, and medical treatment is usually available. Some of the associated diseases, such as colds, influenza, and Legionnaire's disease, are difficult to treat and can lead to complications that may result in death. However, rising medical costs prevent many individuals from seeking appropriate health care, thus increasing the importance of developing measures to limit human exposure to this particular class of pollutant. A number of mitigation processes have been developed and are currently in use. These include filtration, humidity control, and the use of disinfectants.

Next to tobacco smoke, radon is probably the most widely publicized of all the indoor air pollutants, and its presence in homes has created considerable concern about its long-term effects, following exposure to even low-level concentrations. Radon is a product of radium, which is found in rock formations and soils throughout the world. Radon finds its way indoors through the use of water, natural gas, and building materials and directly from the soil and rocks under and around basements and foundations. Radon is present indoors at concentrations much lower than the parts per trillion level but is nevertheless a matter of concern. It is chemically inert and does not, itself, react with the lungs upon inhalation. However, radon is radioactive, and it decays to lead and subsequently to bismuth through the emission of energetic alpha particles. The lead and bismuth ions can attach to aerosol particles and, consequently, lodge in the lungs, where the emission of alpha particles has the potential for causing cancer. Exposure to this odorless, tasteless gas by the general public may already account for a substantial number of cases of lung cancer in the United States and worldwide. Recent estimates by the EPA indicate that from 10,000 to 20,000 cases of lung cancer occur each year among the U.S. population from exposure to radon indoors. The danger of cancer from exposure to radon came into focus as a result of the number of cases of cancer among lead and uranium miners. The health effects of radon are currently predicted by using models of radioactive deposition in the human lung, and these models predict that 9000 deaths occur from lung cancer each year in the United States due to radon exposure (NCRP, 1989).

The development of effective techniques for removing air pollutants is of considerable importance and interest. Mitigation methods, such as increased ventilation, pollutant source control, air cleaners, carbon filters, mechanical filters, and electrostatic precipitators, are currently used to control indoor air pollutants, but they may not be adequate. Some of these can and have been used in conjunction with or as part of heating, ventilating, and air-conditioning (HVAC) systems to reduce pollutant levels. Two of the more important and versatile treatment methods make use of the dehumidification portion of an HVAC system. In addition to the traditional vapor recompression air-conditioning method of removing moisture from air, dehumidification can be carried out by employing either a solid or liquid desiccant. The use of both types of desiccants has been shown to be effective in removing a wide range of indoor air pollutants, including bioaerosols, particulates, organic chemicals, and smoke.

The removal of moisture and pollutants by liquid desiccants, such as lithium chloride and triethylene glycol, is accomplished by contacting the polluted, water-laden air with a liquid desiccant in a gas-liquid contactor in which the water vapor and pollutants are absorbed. As will be discussed in detail in this text, absorption entails the contact of a gas and liquid, with one or more components being taken up homogeneously by the liquid phase. Absorption is a physical process, but chemical reaction can occur. This, however, depends on the nature of the absorbent and absorbate (or solute).

Although proper ventilation can be used effectively to reduce the concentration levels of pollutants found indoors, probably the most versatile and effective means of actually removing pollutants from air is by adsorption. In an adsorption process, a material is transferred from a fluid phase to the surface of a solid, where it is retained by surface forces. The adsorbing solid is defined as the adsorbent, whereas the material adsorbed on the surface is denoted as the adsorbate. While the amount adsorbed per unit area is quite small, the surface area of a variety of adsorbents can exceed 1000 m^2/g. Because of the large surface area, the amount of adsorbate adsorbed on a per gram basis can approach the actual weight of the adsorbent. The selection of an appropriate adsorbent for a particular application depends on its surface properties and the type of adsorbate that is to be removed. Activated carbons are effective in adsorbing nonpolar compounds, such as organic carbons, but are considerably less effective in adsorbing water. Silica gel and molecular sieve (zeolites) are examples of adsorbents that are used to adsorb polar compounds, such as water vapor, but may be less effective in adsorbing nonpolar adsorbates. Both of these adsorbents find application as desiccants in dehumidification processes.

The amount adsorbed on a per weight basis of the adsorbent depends on the concentration of the adsorbate in the fluid phase and the temperature at which the adsorption process occurs. A plot of this information (amount adsorbed per unit weight of the adsorbent) as a function of the adsorbate concentration in the fluid phase for a fixed temperature is referred to as an adsorption isotherm and is different for every adsorbate-adsorbent combination and each temperature. The shapes of the various adsorption isotherms are related to the adsorption mechanism and surface forces and have been classified accordingly. Adsorption has been shown to be effective for removing bioaerosols, as well as a wide range of chlorinated and aromatic hydrocarbons, radon, water vapor, and smoke.

The primary focus of this book is on the health effects, sources, sampling, and specific control strategies of a wide range of air pollutants found indoors. In addition to this general theme, however, detailed discussions of the two most promising mitigation methods—adsorption and absorption—are presented. For completeness, we have included a chapter on risk assessment, with the intent of helping the reader gain a better understanding of the risks associated with various pollutants.

REFERENCES

ANONYMOUS, "Indoor Pollution and Stress Cited as Workplace Hazards," *New York Times*, A-21, September 19, 1984.

BURGE, H. A., and HODGSON, M. H., *ASHRAE J.*, **30** (7), 34 (1988).

BERGLUND, B., BERGLUND, U., and LINDVALL, T., "Assessment of Discomfort and Irritation from the Indoor Air", Proceedings IAQ '86: "Managing Indoor Air for Health and Energy Conservation", American Society of Heating, Refrigerating Air-Conditioning Engineers, Atlanta, Ga., 138, 1986.

HAWTHORNE, A. R., GAMMAGE, R. B., DUDNEY, C. S., HINGERTY, B. E., SCHURESKO, D. D., PARZYCK, D. C., WOMACK, D. R., MORRIS, S. A., WESTLEY, R. R., WHITE, D. A., and SCHRIMSHER, J. M., "An Indoor Air Quality Study for Forty East Tennessee Homes", Report No. ORNL-5965, Oak Ridge National Laboratory, Oak Ridge, Tenn., 1984.

HEI-AR (Health Effect Institute—Asbestos Research), "Asbestos in Public and Commerical Buildings: A Literature Review and Synthesis of Current Knowledge", Cambridge, Mass., 1991.

NAS (National Academy of Science), *Indoor Pollutants*, National Research Council, National Academy Press, Washington, D. C., 1981.

NCRP (National Council on Radiation Protection and Measurements), "Report on Radon", Bethesda, Md., 1989.

PLATTS-MILLS, T. A. E., and CHAPMAN, M. D., *J. Allergy Clin. Immunol.*, **80**, 755 (1987).

SAMET, J. M., MARBURY, M. C., and SPENGLER, J. D., *Am. Rev. Respir. Dis.*, **136**, 1486 (1987).

SAMET, J. M., MARBURY, M. C., and SPENGLER, J. D., *Am. Rev. Respir. Dis.*, **137**, 141 (1988).

SHAH, J. J., *Environ. Sci. Technol.,* **22** (12), 1381 (1988).

SINGH, H. B., SALAS, L. J., SMITH, A. J., and SHIGEISHI, H., *Atmos. Environ.*, **15**, 601 (1981).

2

Risk Assessment

2.1 INTRODUCTION

Compared to the health and life expectancy of previous generations, the modern era is relatively risk free (discounting, of course, threats of global wars). Modern technology has, however, brought a certain homogeneity to power production, transportation, and living styles and introduced new substances (pollutants) into the atmosphere and indoor environments. New analytical techniques have enabled the detection of microscopic amounts of potentially harmful substances that have always been present, but whose existence were not known previously. Longer life expectancy has brought into sharper focus the roles of these harmful substances, because such effects are often not immediate, but are recognizable only after long exposure.

There has always been strong public fear of phenomena or accidents that can cause large disruptions, injuries, and/or deaths. Such accidents are rare and have high consequences associated with them; they can be natural (collision with meteorites, earthquakes, volcanic eruptions, tsunamis, and forest fires) or due to modern technology (dam bursts, plane crashes, nuclear accidents). They can also be spectacular and have always fascinated or terrified the public with ample assistance from the media and, in the case of the United States, from tort lawyers. The net contribution of these accidents to the risk that a person would face in his or her lifetime is, however, not large. The risks faced on an almost daily basis from small hazards, such as automobile travel, are much more significant and are well understood.

In between, there are risks from a spectrum of real, suspected, or conjured hazards related to chemical and biological substances. Generally, the effects of such hazards become evident over very long periods of time, and the quantification of the risks is truly difficult and challenging. The task is complicated by the fact that the effect from a single substance on humans under controlled conditions cannot be studied. Therefore, studies are conducted on test animals that are subjected to massive doses of a substance on a relatively short time scale. However, these conditions are atypical of human exposures. The results of these animal studies are extrapolated for applications to human conditions by the use of mathematical models (that is, equations are formulated and solved for application to animal test conditions). If these exercises predict, or *ex post facto* can be made to describe the animal results, then some trust is placed in their application to human conditions. Since the animals are subjected continuously to high doses of a substance(s) over relatively short periods of time, under, let us say, subanimal conditions, the ability of models to describe the results of such tests have little to do with human exposure and susceptibilities. It is for these reasons that a greater understanding of both the exposure to and effects of indoor pollutants are required. However, it remains doubtful if additional studies will resolve these problems. This has not happened, for instance, in the case of the effects of low levels of ionizing radiation, for which a large number of studies over a long period of time have been conducted in many nations. It is nevertheless useful to discuss here, briefly, the topic of risk assessment, which may also provide perspectives on the relative role of indoor air pollution *vis-a-vis* other risks to which the general population is exposed.

2.2 DEFINITIONS AND QUANTIFICATION

A hazard is referred to as the potential of an entity (or activity) to cause harm to nature, property, or people (McCormick, 1981 and Rasmussen, 1981). The risk from a hazard is defined as

risk (harm/unit time) = frequency (event [exposure]/unit time)
× consequence (harm/event [exposure])

Note that two equally hazardous substances may pose risks that are quite different, since the risk depends not only on the potential (or the consequence) but also on the frequency of exposure. Thus, assessment of risk (R) requires a knowledge of both the frequency (f) and the associated consequence (C). Minimization of the risk then requires reduction of the frequency and/or the consequence. In symbolic form, the above expression is

$$R_i = f_i C_i \tag{2.1}$$

where the subscript i indicates the risk from hazard i. Thus, the total risk R from hazards $i = 1, 2, \ldots, n$ is simply a summation of all risks:

$$R = \sum_i R_i \tag{2.2}$$

There are difficulties associated with the notions of f and C and the definition of R. The frequency f can be estimated from the time people spend indoors where they are exposed to a particular level of pollutants. But social, regional, and national differences should be considered, and it is obvious that no two persons will have the same frequency of exposure.

The above definition is more appropriate for low values of the consequence C. For high values (for example hundreds of fatalities and/or large property losses associated with a dam burst or nuclear reactor accident), a weighting factor might be included. Thus, the risk can be defined as

$$R_i = f_i F_i (C_i) \tag{2.3}$$

where F is a function of C and may have the form

$$F = C^n \tag{2.4}$$

and n is some known number that reflects the roles of disruptions by severe accidents, as well as the public perception. For indoor air pollution, C is low, n might be set equal to one, and the equations above are still applicable. Except perhaps for tobacco smoke, reliable epidemiological data for most indoor pollutants are not available, and we must rely on limited data and their extrapolations by using mathematical, biological, and physicochemical models (the dose models) to actual circumstances. Such extrapolations, of course, lead to major contention among the various parties, such as the public, manufacturers of the building materials and cigarettes, builders, and government agencies. It must be recognized that there is no universal agreement on the risks posed by the various pollutants. Thus, improved knowledge of the frequency of exposure to a pollutant and the consequence of such exposure (note that the C_i's depend on the exposure level) is needed. This alone is hardly sufficient, since the public is exposed to a medley of pollutants. The synergism (that is, the C_i's and R_i's are interdependent), as well as the public perceptions of the exposures and the resulting psychosomatic factors, can play a strong role in the risk assessment.

The applications of the above definitions for the assessment of risks from indoor air pollutants are explained in the following example (see Tancrede et al., 1987). Dropping the subscript i for brevity, we have

d = daily dose, $\mu g/m^3$ of air, of a pollutant to an individual (that is, the amount of the pollutant present in 1 m^3 of air)

R = lifetime *excess* cancer risk (excess cancer per lifetime) to an individual from daily exposure to a dose d of the pollutant

B = breathing rate of the individual, m^3 of air/day

W = weight of the individual, kg

β_a = "potency" of the pollutant for inhalation, $(\mu g/kg\text{-day})^{-1}$; that is, if an animal weighing 1 kg inhaled 1 μg of the pollutant daily, then β_a would be the probability of lifetime excess cancer for this animal (note that the animal can be of any weight, since the value is normalized to unit weight)

K_{ah} = a conversion factor expressing the ratio of the risk to a human to the corresponding risk to an animal based on inhalation toxicity data (potency)

I = a factor relating inhalation data to risk if other pathways (for example, oral) were also available

Then

$$f = \left(\frac{\text{daily exposure over a lifetime of 1 µg of the pollutant}}{\text{person's weight in kg}}\right) = \frac{B}{W}d \tag{2.5}$$

$$C = \left(\frac{\text{lifetime excess cancer risk}}{\text{daily exposure to 1 µg of the pollutant/person's weight in kg}}\right)$$

$$= \beta_a K_{ah} I \tag{2.6}$$

and

$$R = fC \tag{2.7}$$

Thus, the lifetime excess cancer risk from the pollutant to this individual is expressed as

$$R = \left(\frac{B}{W}d\right)(\beta_a K_{ah} I) \tag{2.8}$$

Let us consider the risk to an adult from the exposure to tetrachloroethylene, which is an organic vapor. Thus if,

$$\begin{aligned} B &= 20 \\ W &= 70 \\ d &= 3.5 \\ \beta_a &= 9.2 \times 10^{-6} \\ K_{ah} &= 1 \\ I &= 1 \end{aligned}$$

then, the lifetime excess cancer risk to the individual is $R = 9.2 \times 10^{-6}$, which is not a large risk. If either the exposure or the potency was higher, the risk would increase proportionately. In simple terms, this notion of risk implies that an individual has a 9.2 chance in a million of getting cancer over his or her lifetime because of the daily exposure to tetrachloroethylene at a concentration level of 3.5 $\mu g/m^3$ of air. If the individual had a life span of 70 years, then the chance of getting cancer in any one year from this pollutant is 9.2/70 = 0.13 in a million. Thus, if 100 million people (with life expectancies of 70 years) are exposed to the pollutant, in any one year 13 of them are likely to get cancer because of the exposure. Tancrede et al. used the above framework to compute risks for 52 different organic pollutants and compared their results with those posed by formaldehyde, radon, and environmental tobacco smoke.

Let us explore the case of formaldehyde, a volatile organic compound that has come to be recognized as a highly toxic indoor air pollutant. Formaldehyde has a very low boiling point (−19.5°C) that results in a high vapor pressure at room temperature which leads to a large number of molecules in the air. Its source in indoor air is from emissions by building materials, such as particle board, upholstery, and foam insulation. Typical concentrations of formaldehyde could range from 0.02 to 3.0 ppm (0.02 to 3 molecules per 1 million molecules of air). Formaldehyde has a strong odor (it is a preservative used in biology and zoology labs) that is easily recognized, but precise detection at the parts per million level is possible only with sophisticated instrumentation.

Formaldehyde is of concern for several reasons. It has an unpleasant odor that is intolerable to some, it is irritating to eyes, and it affects the skin. It is also classified as a probable human carcinogen. The evidence for this classification comes from limited animal tests, in which a group of rats were subjected to a range of formaldehyde doses (1 to 15 ppm) under conditions described earlier (Gibson, 1983; Lewis, 1990; Marbury and Krieger, 1991). At higher doses, some rats did develop cancer. However, those doses were quite high and the inhalation was entirely nasal, a circumstance not common with human exposure to formaldehyde. Consequently, there is little proof that even at the 15-ppm level, formaldehyde would be a carcinogen to humans, hence the designation "probable".

Following the procedures of Tancrede et al., the risk from formaldehyde can be estimated as follows. Using typical values for human exposure, we have

$$\begin{aligned} B &= 20 \\ W &= 70 \\ d &= 0.05 \text{ ppm} = (0.05)(1.2 \times 10^3)\ \mu\text{g/m}^3 \text{ of air} \\ \beta_a &= 1.1 \times 10^{-4} \\ K_{ah} &= 1 \\ I &= 1 \end{aligned}$$

Thus, the lifetime excess cancer risk to an individual from formaldehyde exposure is

$$\begin{aligned} R &= \left(\frac{B}{W}d\right)(\beta_a K_{ah} I) \\ &= 1.9 \times 10^{-3} \end{aligned}$$

which is much higher than the value obtained for tetrachloroethylene exposure. This risk estimate indicates that, for a life span of 70 years, the annual excess cancer risk to a person would be $1.9 \times 10^{-3}/70 = 2.71 \times 10^{-5}$. The annual excess cancer risk to the U. S. population (~ 250 million) from formaldehyde is then $(2.71 \times 10^{-5})(250 \times 10^6) = 6800$. The value 1.9×10^{-3} is only an estimate of the median value; the mean value of the individual lifetime excess cancer risk computed by Tancrede et al. is 3.4% or 3.4×10^{-2}, which translates into an individual annual excess cancer risk of 4.8×10^{-4}. Therefore, the annual excess cancer risk to the U.S. population from formaldehyde is $(4.84 \times 10^{-4})(250 \times 10^6) = 120{,}000$. In comparison, annual deaths from auto accidents in the United States are about 50,000. If the above estimate of the risks from formaldehyde is accurate, major efforts to reduce its level would need to be initiated. The estimated individual lifetime excess cancer risk from formaldehyde is actually anywhere from 3.4×10^{-2} to 1×10^{-7} (Tancrede et al., 1987; McCann et al., 1986). Whenever such a wide spread in estimations exists, it is difficult to place a great deal of trust in the values.

A summary of the available risk estimates for various indoor pollutants has been given by Naugle et al. (1989). Figures 2.1 and 2.2 are reproduced from their work. Figure 2.1 provides a summary of the estimates for individual lifetime excess cancer risk, while

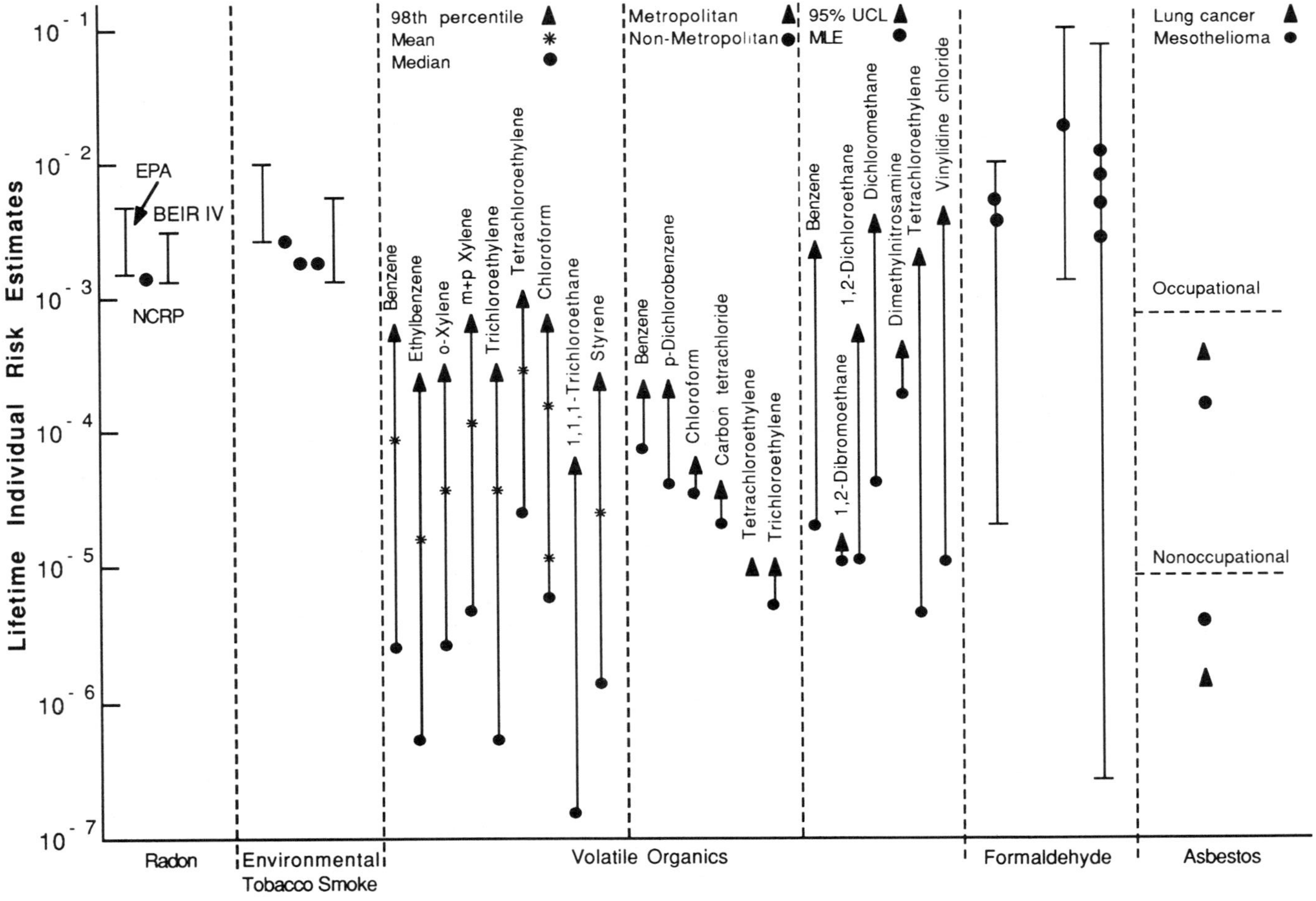

Figure 2.1 *Comparison of individual lifetime cancer risk due to indoor air pollutants. (Source: Naugle et al., 1991.)*

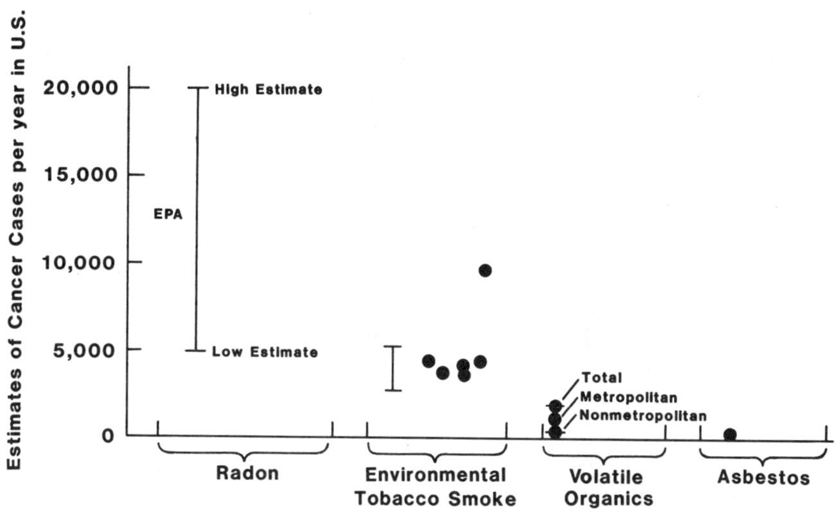

Figure 2.2 *Comparison of annual cancer cases due to indoor air pollutants. (Source: Naugle et al., 1991.)*

Figure 2.2 is an estimate of the annual excess cancer risk to the entire U.S. population from various pollutants. The authors have obviously used their own judgments in arriving at Figure 2.2 from Figure 2.1. Given the ranges of the estimates in Figure 2.1, we must be subjective. In terms of the excess annual cancer risk to the U.S. population, indoor air pollutants are ranked as

Radon (5000 to 20,000)
Environmental tobacco smoke (3000)
Volatile organic compounds (~100)

where the numbers in the parentheses indicate the excess cancers. Note that the tobacco smoke estimate is for passive smoking. Risks to smokers from mainstream smoke (inhaled puffs) are very high (~400, 000 deaths/year), but that is not an indoor air quality and control issue.

Our own estimate is that, in terms of risks, indoor pollutants can be ranked in the following way:

Particulates (tobacco smoke, asbestos, and bioaerosols)
Volatile organic and inorganic compounds
Radon

It is important to emphasize that animal tests do serve a purpose in furthering our understanding, especially when a pollutant causes a distinctive type of malady. This has been the case with asbestos, for example. Much has been learned about the effects of tobacco smoke and radon and the synergistic effects of various pollutants. Still, extrapolations to human conditions are fraught with difficulties.

2.3 UNCERTAINTIES

The difficulties faced in making a meaningful prediction about a particular pollutant are discussed above. The limitations of the point estimations are clear in that it is a rare human who is subjected daily to the same level of a pollutant. Even in a home, the pollutant levels are not uniform, and they vary with respect to location, the season, and time of day. Also, the potency is not known precisely and is based on a questionable and limited set of data. This is a particular case of Bayesian statistics with which we can make estimations using probability values that are known to be inherently in error. These limitations and omissions can be improved upon by appealing to statistics and regarding f, C (and its components), and R as random variables (which in fact they are) and as functions of time. If the distributions of f and C are measured or assumed, the latter being the common case, then the distribution of R, which is the likelihood of a person being susceptible to a certain lifetime excess cancer risk, can be calculated by an applicable statistical method (for example, the Monte Carlo technique). Often it is convenient to assume that the distributions of f and C are log normal with known median or mean values and standard deviations. Then, R also has a log-normal distribution, and its median, mean, and standard deviation are simply related to those of f and C. It is important to know the uncertainties of the results. In the wrong hands, the matters may get sufficiently confused, the central issues get ignored, and the researcher may become more involved in finding uncertainties of uncertainties. It is perhaps sufficient to emphasize that the smaller the size of a sample on which an estimate is based, the less confident we are of the results; and the uncertainties in risks from most pollutants are factors of 10 to 100 of the mean values.

According to the EPA (1989),

> Risk estimates are not available for most pollutants, but available estimates for radon, tobacco smoke, and volatile organic compounds (VOCs) demonstrate that indoor air pollution is among the nation's most important environmental health problems.
>
> The potential economic impact of indoor air pollution is quite high and is estimated in the tens of billions of dollars per year. Such impacts include direct medical costs and lost earnings due to major illness, as well as increased employee sick days and lost productivity while on the job.
>
> Labor costs may be 10 to 100 times greater per square foot of office space than energy and other environmental control costs. Thus, from a profit and loss standpoint, remedial actions to improve indoor air quality (IAQ) where productivity is a concern are likely to be cost effective even if they require an extensive retrofit.

The EPA has emphasized research on

risk assessment methodology, focusing on health and hazard identification, dose-response assessment, exposure assessment, and risk characterization frameworks and methods, especially as they relate to the comparability of results from oral versus respiratory toxicity studies;

exposure assessment and modeling, including methods development and evaluation, measurement studies, development of predictive models, and the management of measured data; and

additive or synergistic effects from multiple chemical contaminants, even where the concentrations of individual compounds are below their known health threshold, and acute reactions in some people and how they exacerbate the chronic effects of pollutants such as environmental tobacco smoke and radon.

More research would help, but it is doubtful if all or even most of the needed answers will emerge from the kind of animal research that is being pursued. Toxicological research appears to be at least five years behind biological research, and perhaps molecular toxicology should be explored to determine the potency factors.

The public interest in controlling the levels of indoor pollutants is very keen. The reader is cautioned, however, that much in the areas of health effects and risk assessment is unknown and uncertain, and the public should neither become unduly alarmed by the media or scientific hype nor become complacent and assume that indoor pollutants do not affect human health.

REFERENCES

EPA (U. S. Environmental Protection Agency), "Report to Congress on Indoor Air Quality," NTIS Document No. PB90-167370, 1989.

GIBSON, J. E., *Formaldehyde Toxicity*, Hemisphere, New York, 1983.

LEWIS, H. W., *Technological Risk*, Norton, New York, 1990.

MARBURY, M. C., and KRIEGER, R. A., in *Indoor Air Pollution: A Health Perspective*, (eds. J. M. Samet and J. D. Spengler), Johns Hopkins University Press, Baltimore, Md., 1991.

MCCANN, J., HORN, L., GIRMAN, J., and NERO, A. V., "Potential Risks from Exposure to Organic Carcinogens in Indoor Air", presented at the EPA Symposium on the Application of Short-term Bioassays, Durham, N. C., 1986.

MCCORMICK, N. J., *Reliability and Risk Analysis*, Academic Press, New York, 1981.

NAUGLE, D. F., PIERSON, T. K., and LAYNE, M. E., "Indoor-Air-Assessment: A Review of Indoor Air Quality Risk Characterization Studies", Report No. EPA/600/8-90/044, Environmental Protection Agency, Research Triangle Park, N. C., 1991.

RASMUSSEN, N., "The Application of Probabilistic Risk Assessment Techniques to Energy Technologies", 123, *Ann. Rev. Energy*, Annual Reviews, Palo Alto, Calif., 1981.

TANCREDE, M., WILSON, R., ZEISE, L., and CROUCH, E. A. C., *Atmos. Environ.*, **21**(10), 2187 (1987).

3

Volatile Organic Pollutants

3.1 INTRODUCTION

Any chemical compound that contains at least one carbon and a hydrogen atom in its molecular structure is referred to as an organic chemical. Organic compounds are further classified into various categories, such as volatile organic compounds (VOCs), semivolatile organic compounds, and nonvolatile organic compounds. This classification is based primarily on the vapor pressure of the particular organic compound at room temperature. A majority of organic compounds are either liquid or solid at room temperature, but their vapor pressures are high enough that there will be a substantial amount of the vapor phase. Volatile organic compounds can be defined as those having a vapor pressure greater than 1 mmHg at room temperature and are present in air as a vapor. Pollutants such as benzene, toluene, and trichloroethane belong to this class. Polycyclic aromatic hydrocarbons (PAHs), pesticides, polychlorinated biphenyl (PCBs), and furans have vapor pressures in the range of 10^{-7} to 1 mmHg and are usually referred to as semivolatile organic compounds. Benzo(a)pyrene and perylene, for example, have vapor pressures below 10^{-7} mmHg and are classified as belonging to the nonvolatile group.

Research directed at the identification and measurement of volatile organic chemicals indoors was initiated in the early 1970s by various national and international organizations, such as the U. S. Environmental Protection Agency (EPA), the Danish Building Association, and the Swedish Building Association. The focus of these early studies was to demonstrate that indoor concentrations of a number of chemicals can exceed their outdoor concentrations, and the sources of many of these pollutants are indoors. Subsequent studies showed that the total nonmethane hydrocarbon concentration indoors was higher than outdoor concentrations in 90% of the measurements. Also, the organic content of a number of indoor particulates was higher than in outdoor air. The most convincing and conclusive study that related the occurrence of organic chemicals in residences to furnishings and building materials was conducted by Mølhave (1979) in Denmark. Numerous studies have been conducted since then to identify and quantify organic chemical pollutants indoors. The EPA also conducted one of the most comprehensive studies on indoor air quality in the United States under the project name of Total Exposure Assessment Methodology (TEAM). This study included approximately 600 residences in the states of New Jersey, North Carolina, North Dakota, and California. Their results showed that in most cases the concentration of 22 targeted chemicals was higher indoors than outdoors, and for some chemicals it was two orders of magnitude greater.

The quantitative and qualitative measurements of chemical mixtures indoors present a challenging and expensive task. Over 800 different chemical compounds have been identified at least once in the air of four office buildings that were studied by the EPA. The organic contaminants identified indoors belong to several organic subgroups, including *n*-alkanes, branched alkanes, alkenes, chlorinated hydrocarbons, aromatics, aldehydes, alkylbenzenes, alcohols, and polycyclic aromatics, to name but a few. Berglund et al. (1986) listed 307 organic chemicals that are present indoors at various times. However, not all of them are present simultaneously in an indoor environment. Indoor concentrations and emission rates from building materials have been measured for only selected organic compounds. Berglund et al. (1986) concluded that even after careful screening, there may be about 70 pollutants that are frequently present indoors.

The EPA targeted 32 volatile organic compounds when characterizing the air and materials in 10 buildings (Sheldon et al., 1989). These targeted chemicals belong to four groups: aromatic, aliphatic, chlorinated, and oxygenated hydrocarbons. Mølhave et al. (1984) prepared a mixture containing 22 organic compounds, which are commonly found in Danish homes, to study the effects of indoor pollutants on healthy but sensitive human subjects. The 22 organics are listed in Table 3.1. According to Amman et al. (1986), the same 22 organics may not be as prevalent in U. S. homes and nonresidential environments.

Pollutants that are known or suspected of being carcinogenic and/or neurotoxic for humans or animals are present in U. S. residences and are identifiable and quantifiable by available analytical techniques. They should be selected for further study. Amman et al. provided a list of 25 organics that satisfy the above criteria. These also are included in Table 3.1. This list of organic chemicals is still too large to study in the laboratory and obtain meaningful results, particularly for the evaluation of samplers and air cleaners. Daisey and Hodgson (1989) chose NO_2 and six representative volatile organic compounds to test the performance of several air cleaners. The six VOCs were dichloromethane, 2-butanone, n-heptane, toluene, tetrachloroethylene, and hexanal. Care must be taken when targeting only certain volatile organic compounds and studying their effect on human health, because the actual pollutants present indoors depend on the source and emission rates from the building materials and furnishings.

TABLE 3.1 Selected Chemicals Used in the Preparation of a Representative Mixture

Chemicals used by Mølhave et al., 1984	*Chemicals suggested by Amman et al., 1986*
n-Hexane	Vinyl chloride
n-Nonane	Methylene chloride
n-Undecane	Chloroform
1-Octene	Carbon tetrachloride
1-Decene	1,1,1-Trichloroethane
Cyclohexane	Trichloroethylene
o-Xylene	Tetrachloroethylene
Ethylbenzene	Benzene
1,2,4-Trimethylbenzene	*o*-Xylene
n-Propylbenzene	*m,p*-Xylene
α-Pinene	*m,p*-Dichlorobenzene
n-Pentanal	Ethylbenzene
n-Hexanal	Styrene
iso-Propanol	1,2-Dichloroethane
n-Butanol	1,1,1,2-Tetrachloroethane
2-Methyl-3-butanone	1,1,2,2-Tetrachloroethane
4-Methyl-2-pentanone	*n*-Butylacetate
1,1,2-Trichloroethane	Hexachloroethane
Ethoxyethyl acetate	Decane
1,2-Dichloroethane	Chlorobenzene
	Ethylphenol
	Acrolein

Formaldehyde has been singled out as one of the more toxic organic chemicals, because of numerous and well-documented health complaints associated with the use of building materials, such as urea formaldehyde foam insulation (UFFI). These materials slowly release free formaldehyde indoors over a long period of time. Because formaldehyde is suspected of being carcinogenic to humans at the present indoor levels, this may become a critical problem.

3.2 HEALTH EFFECTS AND STANDARDS

In indoor environments, particularly in residences, the occupants may be exposed periodically to various pollutants at high concentration levels for short periods of time or continually to low concentrations for a much longer time period. The type of exposure that is more damaging to health is not clear. The health risks posed by air pollutants depend on the exposure time of the individual, on the type of pollutants present, and on the state of health of the individual at the time of exposure; the pollutant concentration is most certainly an important factor. A number of pollutants are colorless, tasteless, and odorless at their indoor concentration levels. Also, humans usually cannot identify a particular chemical in a mixture by its odor or by sensory irritation alone. As a result, the health risks and the inhalation dose-response data of individual chemicals have not been well documented. It should be remembered, however, that these pollutants are present in a mixture that often contains more than 50 compounds at any one time. Frequently the concentration of a single compound or group, such as combustion products or a mixture of chlorinated hydrocarbons, may be an order of magnitude higher than the others, but the synergistic interaction of an individual component with other pollutants should not be ruled out. During short-term exposures, individuals are typically exposed to a single pollutant or a group of specific pollutants, depending on their activity during that time period. Table 3.2 lists some pollutant exposures associated with various activities.

Little is known about the acute and chronic effects of a mixture of volatile organic chemicals on human health, particularly at the low concentrations that are characteristic of nonindustrial environments. It is extremely important that an individual be aware of and understand the potential health effects associated with exposure to individual chemicals. The health effects associated with individual organic chemicals are generally known at concentrations typical of industrial environments, which are often several orders of magnitude higher than in other indoor environments. However, most of these effects were derived from studies with animals. Many manufacturers already specify the type of chemicals that are used in their products, and others are now being required to do likewise. The health hazards associated with a specific group of organic chemicals, such as aldehydes, ketones and ethers, aliphatic, cyclic, aromatic, and chlorinated hydrocarbons, are discussed briefly next. Additional information can be obtained from books on toxicology (for example, Anderson and Scott, 1981).

TABLE 3.2 Pollutants Associated with Various Activities

Activities	*Possible exposure to pollutants*
Cleaning	
Windows	Ammonium hydroxide
Spots/textiles	Tetrachloroethylene, trichloroethylene, methanol, petroleum-derived solvents, benzene
Soaps/detergents	Polyether sulfates, alcohol, sulfonates, alkylsodium isothionates
Oven	Sodium hydroxide, potassium hydroxide
Drain/toilet bowl	Sodium hydroxide, lye
Vacuuming of carpets	Dust
General cleaning	Ammonium hydroxide, chlorine, lye, sodium hypochlorite, sodium peroxide
Painting/varnishing	Toluene, xylene, methylene chloride, heavy metals, pigments, methanol, ethylene glycol, benzene
Application of pesticides	Organophosphates, carbamates, pyrethroids
Gardening, lawn/yard care	Pesticides, herbicides, gasoline, oil, fertilizers
Cooking	Combustion products, formaldehyde
Use of aerosol cans	Propane, butane, methylene propellents, isobutane, fluorocarbon 11 and 12
Disinfectants	Sodium hypochlorite, quaternary ammonium salts, phenols, pine oils
Smoking	Tobacco smoke
Furniture/carpets offgassing	Formaldehyde, VOCs

Source: Sterling et al., 1990.

Aldehydes, Ketones, and Ethers

Formaldehyde is a colorless gas with a distinct, pungent odor that can be detected by most people at concentrations as low as 0.05 ppm. Indoor concentration levels of formaldehyde, particularly in mobile homes and houses with urea formaldehyde foam insulation, can be well above this odor threshold limit. Besides the pungent odor, both acute and chronic health effects have been associated with formaldehyde exposure. Formaldehyde concentrations higher than 1 ppm often produce coughing, wheezing, and shortness of breath. Neuropsychological effects of formaldehyde exposure include headaches, memory lapse, fatigue, and sleep deprivation (Olsen and Dossing, 1982; Schenker et al., 1982). Other health concerns are sensitization and asthma. Table 3.3 summarizes the health effects of formaldehyde at various concentration levels on humans (NAS, 1981).

Symptoms of individuals who have been exposed to formaldehyde are lessened dramatically when they move away from the building in question. Common symptoms associated with formaldehyde exposure in buildings include acute mucous membrane irritation of the eyes, nose, and sinuses and of both the upper and lower respiratory airways. A large quantity of health data has been obtained from surveys of residents of mobile homes and homes with urea formaldehyde foam insulation. These data have been compared with similar data obtained from occupants of other structures to assess the health hazards from formaldehyde exposure; these findings are summarized in Table 3.4. Although the concentration inside buildings is below 0.1 ppm in most cases, formaldehyde is known to act synergistically with other indoor pollutants. Ahlstrom et al. (1984) exposed healthy individuals to 0.082 ppm of formaldehyde in a chamber. When the air flow from a sick building to the chamber was increased from 10 to 100%, while maintaining the formaldehyde concentration at the same level, symptoms of mucous membrane irritation increased four-fold.

There is conclusive evidence that formaldehyde is an animal carcinogen, and, as noted earlier, it is also suspected of being a human carcinogen. Controversy exists over the concentration at which formaldehyde might affect humans. One study indicated cancer occurred in the nasal cavity of about 50% of the rats exposed to a concentration of 14 ppm, but only 2 rats out of 235 showed the same symptoms when exposed to a concentration of 5.6 ppm (Kerns et al., 1983). An epidemiological study by the Du Pont Corporation found

TABLE 3.3 Health Effects from Formaldehyde Exposure at Various Concentrations

Formaldehyde Concentration (ppm)	*Observed Health Effects*
0.0–0.5	None reported
0.05–1.5	Neurophysiologic effects
0.05–1.0	Odor threshold limit
0.01–2.0	Irritation of eyes
0.10–25	Irritation of upper airway
5–30	Irritation of lower airway and pulmonary effects
50–100	Pulmonary edema, inflammation, pneumonia
>100	Death

Source: NAS, 1981.

TABLE 3.4 Sypmptoms Related to Formaldehyde Exposure

Study Population	*Concentration of Formaldehyde*	*Reported Symptoms*
1396 residents of UFFI homes and 1395 of non-UFFI homes (Thun et al., 1982)	Not reported	Wheezing: exposed, 60%; nonexposed, 10% Burning skin: exposed, 70%; nonexposed, 10%
70 employees in 7 mobile homes and 34 nonexposed employees in 3 buildings (Olsen et al., 1984)	Mobile home: 0.24–0.55 ppm Building: 0.05–0.11 ppm	Menstrual irregularities: exposed, 35%; nonexposed, none Excessive thirst: exposed, 60%; nonexposed, 5% Eye irritation: exposed, 55%; nonexposed, 15% Headache: exposed, 80%; nonexposed, 50%
21 workers in mobile homes and 18 workers in buildings (Main and Hogan, 1983)	Mobile home: 0.12-0.16 ppm Building: not reported	Eye irritation: exposed, 81%; nonexposed, 17% Throat irritation: exposed, 57%; nonexposed, 22% Fatigue: exposed, 81%; nonexposed, 22% Headache: exposed, 76%; nonexposed, 11%

no significant cancer-related deaths among occupationally exposed employees of a garment factory as compared to unexposed persons (Levin and Purdom, 1983). Recently, however, Stayner (1986) and Vaughan et al. (1986) reported significant correlations between buccal cavity cancer in garment workers and between nasopharyngeal cancer in mobile home residents. Even at low concentrations, formaldehyde appears to pose a serious health threat. It has been estimated that fewer than 20%, but more than 10%, of the general U. S. population may be hypersensitive to the irritant effects of formaldehyde at any concentration (NAS, 1981). According to another estimate by the EPA, the risk of developing cancer is 1 in 10, 000 for persons exposed to 0.07 ppm of formaldehyde for more than 10 years, and the risk of cancer development increased to 1 in 5000 for persons exposed to 0.1 ppm.

Virtually all ethers, including methyl, ethyl, and butyl ether, have an anesthetic effect on humans. The irritation of mucous membranes can lead to pulmonary edema, vomiting, headaches, and nausea. Exposure to high concentrations of ketones, such as acetone, methyl ethyl ketone, and methyl isobutyl ketone, is known to produce narcosis, nausea, headaches, dizziness, irritation of the mucous membranes, and a loss of coordination. However, most esters, such as ethyl acetate, butyl acetate, and ethyl butyrate, have a pleasant odor and are used as flavors and in perfumes; these are physiologically inert.

Aliphatic and Aromatic Hydrocarbons

Various aliphatic hydrocarbons, including methane, ethane, propane, hexane, and octane, have been identified indoors. Propane and higher hydrocarbons can depress the central nervous system, and their vapors can cause mild irritation of the mucous membranes (Elkins,

1959). Unsaturated hydrocarbons, such as ethylene and isoprene, are usually weak anesthetics. Cyclic hydrocarbons (cyclohexane, methyl cyclohexane, and turpentines) typically have the same effects as the aliphatic hydrocarbons.

Aromatic hydrocarbons, such as benzene, toluene, ethylbenzene, xylenes, styrene, and naphthalene, are not only strong irritants to mucous membranes, but they can also cause pulmonary edema, pneumonitis, and irritation of the eyes, skin, and respiratory system. Polycyclic aromatic hydrocarbons have been related to the same type of chemicals found in the workplace that are known to produce skin cancer (NAS, 1972). Styrene also causes neurotoxic and behavioral effects, such as fatigue, headache, and memory loss and is suspected of being a carcinogen for humans. Benzene has long been suspected of being a human carcinogen, and recent studies appear to confirm that suspicion.

Chlorinated Hydrocarbons and Alcohols

Chlorinated hydrocarbons are frequently used as solvents and as monomers in the manufacture of a number of plastic materials. The high volatility of the compounds leads to substantial danger of exposure to humans by inhalation. A considerable quantity of laboratory animal exposure data as well as field data on humans are available. Henschler (1990) summarized the historic evaluation of some of these chlorinated hydrocarbons and observed that over a 30-year time period the exposure limits have been reduced by a factor of 2.5 for methyl chloroform and by a factor of 10 for chloroform. The associated health effects are continuously being evaluated, and the exposure limits are being appropriately modified. In addition to being irritants to the eyes, skin, and respiratory system, chlorinated hydrocarbons can cause functional and destructive damage to the liver and kidneys (Anderson and Scott, 1981). A number of these chemicals are now suspected of being carcinogenic to humans (see Table 3.5).

Phenol, cresol, pentachlorophenol (PCP), and other phenolic compounds create adverse physiological problems, including vomiting, difficulty in swallowing, diarrhea, tremors, convulsions, and headaches. Prolonged exposure can cause extensive damage to the liver and kidneys. Although alcohols (methanol, ethanol, propanol, and others) are known for their narcotic effects, they can induce many of the same symptoms as the phenolic compounds. The ingestion of methanol, in particular, can damage the optic nerve and even cause blindness.

Mixture of VOCs

Dose-response data for most organics at indoor concentrations are not available. Furthermore, synergistic interactions of pollutants among themselves may complicate the interpretation of data obtained from exposure to certain chemicals. Several studies of employees in office buildings, particularly in those referred to as energy-efficient buildings, demonstrated a direct relationship between a variety of health problems and toxic indoor chemicals. Symptoms associated with these health problems are collectively referred to as *sick building syndrome* (SBS) or *tight building syndrome* (TBS) and are summarized in Table

TABLE 3.5 Changing Exposure Limits of Chlorinated Hydrocarbons

Pollutants	*German MAK List (concentration in ppm)*				
Carbon tetrachloride	25	10			B
Chloroform	100	50		B, 10	
Methylene chloride	500			200	100
Methyl chloride	50				B
Vinyl chloride	500		100	B, A2, A1	
Trichloroethylene	200	100	50	B	
Tetrachloroethane	200	100			50
1,2-Dichloroethane	100		20	B	
Methyl chloroform	500	200			
	1958	1960	1970	1980	1990

A1, human carcinogen; A2, animal carcinogen; B, suspected as human carcinogen.
Source: Henschler, 1990.

3.6. Illnesses that result from exposure to a sick building are generally denoted as *building-related illnesses*.

Most of the information on building-related illnesses was derived from studies conducted by federal and state agencies on health hazards, rather than from epidemiological studies (Samet et al., 1987). Evaluations are typically made on the basis of a questionnaire completed by an individual. As noted by Hedge (1990), who compared eight sets of such questionnaires, there is considerable variation in the range of symptoms covered by these questionnaires, and comparisons of symptom prevalence between studies may be misleading.

The National Institute of Occupational Safety and Health (NIOSH) evaluated several commercial buildings for indoor air quality. Indoor contaminants were found to be the leading cause of building-related illnesses in 15% of the 529 buildings that were studied (Seitz, 1990). These cases were the results of exposure to contamination from copy machines, tobacco smoke, carbonless copy paper, and many others. The emanation of chemicals from building materials such as formaldehyde from carpets, insulation, and furnishings are considered to be responsible in 4% of the cases, and outdoor air (automobile exhaust and dust)

TABLE 3.6 Symptoms Related to Sick Building Syndrome

Irritation of eyes, nose, and throat
Dry mucous membranes and skin
Erythema
Mental fatigue, headaches
Airway infections, coughing
Hoarseness, wheezing
Unspecific hypersensitivity reactions
Nausea, dizziness

Source: Mølhave, 1984.

is responsible in 10% of the cases. One of the most comprehensive studies relating volatile organic compounds to sick building syndrome was conducted by Mølhave et al. (1984) in Denmark. They exposed 62 healthy volunteers, all of whom had previously experienced symptoms related to sick buildings, to various concentrations of a mixture of 22 chemicals (see Table 3.1). The total organic concentrations used were 0, 5, and 25 mg/m^3. These levels corresponded to clean air in a normal new residence and to very contaminated indoor air. They also used a set of questionnaires to evaluate subject response and concluded that, when exposed to either 5 or 25 mg/m3, the reported occurrence of dry mucous membrane irritations, which included eyes, nose, and throat irritations, increased significantly.

Exposure Limits

A large body of literature is now available that relates to the health effects posed by organic chemicals, but many of the findings remain controversial. A number of review articles and several reports and books provide further information for the interested reader and are referenced in a literature survey conducted by Hines et al. (1991). This survey noted that no single indoor air quality guideline has been agreed on by the several agencies interested in the problem. For example, using the threshold limit values (TLV) of toxic chemicals suggested by the American Conference of Governmental Industrial Hygienists (ACGIH) as a basis, the Occupational Safety and Health Association (OSHA) has established one-tenth of the TLV as a guideline for indoor air contaminants. The American Society of Heating, Refrigerating and Air-Conditioning Engineers, Inc. (ASHRAE) Standard 62-1981R committee recommended as a preliminary guideline for residential, office, or retail spaces a concentration of one-tenth of the TLV. The recommendation is based on the assumption that this level would not result in complaints from a nonoccupational population from residential and office environments. If a concentration of one-tenth of the TLV cannot be provided, the standard recommends that the designer or building operator seek expert help in evaluating what level of such a chemical or combination of chemicals would be acceptable.

On the basis of a study by Mølhave et al. (1984), the American Industrial Hygiene Association (AIHA) has adopted a guideline of 5 mg/m^3 for total VOCs. Mølhave et al. found that the majority of people suffered from mucous membrane irritation at this level. Thus, if the one-tenth of the TLV guideline is followed, the majority of people would have some symptoms related to mucous irritation. It also has been suggested that the CO_2 level be used as a criterion for setting the ventilation rate and therefore the indoor air quality. ASHRAE Standard 62-1981 specifies that sufficient outdoor air must be provided to ensure that the level of CO2 indoors does not exceed 2500 ppm. In the revised ASHRAE Standard 62-1981R, it is recommended that the air exchange rate be increased to 15 ft^3/min/person in order to reduce the CO_2 level below 1000 ppm.

3.3 SOURCES AND INDOOR CONCENTRATIONS

Numerous sources of chemical pollutants have been identified in indoor environments. Pollutant types and their concentrations depend, however, on the indoor environment itself. Emissions of pollutants from sources such as carpets and painted walls are continuous, but the rate may slowly decrease over time. Emissions from other sources, such as photocopy machines, laser printers, and FAX machines, are intermittent and occur only during the use of the related equipment.

Formaldehyde

Formaldehyde is a major VOC pollutant. The primary sources of formaldehyde are building materials and furnishings such as particleboard, medium-density fiberboard, plywood, carpet, furniture, tobacco smoke, and to a lesser extent combustion appliances. Formaldehyde is an essential component in the manufacture of several resins: urea formaldehyde (UF), phenolic, acetal, and melamine. It is also used in the production of urea formaldehyde foam insulation (UFFI). This foam, which is created from a mixture of UF resin, a surfactant, and an acid catalyst, is injected into wall cavities to supplement the insulation in existing homes and buildings. The foam hardens within a few minutes and dries completely within a few days. If the mixture is not properly formulated and the proper technique is not used when injecting the foam, the UF resin can break down easily in the presence of heat and moisture, liberating free formaldehyde that eventually diffuses into the living and working areas of buildings.

Formaldehyde-based resins also have superb bonding properties and are inexpensive. As a result, they are used extensively as adhesives in the manufacture of a variety of household products. Examples of formaldehyde use in the manufacture of selected products are listed in Table 3.7. As noted earlier, one prominent use of UF resin is in the manufacture of particleboard, plywood, and chipboard. Several thin sheets of wood are glued together by the UF resin to produce plywood; whereas, particleboard and chipboard are manufactured by mixing wood chips and sawdust with the resin and then pressing the mixture into its final form at a high temperature. Andersen et al. (1975) found that temperature and humidity affected the emission rate of formaldehyde from the particle board. Figure 3.1 shows that the emission rate increases with both temperature and humidity. Both UF and melamine resins are used widely as a base and for the finish coatings for wood products. These wood products are used for subfloors, cabinets, partition walls, paneling, and furniture and can release formaldehyde continuously for several years, although the emission rate decreases slowly with time. The formaldehyde emission rates for a number of materials are listed in Table 3.8.

TABLE 3.7 Formaldehyde Uses and Potential Indoor Sources

Products	*Examples*
Paper products	Grocery bags, wax paper, facial tissues, paper towels, disposable sanitary products
Stiffeners, wrinkle resisters, and water repellents	Floor covering (rugs, linoleum, varnishes, plastics), carpet backings, adhesive binders, fire retardants, permanent-press clothes
Insulation	Urea formaldehyde foam insulation (UFFI)
Combustion devices	Natural gas, kerosene, tobacco smoke
Pressed-wood products	Plywood, particle board, decorative paneling
Other	Cosmetics, deodorants, shampoos, fabric dyes, inks, disinfectants

Formaldehyde is also produced during the operation of combustion appliances, such as wood and gas stoves, and when smoking. Formaldehyde is not a constituent of natural gas, but significant quantities can be produced through complex chemical reactions if appliances are not properly adjusted. Formaldehyde emission rates from selected appliances and from tobacco smoke are given in Table 3.9. The concentration level of formaldehyde indoors can vary widely among mobile homes, conventional homes, homes containing urea formaldehyde foam insulation, and office buildings. A wide range of concentrations, varying from 0.01 to 11 ppm, has been reported. Indoor concentrations depend on several factors, including the type and quantity of the building materials used, the operation of combustion appliances, and occupant smoking. The concentration is highest in mobile homes because of the extensive use of wood products. In homes, the primary sources are urea formaldehyde foam insulation and, to some extent, other building materials and combustion appliances. In office buildings, the primary sources are wood furnishings.

Formaldehyde emission rates can fluctuate significantly with seasonal variations of temperature, relative humidity, and the moisture content of the source material. Godish and Rouch (1986) reported a two-fold increase in the formaldehyde concentration for a 5° to 6°C temperature increase. A combination of temperature and relative humidity may increase the formaldehyde level indoors by as much as a factor of five. Garry (1980) found that formaldehyde levels in homes were the highest during the summer months and were the lowest during the winter. The seasonal variation was associated with the direct impingement of sunlight on a urea formaldehyde foam-insulated wall, which led to the breakdown of the foam resin and to the subsequent release of formaldehyde. The EPA collected data on indoor formaldehyde concentrations for different types of residences. The average formaldehyde concentrations likely to be found indoors are shown in Figure 3.2.

Formaldehyde concentrations will decrease with time as products containing formaldehyde age and cure. A 9-month average of formaldehyde concentration in 40 east Tennessee homes indicated that in conventional homes which were less than five years old, the average formaldehyde concentration was 0.08 ppm. In homes that were more than five years old, the average concentration was 0.04 ppm. Similar results were also reported for

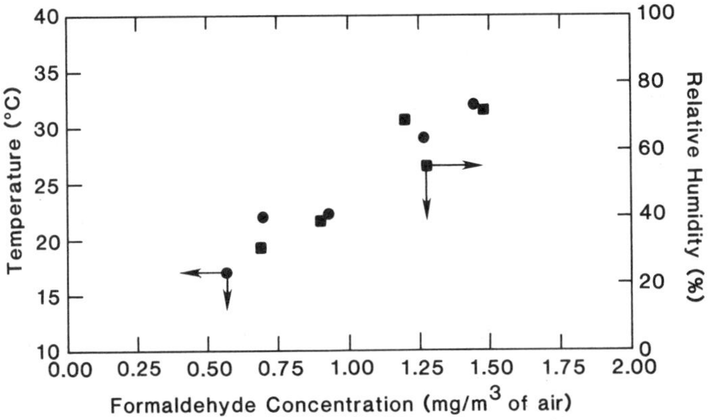

Figure 3.1 *Effect of temperature and humidity on the equilibrium concentration of formaldehyde from a particle board. (Source: Andersen et al., 1975.)*

TABLE 3.8 Formaldehyde Emission Rates of Several Building Materials

Material	*Emission Rates ($\mu g/m^2$ day)*
Medium-density fiberboard	17,600–55,000
Hardwood plywood paneling	1500–36,000
Particle board	2000–25,000
Urea formaldehyde foam insulation	1200–19,200
Softwood plywood	240–720
Paper products	260–680
Fiber–glass products	400–470
Clothing	35–570
Resilient flooring	<240
Carpeting	0–65
Upholstery fabric	0–7

Source: NAS, 1981; Pickrell, 1983; Matthews, 1985.

mobile homes. The average formaldehyde concentration in mobile homes that were two to three years old was 0.23 ppm, while that in unoccupied homes, which were one to two months old, was 0.8 ppm. On the basis of this observation, Hawthorne and Matthews (1985) concluded that most materials release one-half of their formaldehyde content during the first two to five years.

Volatile Organic Compounds

Volatile organic compounds other than formaldehyde are released indoors from combustion by-products, cooking, construction materials, building decorations, paints, varnishes, solvents, adhesives, furnishings, office equipment, bioeffluents, consumer products, home

TABLE 3.9 Formaldehyde Concentrations in Homes Resulting from Various Combustion Appliances[a]

	Winter			*Summer*		
Combustion Sources	*No. of Homes*	*Mean (ppm)*	*Std. Dev.*	*No. of Homes*	*Mean (ppm)*	*Std. Dev.*
None	31	0.046	0.035	34	0.059	0.043
Woodstove	63	0.053	0.032	62	0.082	0.038
Kerosene heater	39	0.055	0.028	42	0.067	0.033
Kerosene heater and woodstove	17	0.050	0.026	13	0.075	0.032
Smoker	33	0.046	0.025	25	0.055	0.031
Smoker and woodstove	41	0.048	0.024	39	0.068	0.032
Smoker and kerosene heater	32	0.042	0.018	32	0.054	0.026
Smoker, woodstove, and kerosene heater	25	0.047	0.027	22	0.060	0.025

a, Seasonal arithmetic means.

Source: Hawthorne and Matthews, 1985; Apte and Traynor, 1986.

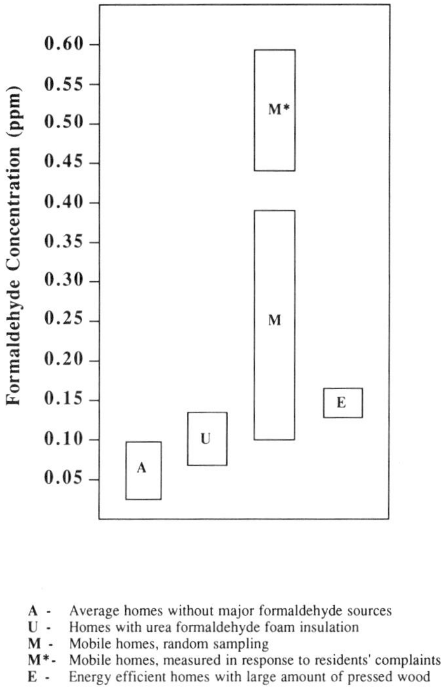

Figure 3.2 *Average formaldehyde concentrations in homes. (*Source: *EPA, 1985.)*

and personal health care products, and pesticides. The relative contribution of these products to indoor air quality depends on factors such as personal habits, the activities taking place in the particular indoor environment, the number of occupants, and the frequency or volume of use of the products. Therefore, the types of organics present and their indoor concentrations vary widely.

VOCs from combustion products. Organic compounds that are emitted as a result of the incomplete combustion of natural gas in gas stoves include aliphatic hydrocarbons such as methane, ethane, propane, and hexane. Cooking with oils and butter also introduces organic compounds indoors. The burning of wood, on the other hand, releases polynuclear aromatic hydrocarbons (PAH), whose emissions depend on stove characteristics. The concentrations of PAHs from airtight and nonairtight woodstoves are shown in Figure 3.3. Tobacco smoking usually contributes only a small fraction of the volatile organics found indoors. Depending on the number of cigarettes smoked, however, the contribution can be significant, particularly in public places such as restaurants, cafeterias, and airports. The types of organic compounds contained in tobacco smoke are shown in Table 3.10. There may be some variation in the amounts released because of differences in cigarette brands, but the basic types of organics are the same.

VOCs from building materials. Studies in laboratories, homes, and other buildings have been conducted to identify the sources and concentrations of the VOCs. Mølhave

(1982) measured the emission rates from 42 commonly used building materials and the total concentrations of the VOCs in an 0.8-m3 test chamber. As pointed out by Mølhave, emission rates differ significantly from one product to another, which in turn can influence indoor concentrations. The total concentrations of pollutants and emission rates of some selected building materials are shown in Table 3.11.

Most recent studies have been concerned with measuring the emissions of selected organics from building materials. Ozkaynak et al. (1987) calculated the emission rates of 10 common VOCs that are present in household products, such as fabrics, adhesive tapes, deodorants, ink pens, paper, photo equipment, and electrical equipment, by using their own test results and information from a database developed for NASA's space shuttle program. The VOC emission rates of 5000 materials are listed in the NASA database. The emissions of the 10 VOCs reported by Ozkaynak et al. are presented in Table 3.12.

Basically, every household product contains organic compounds that are released indoors during their use. It is now customary to analyze every building material during the investigation of the air quality in a sick building. The objective is to identify the source of the chemicals so that the building material can be removed or replaced by an equivalent product that has a lower emission rate and is less harmful. Gebefuegi and Korte (1990) investigated the sources of organics in an office building and analyzed a variety of cleaning fluids, including floor wax and carpet and glass cleansers. The major organic pollutants identified in these cleansers are shown in Table 3.13. The elevated level of limonene was attributed to the daily use of these cleansers.

Depending on the chemical formulation of the product, adhesives that are commonly used to seal fresh air infiltration paths or to glue carpets and floor tiles can be significant

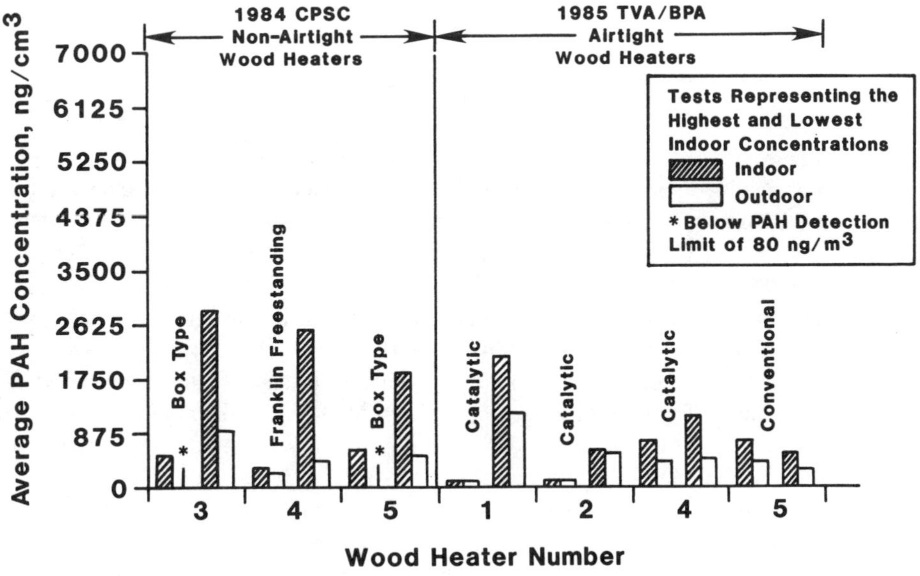

Figure 3.3 *PAH concentrations resulting from combustion in airtight and non-airtight woodstoves. (Source: Knight et al., 1986.)*

TABLE 3.10 Emission of Chemicals from Mainstream and Sidestream Smoke

Chemicals	*Mainstream (μg/cigarette)*	*Sidestream (μg/cigarette)*
Gas and Vapor Phase		
Carbon monoxide	1000–20,000	25,000–50,000
Carbon dioxide	20,000–60,000	160,000–480,000
Acetaldehyde	18–1400	40–3100
Hydrogen cyanide	430	110
Methyl chloride	650	1300
Acetone	100–600	250–1500
Ammonia	10–150	980–150,000
Pyridine	9–93	90–930
Acrolein	25–140	55–300
Nitric oxide	10–570	2300
Nitrogen dioxide	0.5–30	625
Formaldehyde	20–90	1300
Dimethylnitrosamine	10–65	520–3300
Nitrosopyrolidine	10–35	270–945
Particulates		
Total suspended particles	36,200	25,800
Nicotine	100–2500	2700– 6750
Total phenols	228	603
Pyrene	50–200	180–420
Benzo (a) pyrene	20–40	68–136
Naphthalene	2.8	4.0
Methyl naphthalene	2.2	60
Aniline	0.36	16.8
Nitrosonornicotine	0.1–0.55	0.5–2.5

Sources: Wakeham, 1972; HEW, 1979; Hoegg, 1972.

sources of volatile organics. Adhesives can be either solvent or water based. Solvent-based products incorporate either toluene, methylene chloride, or cyclohexane as the base chemical to dissolve various other ingredients when formulating the adhesive, whereas water is used as the solvent in the water-based adhesives. Water-based adhesives were found to emit a greater quantity of VOCs than solvent-based products (Girman et al., 1986). The emission rates of eight adhesives, which included five solvent-based and three water-based products, were measured by Girman and his co-workers, who observed that the water-based adhesives not only took longer to dry (approximately 14 days) than solvent-based adhesives (9 to 13 days). The total quantity of VOCs emitted was in the range of 610 to 780 mg/g·h, compared to a range of 0.59 to 48 mg/g·h for the solvent-based adhesives. The water-based adhesives emitted mostly alkanes, whereas solvent-based products released aromatic hydrocarbons.

VOCs from carpets and paints. The two predominant sources of organic compounds found indoors are paints and carpets. About 95% of all interior surfaces are either painted or covered with carpet, and extreme care must be taken in the selection of the covering to be used. The emission of VOCs from water-based paints is significantly lower than

TABLE 3.11 Total Concentration and Emission Rates of VOCs from Selected Building Materials

Material	*Description*	*Concentration* *(mg/m^3)*	*Specific Emission* *($mg/m^2 \cdot h$)*
Particleboard	Urea formaldehyde glue	1.56	0.12
Particleboard	Urea formaldehyde glue	1.73	0.13
Plasterboard	12-mm Paper board	0.66	0.026
Calcium-silicate board	22.8-mm Board	1.69	0.064
Sealing agent	Plastic compound	169.00	72.0
Sealing agent	Plastic silicone compound	77.90	26.0
Sealing agent	Putty, strips 5 × 7 mm	1.38	0.34
Insulation batch	Mineral wool	0.38	0.012
Particleboard	Urea formaldehyde glue	3.56	0.14
Plywood lining	Teak	1.07	0.044
Woodfiber board	12-mm Board	2.96	0.12
Tightening filet	Neoprene/polyethylene	0.81	0.16
Tightening filet	Plasticized PVC	1.05	0.056
Felt carpet	Synthetic fiber/plastic	3.15	0.11
Felt carpet	Synthetic fiber	1.95	0.08
Wallpaper	Vinyl and paper	0.95	0.04
Wallpaper	Vinyl and glass fiber	7.18	0.30
Wallpaper	Printed paper	0.74	0.031
Floor covering	Linoleum	5.19	0.22
Wall and floor glue	Water-based EVA	1410.00	271.0
Texture glue	Water-based EVA	9.81	2.1
Filler	PVA glue/cement	57.80	10.2
Filler	Sand, cement	3.95	0.73
Wall covering	Hessian	0.09	0.0054
Floor covering	Synthetic fiber/PVC	1.62	0.12
Floor covering	Rubber	28.40	1.4
Wallpaper	PVC foam	5.50	0.23
Tightening fillet	Heat-expanding neoprene	0.35	0.016
Fiberboard	Glass fiber/polyester	0.40	0.017
Paint	Acryl latex	2.00	0.43
Floor varnish	Epoxy, clear	5.45	1.3
Floor varnish	Isocyanate	28.90	4.7
Floor varnish	Acid hardener	3.50	0.83
Wall covering	PVC	2.43	0.10
Laminated board	Plastic	0.01	0.0004
Floor covering	Soft plastic	3.84	0.59
Insulating foam	Polystyrene	40.50	1.4
Insulating foam	Polyurethane	3.59	0.12
Floor covering	Homogeneous PVC	54.80	2.3
Floor/wall covering	Textile	39.60	1.6
Floor/wall covering	Textile	1.98	0.083
Cement flag	Concrete	1.45	0.073

Source: Mølhave, 1982.

TABLE 3.12 Emissions of Organic Chemicals from Household Products

	Emission (μg/g)								
Compound	*Adhesives*	*Coatings*	*Fabrics*	*Foam*	*Lubricants*	*Paints*	*Rubber*	*Tape*	*Cosmetics*
1,2-Dichloroethane	0.80	—	—	0.75	—	—	—	3.25	—
Benzene	0.90	0.60	—	0.70	0.20	0.90	0.10	0.69	—
Carbon tetrachloride	1.00	—	—	0.18	—	—	4.20	0.75	—
Chloroform	0.15	—	0.10	0.04	0.20	—	0.90	0.05	—
Ethyl benzene	—	—	—	—	—	527.80	—	0.20	—
Limonene	—	—	—	—	—	—	—	—	—
Methyl chloroform	0.40	0.20	0.07	1.00	0.50	—	0.10	0.10	0.20
Styrene	0.17	5.20	—	0.02	12.54	33.50	0.15	0.10	1.10
Tetrachloroethylene	0.60	—	0.30	65.0	0.10	—	0.20	0.08	0.70
Trichloroethylene	0.30	0.09	0.03	0.10	0.10	—	0.07	0.09	1.90

TABLE 3.12 Continued

	Emission (µg/g)							
	Deodorants	*Health and Beauty Aids*	*Electrical Equipment*	*Miscellaneous Housewares*	*Ink Pens*	*Paper Equipment*	*Photo Film*	*Photo Equipment*
1,2-Dichloroethane	—	—	0.06	—	—	—	—	
Benzene	—	1.85	0.02	1.10	0.40	0.03	1.51	0.04
Carbon tetrachloride	—	—	0.00	0.04	0.20	—	2.50	—
Chloroform	—	—	0.23	4.85	10.00	0.10	2.50	0.10
Ethyl benzene	—	—	0.80	—	—	—	10.50	0.13
Limonene	0.40	1.00	—	1.80	—	—	—	—
Methyl chloroform	—	0.01	0.03	0.19	0.10	0.26	0.08	1.90
Styrene	0.15	0.17	0.05	0.02	0.30	—	0.04	0.10
Tetrachloroethylene	—	—	0.05	—	2.00	0.42	—	—
Trichloroethylene	—	0.11	0.01	0.06	0.07	0.10	0.03	0.13

Source: Ozkaynak et al., 1987.

TABLE 3.13 Volatile Organic Compounds in Cleaning Fluids

Product	*Main Pollutants*
Floor wax	1,4-Diethylbenzene, butylbenzene, decane, 1,2,5-trimethylbenzene, 1-nonene, ethylbenzene, xylene, limonene
Carpet cleanser	Ammonia
Ceiling cleanser	Limonene, *p*-cumene, undecane, α-pinene
Ceramic floor cleanser	Limonene, *p*-cumene
Stone floor cleanser	Heptane, undecane, nonane, decane
Desk cleanser	Limonene, undecane, *p*-cumene
Glass cleanser	Heptane, ammonia

Source: Gebefuegi and Korte, 1990.

from solvent-based paints. A water-based paint normally contains 30% to 40% water, 40% to 50% nonvolatile matter, and only 6% to 10% solvents. Although the concentration of organic compounds near the painted surface can be in the range of 200 to 300 mg/m^3 immediately after painting, the concentration drops to background levels after about two weeks. Bayer and Papanicolopoulos (1990) analyzed 30 different carpet samples in an environmental chamber. A list of the VOCs most frequently identified in carpets is given in Table 3.14. In recent years the number of studies to identify and quantify the types of volatile organic compounds present in building materials and household products has increased. The results of these studies are summarized in the appendix.

The Research Triangle Institute, working with the EPA, has developed a PC-based database for indoor air pollution sources (DIAPS). The database provides emission rates of numerous pollutants, the total emission from the source itself, and the experimental conditions employed during the tests (Lawless et al., 1990).

VOCs from aerosol cans. A number of products, such as air fresheners, antiperspirants, hairsprays, and dusting aids, are available in spray cans. These sources deliver VOCs indoors in the form of aerosols. The aerosol propellants include propane, isobutane, trichlorofluoromethane, and dichlorodifluoromethane. Mokler et al. (1979a, 1979b) measured the total aerosol concentration and size distribution from several spray cans, and found that spray paints and antiperspirants had the greatest aerosol content (see Table 3.15). Immediately after spraying, the concentration of an individual component can increase up to several hundred parts per million. The concentrations of vinylchloride and Freon 12 from insect sprays were 466 and 380 ppm, respectively, in the breathing zone, even after one minute following release of the spray. The spray was released over a 30-second period.

VOCs from humans. Humans also constitute a major source of organic and inorganic compound emissions. Table 3.16 lists the emission rates of 12 organic and 4 inorganic chemicals that were found by Wang (1975) in a college classroom during regular class periods and during examinations. Human breath generally contains an average of

TABLE 3.14 Volatile Organic Compounds Frequently Detected in Carpets

Benzene	Ethylmethylbenzene
4-Phenylcyclohexene	Trimethylbenzene
Ethanol	Chlorobenzene
Carbon disulfide	Chloroform
Acetone	Benzaldehyde
Ethyl acetate	Styrene
Ethylbenzene	Undecane
Methylene chloride	Xylenes
Tetrachloroethene	Trichloroethene
Toluene	Phenol
1,1,1-Trichloroethane	Dimethylpentane
Hexane	Butylbenzyl phthalate
Octanal	1,4-Dioxane
Acetaldehyde	Pentanal
Methylcyclopentanol	Methylcyclopentane
Hexene	

1200 $\mu g/m^3$ of acetone and 240 mg/m^3 of ethanol. As a result, the concentration of these two chemicals increased by a factor of two when students were present in the classroom.

VOCs from pesticides. Pesticides and insecticides sprayed in homes can be a potential source of organic pollutants, although they are generally present at very low concentrations. Still, illnesses of residents from the exposure to pesticides have been frequently reported to local hospitals. The EPA initiated a nonoccupational pesticide exposure study (NOPES) to address the variability of exposure and associated health effects (Immerman and Schaum, 1990). Approximately, 32 household pesticides were evaluated in this study. Although a number of pesticides have been banned, some can still be detected in indoor air samples.

One of the more frequently identified pesticides indoors is chlordane. Air samples from the living area of 157 residences in New Jersey had an average chlordane content of 1.72 mg/m^3 (Fenske and Sternbach, 1987), and 50% of the homes had chlordane levels above the interim guideline of 5 mg/m^3 set by the National Academy of Science. Professional sprayers of pesticides or insecticides are usually aware of the associated health hazards. Standard guidelines as well as protective equipment are available to minimize exposure. However, the greatest health threat is probably to the occupants of homes or to workers in office buildings where these pesticides are sprayed. A significant quantity of these chemicals remains airborne after their application, even after one to two days, particularly if the space is not well ventilated. These chemicals then settle on floors, carpets, furniture, or any exposed surface. As a result, the surface concentration is expected to be higher after one to two days than it is immediately after spraying (see Table 3.17).

In commercial buildings, spraying is generally done on a Friday or over the weekend. Since the HVAC systems in many buildings are shut down over the weekend to save en-

TABLE 3.15 Total Concentration of Aerosols in Household Products and Their Size Distribution

Product	*Total Aerosol Concentration (mg/m^3)*	*Mass Median Aerodynamic Diameter (μm)*
Air freshener	27	5.2–6.3
Antiperspirant	246	5.9–7.3
Dusting aid	86	6.4–7.5
Fabric protector	9	2.6–4.0
Furniture wax	22	3.0–4.9
Hair spray	30	5.8–6.4
Paint	189	7.2–8.7
Wood panel wax	15	1.4–1.5

Source: Mokler et al., 1979a.

TABLE 3.16 Emission Rates of Organics by Humans

		Emission Rate (mg/day per person)	
Organic Compounds	*Typical Concentration, 389 People in Class (ppb)*	*Lecture Class 225 People*	*During Examination*
Acetone	20.6 ± 2.8	50.7 ± 27.3	86.6 ± 42.1
Acetaldehyde	4.2 ± 2.1	6.2 ± 4.5	8.6 ± 4.6
Acetic acid	9.9 ± 1.1	19.9 ± 2.3	26.1 ± 25.1
Allyl alcohol	1.7 ± 1.7	3.6 ± 3.6	6.1 ± 4.4
Amyl alcohol	7.6 ± 7.2	21.9 ± 20.8	20.5 ± 16.5
Butyric acid	15.1 ± 7.3	44.6 ± 21.5	59.4 ± 52.5
Diethylketone	5.7 ± 5.0	20.8 ± 11.4	11.0 ± 7.7
Ethyl acetate	8.6 ± 2.6	25.4 ± 4.8	12.7 ± 15.4
Ethyl alcohol	22.8 ± 10.0	44.7 ± 21.5	109 ± 31.5
Methyl alcohol	54.8 ± 29.3	74.4 ± 5.0	57.8 ± 6.3
Phenol	4.6 ± 1.9	9.5 ± 1.5	8.7 ± 5.3
Toluene	1.8 ± 1.7	7.4 ± 4.9	8.0
Carbon monoxide		48,400 ± 1200	
Ammonia		32.2 ± 5.0	
Hydrogen sulfide		2.73 ± 1.32	2.96 ± 0.68
Carbon dioxide		642,000 ± 34,000	930,000 ± 52,000

Source: Wang, 1975.

ergy, workers are likely to be exposed to high concentrations of these pesticides and insecticides following their return to work. In homes, the situation may be even worse; occupants typically return to their homes within four to five hours after application of the pesticide. Chemicals that settle on surfaces can be easily transferred to the body through ingestion and skin absorption. Small children may be more vulnerable to exposure since they put their hands to their mouths quite frequently.

VOCs from water. Water supplies that are to be used by the general public usually undergo chlorine treatment, resulting in a number of chlorinated hydrocarbons being formed during the chlorination process as by-products. Some of the common by-products are chloroform, trichloroethylene, carbon tetrachloride, 1, 2, 3-trichloropropane, and tetrachloroethylene. Volatilization of these organics from water into indoor air takes place anytime water is used. Additional exposure to VOCs from water can take place by inhalation and direct absorption through the skin when showering, bathing, or washing dishes. The concentration of the VOCs in water, the water temperature, and the duration of use are the primary factors that determine the extent to which an individual is exposed. The concentration of chloroform in exhaled air prior to a shower has been shown to be less than the minimum detection level of 0.9 mg/m^3, but ranged from 3.8 to 13.0 mg/m^3 after a normal shower (Weisel et al., 1990). It has been estimated that chloroform dose and cancer risk from a 10-minute shower with water containing 24.5 mg/L of choloroform is equal to or greater than that from the daily ingestion of the same water.

Indoor concentrations. In the late 1970s the measurement of indoor concentrations of individual volatile organic compounds was confined primarily to public buildings because of better access and convenience. Due to the lack of a standard protocol, a wide variation was observed in the types of organics and in total concentrations. Mølhave (1979) targeted 29 chemicals in 14 office buildings in Denmark. The majority of the organics identified were alkylbenzenes and had concentrations in the range of 0.03 to 2.8 mg/m^3.

Straight chain and cyclic alkanes were targeted for measurement by Miksch et al. (1982), who investigated several public buildings in California. Their study showed that aliphatic hydrocarbons dominated, followed by aromatic and chlorinated hydrocarbons. Air samples from four office buildings in California showed very low concentrations of organics, in the range of 3 to 320 $\mu g/m^3$.

The EPA investigated the indoor air quality in 10 public buildings (Sheldon et al., 1989). Initially, four buildings were selected with the intent of identifying the VOCs in the air samples and the emissions from the building materials. Approximately 500 different types of VOCs were identified in these buildings. The indoor concentrations of more than 100 VOCs were several times higher than outdoors. Thirty-two VOCs were selected for study in the remaining six buildings. The indoor concentrations of these 32 VOCs are summarized in Table 3.18. The VOC concentrations in the newer buildings were extremely high during the first two weeks of the test but decreased slowly over time.

Identification of VOCs and the measurement of their concentrations in homes and residences were initiated by Mølhave and Moller (1979). They characterized 7 unoccupied new homes and 39 old apartments and found that the concentration of individual compo-

TABLE 3.17 Concentrations of Various Pesticides in Air and on Surfaces after Spraying

	Concentration in Air (μg/m^3)				Concentration on Furniture and Floor (ng/cm^2)		
	Immediately after Spraying	*2 Hours after Spraying*	*4 Hours after Spraying*	*24 Hours after Spraying*	*1–2 Hours after Spraying*	*24 Hours after Spraying*	*48 Hours after Spraying*
Diazinon							
Office 1	15.0	35.0	27.5	27.5	22.0	15.8	13.3
Office 2	47.0	90.0	156.0	72.0	18.0	16.0	15.9
Office 3	95.0	93.0	162.5	122.0	28.1	32.7	NM
Chlorpyrifos							
Office 4	6.0	3.0	2.0	1.7	<0.3	4.3	5.9
Office 5	9.6	11.0	14.0	14.0	<0.3	4.4	4.2
Office 6	9.6	18.5	27.5	19.0	3.33	3.0	NM
Bendiocarb							
Office 7	2.65	0.7	0.6	0.25	22.3	17.7	NM

NM, concentration was not measured.

Source: Currie et al., 1990.

nents ranged from 0.02 to 19 mg/m^3. The total VOC concentration varied from 0.48 to 18.7 mg/m^3 in the new houses and from 0.02 to 1.7 mg/m3 in the apartments. Aromatic hydrocarbons accounted for over 40% of the total hydrocarbon concentration. The occurrence of VOCs in the air of 500 homes in Germany was investigated by Krause (1987). The average concentration of 57 organics was about 4 mg/m^3. Aromatic hydrocarbons were the most prevalent and accounted for over 45% of the total concentration, followed by the normal alkanes, chlorinated hydrocarbons, and terpenes. Hawthorne et al. (1985) measured indoor concentrations of 18 VOCs in 40 homes. The concentration level of the VOCs during the summer months was almost twice that found during the winter.

The EPA carried out an interesting study as a part of their Total Exposure Assessment Methodology (TEAM) study. The VOCs were measured in the personal air, in exhaled air (breath), and in outdoor air. Personal air is defined as the air that is within the breathing zone of a person. The concentrations of some selected VOCs measured under the TEAM study are presented in Table 3.19. A summary of the sources, their associated health effects, the average concentration range in indoor air, and the concentration level of concern indoors of the most commonly found indoor VOCs is provided in Table 3.20.

3.4 SAMPLING AND MEASUREMENT

Many techniques exist for the sampling and subsequent analysis of volatile organic compounds. In indoor air quality monitoring, three types of analytical instruments are used:

1. Personal monitors: These light weight monitors may be conveniently carried or worn by a person.
2. Portable monitors: These may be moved during sampling.
3. Stationary monitors: These must be operated from a fixed location.

Either active or passive sampling can be used for many of the monitors. During active sampling, contaminated air is drawn into a sensor or collector, whereas in passive sampling air diffuses into the collector. A longer period of time must be allowed for passive sampling to obtain accurate results. The active sampling method can providc a rcal-timc analysis of the sample, depending on the type of instrument used.

Monitoring instruments can be categorized as analyzers or collectors. An analyzer produces an instantaneous signal, generally electrical, that is dependent on the pollutant concentration. A collector is used to gather the pollutant in a suitable medium for subsequent analysis.

Sampling Methods

The collectors used for obtaining air samples are based on one of four principles: (1) air displacement, (2) condensation (3) gas washing or absorption, and (4) adsorption. Air displacement collectors are basically evacuated flasks or plastic bags. Samples are collected by opening the inlet to the contaminant-laden air. The sample is subsequently analyzed using a suitable analytical method.

TABLE 3.18 Average Concentrations of Volatile Organic Compounds in Various Indoor Environments (ng/L)

	Martinsburg, W.V.			*Fairfax, Va.*		*Worcester, Mass.*		*Washington, D.C.*	*Cambridge, Mass.*	*Martinsburg, W.V.*
	Hospital (new)			*Office (new)*		*Nursing Home (new)*		*Office (old)*	*Office/ School (Old)*	*Nursing Home (old)*
Compound	*Trip 1*	*Trip 2*	*Trip 3*	*Trip 1*	*Trip 2*	*Trip 1*	*Trip 2*	*Trip 1*	*Trip 1*	*Trip 1*
Aromatic Hydrocarbons										
Benzene	1.55	2.13	2.88	2.74	4.95	1.70	2.44	5.61	4.50	3.13
m-Xylene	6.88	3.13	9.91	41.53	15.05	23.80	5.33	27.11	8.72	2.95
o-Xylene	3.05	0.92	3.07	18.40	3.67	8.92	2.07	9.28	3.43	0.99
Styrene	1.00	1.07	1.33	2.52	2.87	2.99	1.27	2.36	1.32	1.19
Ethylbenzene	1.94	1.01	2.88	51.26	5.37	7.90	2.15	10.15	2.69	0.97
Isopropylbenzene	0.31	ND	0.33	3.94	0.67	2.27	0.33	0.79	0.36	ND
n-Propylbenzene	ND	ND	ND	5.00	1.13	2.99	0.70	1.22	0.56	ND
m-Ethyltoluene	1.11	0.86	1.48	27.41	5.57	12.38	2.62	6.07	2.62	0.90
o-Ethyltoluene	ND	ND	0.66	8.89	2.08	4.01	0.73	1.60	0.74	ND
1,2,3-Trimethylbenzene	0.63	0.43	0.76	15.10	2.91	5.32	0.72	1.80	1.06	0.79
1,2,4-Trimethylbenzene	1.48	0.98	1.82	73.51	7.27	13.95	2.52	6.28	2.80	0.98
1,3,5-Trimethylbenzene	ND	ND	0.75	16.97	2.75	6.83	0.92	1.83	1.14	ND
Aliphatic Hydrocarbons										
α-Pinene	ND	ND	ND	14.13	24.64	5.19	ND	ND	2.65	ND
n-Decane	3.65	2.73	2.71	436.38	15.24	68.27	3.81	2.26	5.98	1.87
n-Undecane	3.31	1.96	2.34	210.80	33.93	68.51	3.48	2.85	6.77	ND
n-Dodecane	ND	ND	ND	152.69	23.74	31.42	ND	ND	2.23	ND
Chlorinated Hydrocarbons										
1,2-Dichloroethane	2.06	1.49	2.21	ND	4.51	ND	ND	ND	ND	ND
1,1,1-Trichloroethane	4.98	4.50	15.54	12.54	38.85	4.03	1.76	40.98	10.69	3.09
Trichloroethylene	1.05	ND	ND	ND	7.93	2.58	0.57	0.61	10.89	ND
Tetrachloroethylene	ND	ND	1.79	ND	1.64	1.13	0.96	3.97	4.11	0.99
p-Dichlorobenzene	ND	ND	6.61	ND	2.64	2.17	0.62	0.60	ND	ND
Oxygenated Hydrocarbons										
n-Butylacetate	ND	ND	ND	ND	6.34	ND	1.22	2.63	1.48	ND
2-Ethoxyethyl acetate	1.31	ND	ND	ND	2.16	9.58	ND	1.67	ND	ND

ND, not detected

Source: Sheldon et al., 1989.

TABLE 3.19 Concentration of Volatile Organic Compounds in Personal Air, Outdoor Air, and Breath

	Bayonne and Elizabeth, N.J. Fall 1981 Population: 128,000			*Greensboro, N.C. Fall 1981 Population: 130,000*			*Devils Lake, N.D. Fall 1981 Population: 7000*		
Pollutants[a]	*Personal Air (344)*[b]	*Outdoor Air (86)*	*Breath (322)*	*Personal Air (24)*	*Outdoor Air (6)*	*Breath (23)*	*Personal Air (24)*	*Outdoor Air (5)*	*Breath (23)*
1,1,1-Trichloroethane	17.0	4.6	6.6	32.0	60.0	—	25.0	0.05	9.3
Benzene	16.0	7.2	12.0	9.8	0.4	15.0	—	—	56.0
m, *p*-Xylene	16.0	9.0	6.4	6.9	1.5	3.8	6.2	0.05	6.5
Carbon tetrachloride	1.5	0.87	0.69	—	—	—	—	—	—
Trichloroethylene	2.4	1.4	0.88	1.5	0.2	0.54	0.50	0.08	0.89
Tetrachloroethylene	7.4	3.1	6.8	3.3	0.7	3.9	5.0	0.69	8.0
Styrene	1.9	0.66	0.79	1.4	0.1	0.4	—	—	0.52
p-Dichlorobenzene	3.6	1.0	1.3	2.6	0.4	1.2	1.7	0.07	0.82
Elthylbenzene	7.1	3.0	2.9	2.5	0.3	1.5	2.1	0.03	1.4
o-Xylene	5.4	3.0	2.2	3.6	0.6	1.2	2.7	0.05	2.7
Chloroform	3.2	0.63	1.8	1.7	0.14	0.67	0.38	0.05	2.9

[a]concentrations are in $\mu g/m^3$; [b]number of samples.

Source: Wallace et al., 1987.

TABLE 3.20 Sources of Pollutant, Related Health Effects, Average Indoor Concentration, and Acceptable Concentration Level

Pollutants	*Sources*	*Health Effect*	*Average Concentration*	*Acceptable Concentration*
Formaldehyde	Particleboard, carpets, insulation, plywood, ceiling tile, tobacco smoke	Eye, nose, and throat irritation, rashes	0–0.01 ppm	0.05–0.10 ppm
Dichloromethane	Paint strippers and thinners	Nerve disorders, diabetes, cancer, respiratory irritation	0.005–1 mg/m^3	350 mg/m^3
Chloromethane	Solvents, aerosol sprays	Nerve disorders, possible cancer	—	50 ppm (ACGIH)
Trichloromethane	Chlorine-treated water in hot showers	Anesthetic, eye irritation, cancer	0.0001–0.02mg/m^3	270 mg/m^3
Carbontetrachloride	Ink pens, photo equipment, rubber products, tapes	Headache, cancer, dizziness, burning irritation of eyes and lacrimation	—	—
1,1,1-Trichloroethane	Aerosol sprays, cleaning product	Dizziness, cancer, eye irritation, irregular breathing	—	5 ppm
Tetrachloroethylene	Dry-cleaning fluid fumes on clothes, paneling	Nerve disorders, possible cancer, damage to liver and kidneys	0.002–0.05 mg/m^3	335 mg/m^3
1,2-Dichloroethane	Adhesives, foams, tapes, electrical equipment	Eye, nose, and throat irritation, liver and kidney damage	—	—
Styrene	Photocopiers, plastics, and synthetic rubber products	Fatigue, headache, poor memory, skin, eye, and throat irritation	—	5 ppm
Benzene	Smoking, particleboard, latex caulk, paint	Acute and chronic toxic effect on skin, eyes, and mucous irritation, carcinogenic	0.01–0.04 mg/m^3	1 ppm
1,4-Dichlorobenzene	Air fresheners, mothball crystals	Cancer	0.005-0.1 mg/m^3	450 mg/m^3
Benzo-α-pyrene	Woodstoves, tobacco smoke	Lung cancer	9.9 ng/m^3 (smoking) 0.7 ng/m^3 (nonsmoking)	—
Trimethylbenzene	Wallpaper, needle felt, adhesives, floor varnish	Skin irritation, tension, nervousness, pneumonitis	—	—

TABLE 3.20 Continued

Pollutants	*Sources*	*Health Effect*	*Average Concentration*	*Acceptable Concentration*
Toluene	Needle felt, wallpaper, wall covering, paint printers, floor covering, chipboard	Cancer, fatigue, eye irritation, headache	0.015-0.07 mg/m^3	375 mg/m^3
o, *m*, *p*-Xylene	Needle felt, wallpaper, adhesive, floor covering	Eye irritation, dry throat, headache	0.01–0.05 mg/m^3	435 mg/m^3
n-Nonane	Wallpaper, adhesive, floor covering	Nausea	0.001–0.03 mg/m^3	1050 mg/m^3
n-Decane	Wallpaper, painting	—	0.002–0.04 mg/m^3	—
Limonene	Air fresheners, furniture polish	—	0.01–0.1 mg/m^3	560 mg/m^3
Nitro-dimethylamine		—	$(1–50)10^{-6}$ mg/m^3	
iso-Octane	Floor covering, floor varnish	—	—	
Cyclohexane	Floor covering	Mild conjunctivitis, fissured dermatitis	—	30 ppm
Methyl cyclohexane	Floor wax	Skin, nose, and throat irritation, drowsiness, lightheadness	—	40 ppm
Acetone	Particleboard	Eye, nose, and throat irritation, headache, dizziness, dermatitis	—	75 ppm
n-Propanol	Chipboard, particleboard, lacquers, cosmetics, cleaner	Drowsiness, headache, nausea, diarrhea	—	20 ppm
n-Butanol	Floor lacquer, jointing compounds, cement flagstone, paints	Skin, eye, and throat irritation, headache	—	50 ppm (ACGIH)
Methanol	Window cleaner, paints, thinners, cosmetics, adhesives, human breath	Mild dermatitis, prolonged exposure can cause optic nerve damage, blindness	—	200 ppm (ACGIH)
Ethanol	Cleaners, perfumes, polishes, cosmetics	Headache, drowsiness, tremors, fatigue	—	1000 ppm (ACGIH, TWA)

(*continued*)

TABLE 3.20 Continued

Pollutants	*Sources*	*Health Effect*	*Average Concentration*	*Acceptable Concentration*
Hexane	Fuel, aerosol propellants, perfume, cleaners	Dermatitis, irritation of mucous membranes	—	50 ppm
Ethylacetate	Lacquer solvent, perfume	Eye, nose, and throat irritation, narcosis, dermatitis	—	40 ppm
Heptane	Fuel, aerosol propellants, perfume, cleaners	Dermatitis, irritation of mucous membranes, narcosis	—	40 ppm
1,4-Dioxane	Wax, paint, varnish	Eye, nose, and throat irritation	—	2.5 ppm
Butylacetate	Perfume	Headache, drowsiness, eye, and skin irritation	—	15 ppm
Acetaldehyde	Tobacco smoke	Eye and skin irritation, headache	—	10 ppm
Octane	Fuel, aerosol propellants, perfume, cleaners	Chemical pneumonia, eye, nose, and throat irritation, drowsiness, dermatitis	—	30 ppm
Freon	Refrigeration units	Suspected carcinogen[a]	—	50 ppm

[a]Teichman and Woods, 1987; ACGIH, American Conference of Governmental Industrial Hygienists (one-tenth of threshold limit values); TWA, time weighted average.

Source: Hines et al., 1991.

In the condensation type of collector, air is passed through a U-tube or a suitable container, followed by subsequent cooling of the sample below the boiling point of the pollutant. In most cases, a liquid nitrogen cryogenic trap is used. The condensed liquid is then analyzed.

In gas washers (absorbers), air containing the pollutants is bubbled through a liquid contained in an impinger. Pollutants either dissolve in the liquid or react with it. After sampling is completed, an aliquot of the liquid is analyzed to determine the characteristics and concentrations of the pollutants. Distilled water is commonly used for readily soluble gases, but a number of special reagents have been developed for specific pollutants. Often a liquid medium that can absorb all the pollutants of concern is difficult to find. Therefore, this method is used only for a few specific pollutants, such as formaldehyde, phenol, and ethylene oxide. Care must be taken to avoid using a liquid medium that might react with the pollutants and interfere with the subsequent laboratory analysis. The collection efficiency can be improved by either decreasing the flow rate, improving the distribution of the gas phase in the liquid medium, increasing the residence time, or using two or more collectors in series.

Passive samplers are used mostly to test for occupational exposure and can be either diffusion or permeation controlled. In diffusion-controlled samplers, air flows through a small opening into the collecting medium. In permeation-controlled devices, however, a membrane is used to permit only selected pollutants from the air to enter the sampling medium. The difficulties encountered with diffusion-controlled samplers include convective currents, concentration-dependent diffusivities, sampler geometry, sampling time, and concentration variations during the sampling period. Factors of concern with the permeation-controlled samplers are membrane thickness, possible reaction with the pollutants, and moisture sorption. A few passive samplers are available commercially for specific organic compounds. Passive samplers are used to measure a particular pollutant and, therefore, may not be convenient for use in indoor environments.

To fully characterize indoor air, the emission rates from various building materials should also be measured. Emissions from materials are measured either in an environmental chamber or through a head-space analysis. In chamber studies, the material is placed in a chamber where various environmental parameters, including temperature, humidity, and air exchange rates, are controlled precisely. Air samples from the chamber are collected in solid adsorbent tubes for later analysis with a gas chromatograph. In a head-space analysis, small samples of the material are placed in a container, and the head-space gases are collected and analyzed by a gas chromatograph. A variation of this technique is to gently heat the building materials (30° to 34°C) and collect the vapors directly in solid adsorbent cartridges for subsequent analysis. The emission rate of a particular pollutant is usually expressed as $mg/m^2 \cdot h$. The sampling time depends on the emission rate from the material. While a sampling period of several hours may be required for some solids, a sampling time of 30 minutes to 1 hour is usually sufficient for solvent-based materials.

Solid adsorbing media are becoming more popular because of their convenience. With this method, pollutants are allowed to adsorb for a predetermined period of time on a solid that is contained in a tube or canister. The most frequently used solid adsorbent is activated carbon, but silica gel and molecular sieves are also used. Recently, a number of polymeric adsorbents, such as Tenax-GC, Tenax-TA, Ambersorb XE-340, XAD-2, XAD-4, and Chromosorb 101, 102, and 103 (these are trade names), have been suggested and are rapidly replacing activated carbon. The adsorbed organics may be analyzed after they are extracted from the solid adsorbent either by using an organic solvent or by thermally desorbing them directly into a gas chromatograph.

The two key factors for successful use of adsorbent cartridges are (1) the careful cleaning of the sorbent tube to remove background contamination and (2) accurate determination of the sampling rate. Cartridges can be cleaned by heating them at a high temperature under vacuum or by flowing a small quantity of an inert gas (helium or nitrogen) through the cartridge while heating. The flow of inert gas must be continued while cooling the tube. In the second method, cartridges are washed with a solvent and then dried at 100° to 150°C under vacuum. The cleaned tubes must be plugged at both ends until used. The EPA used Tenax cartridges extensively in their TEAM study and in other studies involving the measurement of VOCs. Table 3.21 gives a blank analysis of Tenax cartridges and the recoveries of various VOCs. Surprisingly, for a number of organics the recovery was greater than 100%. Either the polymeric sorbent is acting as a catalyst, converting some of the pollutants to other chemicals, or the polymer itself breaks down during thermal desorption. Therefore, care must be taken when analyzing the data.

The sampling rate depends on the adsorption-desorption characteristics of the organic pollutants on the adsorbent. The objective is to collect as much of the pollutants as possible without saturating the adsorbent and reaching the breakthrough volume for the targeted chemical. Breakthrough volumes are pollutant and sorbent specific and depend on the sampling temperature, sample flow rate, and composition of the gas mixture being sampled. Multicomponent gas adsorption is a complex problem and is currently receiving a great deal of attention.

Several investigators have been used to determine the breakthrough volume in the laboratory, and this information has then been employed in field sampling. When several adsorbable compounds are present in air, initially all of them will be adsorbed if a sufficient amount of adsorbent and an adequate residence time are provided. Depending on the molecular structure of the pollutant, one compound will have a greater affinity for the adsorbent and will be more strongly adsorbed than the other. As sampling progresses, the strongly adsorbed compounds will displace the weakly adsorbed ones from the adsorbent surface. Even when chemicals with similar properties, such as methane, ethane, and propane, are present in a mixture, the high molecular weight compound typically displaces the lower molecular weight material. The adsorption-desorption characteristics can change dramatically with changes of temperature and mixture composition. Therefore, for accurate measurements,

TABLE 3.21 Recoveries, Blank Levels, and Accuracy for Targeted Volatile Organics in a Tenax Cartridge

Chemical	*Recoveries*[a] *(%)*	*Blank Levels*[a] *(ng/Cartridge)*	*Median RSD*[b] *(%)*
Chlorinated Hydrocarbons			
Chloroform	96	5	31
1,2-Dichloroethane	102	ND[c]	7
1,1,1-Trichloroethane	104	12	31
Trichloroethylene	99	2	20
Tetrachloroethylene	95	ND	16
p-Dichlorobenzene	109	2	20
Carbon tetrachloride	97	ND	23
Aromatic Hydrocarbons			
Benzene	80	36	35
Styrene	109	6	40
Ethylbenzene	111	2	23
o-Xylene	104	1	25
m, *p*-Xylene	104	4	25
Aliphatic Hydrocarbons			
Decane	120	ND	30
Undecane	105	4	22
Dodecane	98	2	22
Mean for all chemicals	102 ± 9	5 ± 9	25 ± 8

[a]Median of 18 triplicate determinations.

[b]Relative standard deviation.

[c]Not detected.

Source: Sheldon et al., 1989.

the air samples must be collected at different flow rates; that is, different volumes of air must be sampled.

Analysis Methodologies

Instruments such as colorimeters and spectrophotometers may be used to quantify a specific pollutant. However, identification and quantification of the individual components in a mixture may require a gas chromatograph equipped with a mass selective detector.

Three types of direct-reading colorimetric instruments are in use for the determination of indoor pollutant concentrations. They utilize liquid reagents, chemically treated papers, and glass indicating tubes that contain solid chemicals, which are coated or impregnated with a special reagent. Glass indicating tubes are used primarily for identifying and measuring inorganic gases.

A colorimetric method that employs a liquid reagent is most frequently used for determining the concentration of formaldehyde. When testing for formaldehyde, a measured volume of sample air is drawn through distilled water into an absorber. The air flow rate and sampling time depend on the anticipated formaldehyde concentration. A formaldehyde collection efficiency greater than 95% can be achieved by using two absorbers in series. This technique, which is referred to as the chromotropic acid method, has very little interference from other aldehydes or indoor air pollutants and is used as a standard reference method.

Another method, typically referred to as the acid bleached pararosaniline method, is equivalent to the chromotropic acid method and can also be used for quantification of formaldehyde in air. The pararosaniline method is more than twice as sensitive as the chromotropic acid method and has been shown to have greater reproducibility at low concentration levels (Miksch et al., 1981).

Components in a mixture may be analyzed by using a gas chromatograph (GC) equipped with an appropriate chromatographic column. Various types of detectors are available that can be incorporated in a GC. The detectors most frequently used for analysis include the following:

Thermal conductivity detector (TCD): measures changes in the thermal conductivity and is most suitable for inorganic gases.

Hot wire detector (HWD): also measures changes in the thermal conductivity and is suitable for inorganic gases.

Flame ionization detector (FID): measures differences in flame ionization due to combustion of the sample in hydrogen and is suitable for aliphatic and aromatic hydrocarbons.

Electron capture detector (ECD): measures current flow between two electrodes due to ionization of the gas by a radioactive source and is suitable for chlorinated hydrocarbons.

Flame photometric detector (FPD): measures light emitted from the excited states of sulfur and phosphorus compounds in a hydrogen flame.

Mass selective detector (MSD): suitable for most indoor pollutants. It not only provides a quantitative analysis, but also can be used to identify unknown chemicals in a mixture.

Often a single detector is not capable of quantifying all the components in a mixture, but can be used in series to enhance the capability of the analysis. The most common configuration is a TCD with either an FID, ECD, or FPD in series. For qualitative analysis, the GC response must be calibrated with a mixture of known composition. Because the concentrations of most indoor air pollutants are in the range of a few parts per million, care must be taken when preparing sample gas mixtures and in the calibration of the instruments prior to analysis.

3.5 SPECIFIC CONTROL STRATEGIES

Techniques that are currently used to improve the quality of indoor air may be grouped into three categories: source control, ventilation, and air cleaning.

Source Control

Indoor concentrations of VOCs can be reduced by selecting products that have low emission rates. Although no standard exists, various state governments and private companies are attempting to incorporate design and operational features into the selection of materials and furnishings to improve indoor air quality. As shown in Table 3.11, products that are used to accomplish the same task can have different emission rates.

Building materials should be selected in such a way that indoor formaldehyde concentration remains below 0.05 ppm and the total concentration of volatile organics will not exceed 5 mg/m^3. To meet these criteria, building materials and furnishings should be limited to specific emission rates. Tucker (1990) suggested allowable emission rates of various materials (see Table 3.22) that may help to achieve a total organic concentration in the range of 1 to 5 mg/m^3 of air. This range was selected on the basis of the work of Mølhave (1979), who suggested that human mucous membrane irritation may occur above the upper value.

Bayer and Black (1988) investigated the indoor air quality of a new office building that employed new construction materials and was built using new building techniques. The office floor was constructed of hardwood covered in places with wool rugs, and there were no carpets or office partitions. Furnishings and furniture were primarily of hardwood and/or metal construction; the use of paints and wallpapers was minimized throughout the building. The indoor air quality of this special building was compared with that of two other similar buildings that had been constructed with traditional materials. The concentrations of 23 organic chemicals were measured in the three buildings. The total VOC concentration in the special building was 137 μg/m^3, compared to 264 and 821 μg/m^3 in the other two buildings. It was noted that materials alone may not be adequate to keep the total organic concentration below the targeted level; the building must also be properly ventilated.

A substantial reduction in the indoor formaldehyde concentration can be achieved by source control and the application of a coating as a physical barrier. Godish and Rouch

(1984) found from their study in a house with a particleboard floor and plywood paneling that formaldehyde levels decreased by 49% and 29% when 75% of the panel board and 50% of the floor, respectively, were removed from the house. However, in other instances, the removal of the urea formaldehyde foam insulation from houses had little effect on indoor formaldehyde levels. Using a mathematical model, Matthews et al. (1983) predicted that a reduction in formaldehyde levels as great as 89% can be obtained if a vinyl rather than particleboard floor is used. They also pointed out that the removal of selected sources might have a limited impact because of the increased emission rate from the remaining sources, which would result from an increased concentration difference between the material and air. The avoidance of formaldehyde-containing materials would no doubt constitute the best control policy, but most of these materials are inexpensive and have a high performance rating. Because complete elimination of these materials is not a practical solution, a number of studies has been reported on the use of coatings, barriers, and chemical treatments. Chemicals such as urea, polyurethane, ammonium sulfite, and sulfur compounds of sodium have been tried as an after-treatment of plywoods. Urea was found to be the most effective. Placing a solid barrier over the particleboard has also been found to reduce formaldehyde emissions. Efforts have been made to modify the chemical formula of resins to reduce formaldehyde emissions. Many manufacturers of wood products are working in this area.

Ventilation

Ventilation does not discriminate among pollutants; the indoor concentration of any number of pollutants can be reduced as long as their outdoor concentrations remain lower than those indoors. According to a study by the U. S. Department of Energy (DOE, 1979), residential and commercial buildings consume approximately one-third of the total energy

TABLE 3.22 Suggested Emission Rates of Products in Buildings

Materials	*Emission Rates (ng/h·m²)*
Flooring material	0.6
Floor coating	0.6
Wall material	0.4
Wall coating	0.4
Movable partitions	0.4
Office furniture	0.25 mg/h/workstation
Office machines (central)	0.25 mg/h/m^3 of space
Ozone emission	0.01 mg/h/m^3 of space
Office machine (personal)	2.5 mg/h/workstation
Ozone emissions	0.1 mg/h/workstation

Assumption: Indoor air is well mixed, ventilation rate is 0.5 exchanges of fresh air per hour. Maximum prudent increment in indoor concentration of organic vapor from any single source type is 0.5 mg/m^3 and that of ozone is 0.02 mg/m^3. Volume of concern for dispersion of emissions from furniture and machines at workstations is 10 m^3. Emission rates from floor coatings and wall coatings may be higher immediately after application, but their steady-state emissions should be considered.

Source: Tucker, 1990.

used in the United States, and ventilation systems consume more than 50% of the total energy used in a building. Therefore, in the last two decades, particularly after the energy crisis of the 1970s, energy conservation became the central theme in the design and retrofitting of building ventilation systems. Reduction of the infiltration of fresh air by making the space airtight, recirculation of up to 80% of the indoor air, and the use of better insulation, such as urea formaldehyde foam insulation, are some of the measures employed to reduce energy costs.

Klauss et al. (1970) studied the historical development of ventilation standards in the United States. The ventilation rate early in the twentieth century was about 30 ft^3/min/person, which was reduced to 5 ft^3/min/person in 1973 (see Figure 3.4). From 1978 through 1988, the National Institute of Occupational Safety and Health (NIOSH) identified 380 commercial buildings with indoor air quality problems. Inadequate ventilation was found to be the cause of poor air quality in approximately one-half of these buildings. The American Society of Heating, Refrigerating and Air-Conditioning Engineers, Inc. (ASHRAE) realized that ventilation can provide a practical means to improve air quality and have recently proposed to increase the ventilation rate to 15 ft^3/min/person. The ventilation requirements under this voluntary standard (ASHRAE, 1989) for building environments are shown in Table 3.23.

Generally, four types of ventilation strategies can be used in residences or commercial buildings to reduce indoor pollutant concentrations. These include infiltration of outdoor air, natural ventilation, mechanical ventilation, and local ventilation.

Infiltration of outdoor air. One common method of reducing energy costs is to seal the infiltration paths of fresh air which, unfortunately, results in the deterioration of indoor air quality. Infiltration of outdoor air into the indoor environment occurs through openings in the structure due to pressure and temperature gradients. The amount of infiltrating air can be incorporated in the building design by considering building materials, the climate, and surrounding terrain. The natural path for infiltration includes cracks in the wall and in the basement, joints between two walls, openings around drain pipes, windows, doors, and heating systems, fireplaces, and vents. Indoor air quality can be improved substantially if these paths remain open.

Natural ventilation. Natural ventilation occurs when doors and windows are open. This method is extremly effective in reducing pollutant levels when short-term activities, such as cooking or cleaning, are taking place. The air flow rate through the house is often increased by opening doors or windows for cross ventilation. The main disadvantage of natural ventilation is the increased energy cost associated with the extra heating or cooling load.

Mechanical ventilation. Mechanical ventilation is often considered an integral part of an HVAC system and is usually more suitable for nonresidential applications, such as commercial buildings, schools, apartment complexes, and hospitals, where large volumes of air are processed. The advantage of a mechanical ventilation system over infiltration or natural ventilation is that the energy required to heat or cool the outdoor air can be reduced by recovering some energy from the exhaust air. This is achieved by using an air-to-air heat exchanger, which is available for both residential and nonresidential applica-

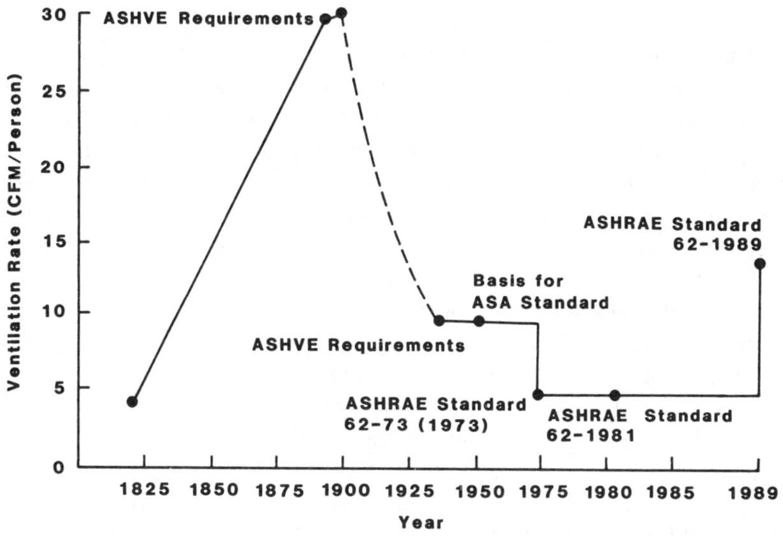

Figure 3.4 *Historical change of guidelines for ventilation rate in buildings.* (Source: *Klauss et al., 1970.*)

tions. Air-to-air heat exchangers for large commercial and public buildings are described in the ASHRAE Handbook (1988).

Figure 3.5 shows a generalized schematic of the heat exchanger. Most residential models are designed to be installed in a ductwork system. Air is withdrawn from locations throughout the structure, exchanges heat with the incoming outdoor air, and is then distributed throughout the structure via the duct system (see Figure 3.6). Air-to-air heat exchangers can also be installed in windows or walls. The efficiency of such a unit depends on the temperature difference between the indoor and outdoor air. Heat recovery of up to 85% can be achieved.

TABLE 3.23 Ventilation Requirements for Building Environments

	Ventilation Rate(ft^3/min/person)	
Space Type/Use	*Smoking*	*Nonsmoking*
Restaurant dining	35	20
Bars/cocktail lounges	50	30
Hotel bedrooms	30	30[a]
Hotel lobbies	15	15
Office space	20	20
Ballrooms and discos	35	25
Theater lobbies	35	20
Classrooms	25	15

[a]cfm/room

Source: ASHRAE, 1989.

Evaporative coolers also may be used as a part of the mechanical ventilation system to adjust the temperature and humidity of outdoor or recirculating air. Air is forced through either a water-saturated filter, or water is sprayed onto the moving air stream. In either case, water absorbs heat from the air and evaporates, thus providing both cooling and humidity adjustment. One major disadvantage of evaporative coolers is the growth of bacteria, fungi, and viruses in the water, which may be carried indoors. Although biocides and fungicides can be used to prevent microbial growth, these chemicals are toxic in nature and can be carried away by the air. This subject is discussed in detail in Chapter 7.

Local Ventilation. Local ventilation is typically achieved by using a blower fan that is capable of producing a low-pressure zone in the area surrounding the pollutant source. Often these systems are installed in copy rooms, blueprint rooms, bathrooms, and other areas of high pollutant concentrations.

In general, lower air exchange rates will result in higher concentrations of any pollutant that has indoor sources. If the indoor sources are continuous, like those of formaldehyde and other volatile organic compounds, the impact of increased ventilation may not be realized immediately. In new structures, an increase in ventilation may initially increase emission rates due to a greater concentration difference between the source and the surrounding air. After a certain period of time, however, the source attains a steady emission rate and the indoor concentration level will start to decrease. As a result, measurements at various time intervals are necessary to evaluate the full impact of increased ventilation.

Shields and Weschler (1989) measured the VOC concentrations in an office building for several ventilation rates and made a second measurement after two months. A substantial reduction in the pollutant concentration was observed during the second measurement. Recently Nagda et al. (1990) studied the impact of increased ventilation rates in a 20-story office building having a total floor area of approximately 9300 m^2. The air exchange rate was increased from 0.84 air changes per hour (ach) to 1.08 ach, which resulted in an increase of ventilation rate from 16.5 to 38.0 ft^3/min/person. Measurements were conducted at 1-week intervals during the summer months for formaldehyde, nicotine, respirable particles, CO_2, and CO. A substantial reduction in the concentration of these pollutants was not

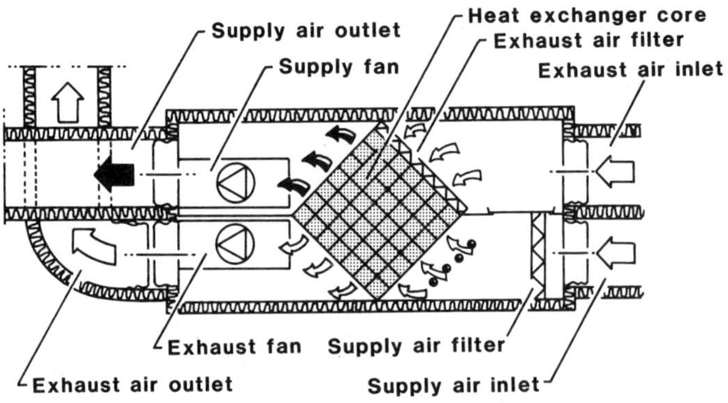

Figure 3.5 *Schematic diagram of an air-to-air heat exchangers.* (Source: *Fisk et al., 1987.*)

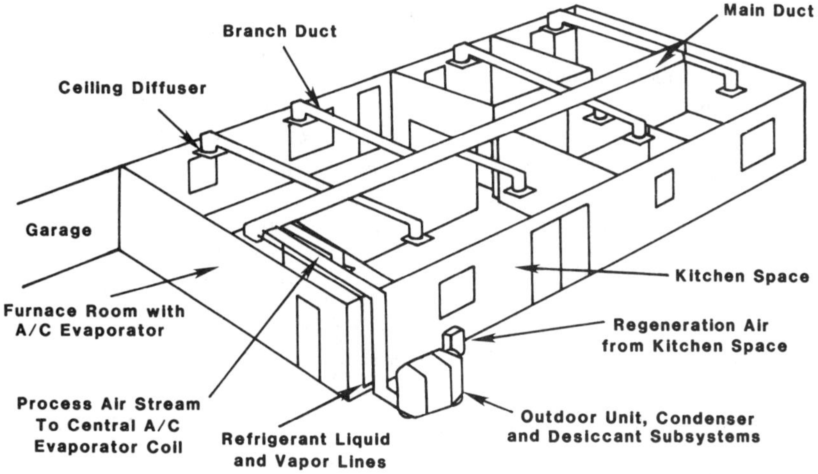

Figure 3.6 *Schematic of a typical residential integrated AC installation.* (Source: *Novosel et al., 1988.)*

observed; on the contrary, formaldehyde concentrations increased. However, mechanical ventilation has been found to be very effective in reducing formaldehyde concentrations in mobile and conventional homes that have urea formaldehyde foam insulation and particle-board subflooring. A significant reduction (in the range of 35% to 70%) was reported when the air exchange rates were increased fourfold (see Table 3.24).

Air Cleaning

Removal by catalytic conversion. Some selected contaminants found in indoor air can be converted to less harmful gases by catalytic conversion. In this process, the polluted air is brought into contact with a solid catalyst under specific conditions. This contact results in a chemical reaction of the pollutant to produce reaction products. Goldsmith et al. (1974) proposed a regenerable platinum-impregnated, activated-carbon system for removing contaminants from a spacecraft atmosphere. The removal capacity of the catalyst, its regeneration ability, and the effects of the repetitive adsorption-desorption cycling on removal capacity were evaluated for seven different spacecraft contaminants, which included caprylic acid, diisobutyl ketone, acrolein, tetrachloroethane, Freon-114, vinyl chloride, and thiophene. These pollutants were removed from air by adsorption on carbon, but were converted to less harmful chemicals during high-temperature desorption with the aid of the catalyst impregnated in the carbon. Oxygenated hydrocarbons were oxidized rather easily, whereas chlorinated and sulfur containing hydrocarbons were not oxidized. A temperature higher than 400°C was required for complete destruction of caprylic acid and diisobutyl ketone. During repetitive cycling, the platinum-impregnated carbon was found to be superior to plain carbon, although both carbons were equally effective during the first few cycles.

Several studies have been carried out (Carhart and Thompson, 1975) to identify methods for removing contaminants from a submarine atmosphere. In a submarine atmosphere, a catalytic burner using a Hopcalite catalyst at 315°C effectively converted H_2, CO, most hydrocarbons (except methane), and oxygenated organic compounds to CO_2 and H_2O. The Hopcalite catalyst was found to be less effective in oxidizing compounds that contained nitrogen and chlorine. Another problem experienced with the catalytic burner was that some chlorinated compounds were converted to compounds that were more hazardous than the original pollutants. For example, some refrigerants were converted to halogen acids, and methyl chloroform was converted to the more toxic vinylidene chloride and trichloroethylene. Frequent use of an alkaline adsorbent (lithium carbonate) was necessary to clean the effluent from the catalytic burner. Although Hopcalite catalytic burners are in use in submarines, the operating temperature is too high for use in homes or public buildings.

Jewell (1980) used several automobile catalytic converters at room temperature to remove formaldehyde from mobile homes. During operation, the converter produced an unknown but irritating vapor. Formaldehyde can also be removed by using a bed of activated alumina impregnated with potassium permanganate via the chemical reaction $HCHO + KMnO_4 = H_2O + CO_2 + KMnO_2$. The bed is discarded when it loses its activity. A bed containing 36 kg of alumina catalyst and operating at an air flow rate of 130 ft^3/min reduced the formaldehyde level in a mobile home by 80% (0.5 ppm to 0.1 ppm). This level was maintained over a period of 140 days. However, Godish (1989) did not find the method to be as effective when a thin filter bed of the same catalyst was used in a mobile home and a home containing urea formaldehyde foam insulation. At an air flow rate of 130 ft^3/min through the bed, the formaldehyde concentration was reduced by 25% to 30% in the mobile home and by 35% to 45% in the foam insulated home. A comparison of the catalytic converter studied by Jewell (1980) with a filter bed containing activated alumina impregnated with potassium permanganate showed that a converter containing platinum-palladium-rhodium-cerium oxide was as effective as the activated alumina.

A number of catalysts have been developed for removing formaldehyde. These include Purafil, activated carbon and alumina oxide, activated carbon and alumina oxide impregnated with urea and ammonium sulfate, and a ceramic material impregnated with $KMnO_4$. Although Purafil and the ceramic material were more effective in removing formaldehyde, the capacity of Purafil declined significantly when tested in homes. The removal capacity of Purafil dropped from 16 g/kg in the laboratory study to 4 g/kg when employed in homes. Other air pollutants were suspected to have reacted with the KMnO4, thereby reducing the capacity for formaldehyde. The removal efficiency ranged from 17% to 72%.

Arthur D. Little, Inc. (1980, 1981) tested several adsorbents, including Purafil, Calgon ASC carbon, Barnebey-Cheney CI carbon, Hopcalite, alumina impregnated with chromia, urea, and ammonium chloride, and molecular sieve 3A, to determine their capacity to remove formaldehyde. Purafil, Calgon ASC carbon, Hopcalite, and other alumina-based materials catalytically converted formaldehyde to carbon dioxide and water, whereas Barnebey-Cheney CI carbon, and molecular sieve 3A removed it by physical adsorption. Purafil, Calgon ASC, and Barnebey-Cheney CI had a higher capacity than the other adsorbents. Following the laboratory tests, filter beds containing 8.6 kg of Purafil, 6.8 kg of Calgon ASC, and a mixed bed of Purafil and Calgon ASC were tested separately in a mobile home using an air flow rate of approximately 575 ft^3/min. The formaldehyde concentration dropped from 1.3-1.8 ppm to 0.2-0.3 ppm within a few hours of operation, but the concentration returned to its original level after one to five days of operation. Other air contaminants were suspected of reducing the uptake capacity of the adsorbent.

TABLE 3.24 Effectiveness of Mechanical Ventilation in Reducing Formaldehyde Levels in Residential Structures

Investigators	*Residence Type*	*Air Exchange Rate (ach)*	*Formaldehyde Concentration (ppm)*	*Reduction (%)*
Jewell (1984)	Mobile home	0.56	0.21	
		1.20	0.11	48
		2.00	0.08	62
	Mobile home and added particleboard	0.56	0.28	
		1.20	0.15	46
		2.00	0.13	54
Godish and Rouch (1984)	Mobile home			
	Series 1	0.20	0.20	
		0.45	0.11	45
	Series 2	0.17	0.19	
		0.77	0.09	53
	Mobile home and added particleboard			
	Series 1	0.19	0.18	
		1.22	0.06	67
	Series 2	0.21	0.21	
		1.33	0.09	57
	Urea-formaldehyde foam-insulated (UFFI)			
	Series 1	0.20	0.25	
		1.45	0.06	76
	Series 2	0.18	0.10	
		1.36	0.03	70
	Series 3	0.20	0.12	
		0.85	0.03	70
	UFFI and particleboard Subflooring			
	Series 1	0.21	0.22	
		0.70	0.10	55
	Series 2	0.24	0.19	
		0.48	0.08	58
Matthews (1985)	Conventional-particle-board subflooring			
	Series 1	0.20	0.091	
		0.50	0.085	7
		0.94	0.059	35
	Series 2	0.25	0.099	
		0.46	0.082	17
		0.98	0.053	47

Eian (1984) developed 33 sorbents for the removal of formaldehyde from air, including activated carbon impregnated with 31 salts, and silica gel and activated alumina impregnated with sodium sulfamate. Three commercially available adsorbents, Purafil, Barnebey-Cheney CI, and a Norton Company formaldehyde cartridge, were also used to remove formaldehyde from air. These were evaluated by passing an air stream at a flow rate of 64 L/min, containing 100 ppm formaldehyde, through a bed that had a cross-sectional area of 61 cm2 and a depth of 3.2 cm. The best service life (defined as the time elapsed before the formaldehyde concentration in the outlet stream reached 1 ppm) was observed to be 575 minutes with activated carbon impregnated with 40% sodium sulfamate solution.

Ammonia fumigation is another technique that has been examined for formaldehyde removal, particularly from mobile homes. A substantial reduction in formaldehyde concentration in ammonia-fumigated homes was reported by Jewell (1984). Formaldehyde levels were reduced by as much as 81% after 40 to 60 weeks, and they remained 60% lower than the initial level even after three to five years. Muratzky (1987) also observed a reduction of about 75% in the formaldehyde level over a period of 1 to 48 months. He found that the house was not habitable during the initial 2- to 4-week period after fumigation and that an ammonia odor remained for four months.

Removal by absorption processes. The absorption process offers promise as a method for removing organic pollutants from indoor air, but this technique has not been explored fully. The design of an absorption process and other aspects related to its application indoors are discussed further in Chapter 9.

Pedersen and Fisk (1984) tried to remove formaldehyde from air by washing the air with water in a system that incorporated a refrigeration cycle to control the humidity of the outlet air stream. Two dehumidifiers were tested; one had a simple coiled tube, and the other was finned to provide extra surface area. The air washers were tested using air that contained 0.07 to 0.4 ppm of formaldehyde and flow rates that ranged from 100 to 160 L/s. Formaldehyde removal efficiencies varied from 30% to 63%. Although their experimental results demonstrated that washing with water can be an effective method for removing formaldehyde, the system was not tested in field studies where other air contaminants can interfere with the removal efficiency.

Removal by adsorption processes. Selected solid materials can attract molecules from the gas phase and retain them on the solids surface until the process is reversed by applying either heat, vacuum, or a combination of the two. The attractive energy varies both with the solid and the gas. The capacity for adsorbing certain pollutants by these solid materials can be enhanced by increasing the internal surface area of the solid (by activation) or by changing the surface characteristics. Before choosing an activated solid material for adsorbing a particular pollutant or a group of pollutants, the amount that can be adsorbed on the solid must be known. The amount of a pollutant that a solid material can adsorb at a certain temperature and pressure is available in the literature for a number of pollutants. With proper understanding and interpretation, these data can be used to design a system following the procedures discussed in Chapter 10. A number of systems have already been developed and are currently being used to remove selected pollutants from indoor air.

The adsorptive property of solids (wood charcoal) was recognized as early as 1773 by

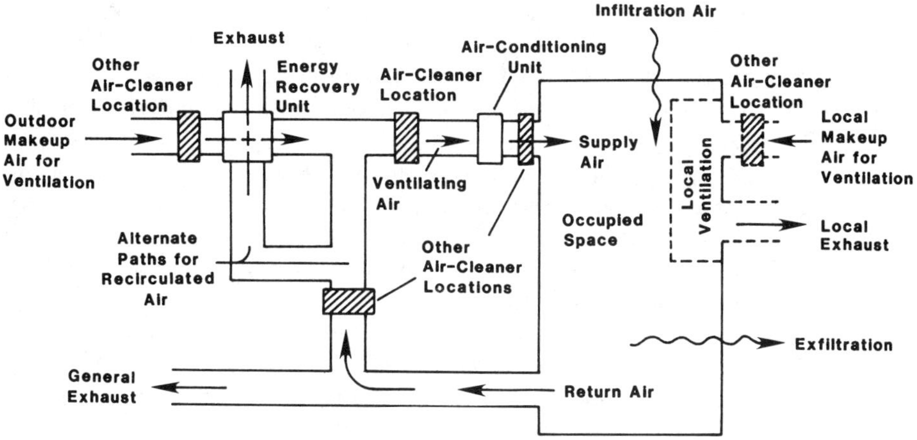

Figure 3.7 *Schematic of a ventilation system with solid-adsorbent air cleaners.* (Source: *ASHRAE Standard 62-1989.*)

Scheele, but the adsorption method was not used commercially until the middle of the nineteenth century. Even then it was used primarily in air filters in ventilation systems and for the treatment of sewage. The rapid development of activated carbons for gas adsorption occurred during World War I, when it was used in gas masks to adsorb poisonous gases. Since then the removal of contaminants by activated carbon from air and other gas streams, both in industrial and other indoor environments, is by far the most prevalent application.

Adsorption processes received a big boost from the space shuttle program. Adsorption processes were considered as one of the options for controlling humidity and CO_2 levels in a spacecraft cabin and for the removal of trace contaminants. Silica gel and molecular sieve were leading candidates for humidity control, while adsorption on activated carbon was the primary technique proposed for the removal of organic pollutants.

An adsorption process is most economical if the adsorbent can be regenerated for repeated use. Therefore, most systems undergo repeated adsorption-regeneration cycles. Carbon filters are more frequently used in conjunction with other air cleaners, such as panel filters, extended surface filters, and electrostatic precipitators. Carbon filters are also used in HVAC systems to clean the intake as well as the recirculated air. A schematic diagram of a ventilation system that uses solid adsorbent based air cleaners is shown in Figure 3.7. A review of these is available in the ASHRAE Handbook (ASHRAE, 1988). In residential air-conditioning applications, thin bed filters oriented perpendicular to the air flow are employed.

REFERENCES

ACGIH (American Conference of Governmental Industrial Hygienists), Cincinnati, O., 1989.

AHLSTROM, R., BERGLUND, B., BERGLUND, U., and LINDVALL, T., "Odor Interaction Between Formaldehyde and the Indoor Air of a Sick Building", Proceedings 3rd International Conference on Indoor Air Quality and Climate, Stockholm, Sweden, **3**, 461, 1984.

AMMAN, H. M., BERRY, M. A., CHILDS, N. E., and MAGE, D. T., "Health Effects Associated with

Indoor Air Pollutants", Proceedings IAQ '86: Managing Indoor Air for Health and Energy Conservation, American Society of Heating, Refrigerating and Air-Conditioning Engineers, Atlanta, Ga, 70, 1986.

ANDERSEN, I. B., LINDQUIST, G. R., and MØLHAVE, L., *Atmos. Environ.*, **9,** 1121 (1975).

ANDERSON, K., and SCOTT, R., *Fundamentals of Industrial Toxicology,* Ann Arbor Science, Mich., 1981.

APTE, M. G., and TRAYNOR, G. W., "Comparison of Pollutant Emission Rates from Unvented Kerosene and Gas Space Heaters", Proceedings IAQ '86: Managing Indoor Air for Health and Energy Conservation, American Society of Heating, Refrigerating and Air-Conditioning Engineers, Atlanta, Ga., 405, 1986.

ARTHUR D. LITTLE, Inc., "Formaldehyde Concentration Level Control in the Air of Mobile Homes", Arthur D. Little, Inc., Cambridge, Mass., 1980.

ARTHUR D. LITTLE, Inc., "Formaldehyde Concentration Level Control in the Air of Mobile Homes; Phase I and II Report to the Formaldehyde Institute", Arthur D. Little, Inc., Cambridge, Mass., 1981.

ASHRAE (American Society of Heating, Refrigerating and Air-Conditioning Engineers), *ASHRAE Handb. Equipment,* The Society, Atlanta, Ga., 1988.

ASHRAE, "Ventilation for Acceptable Indoor Air Quality", ASHRAE Standard 62-1989, The Society, Atlanta, Ga., 1989.

BAYER, C. W., and BLACK, M. S., "Indoor Air Quality Evaluations of Three Office Buildings", Proceedings IAQ '88: Engineering Solutions to Indoor Air Problems, American Society of Heating, Refrigerating and Air-Conditioning Engineers, Atlanta, Ga., 294, 1988.

BAYER, C. W., and PAPANICOLOPOULOS, C. D., "Exposure Assessments of Volatile Organic Compound Emission from Textile Products", Proceedings 5th International Conference on Indoor Air Quality and Climate, Toronto, Canada, **3,** 725, 1990.

BERGLUND, B., BERGLUND, U., and LINDVALL, T., "Assessment of Discomfort and Irritation from the Indoor Air", IAQ '86: Managing Indoor Air for Health and Energy Conservation, American Society of Heating, Refrigerating and Air-Conditioning Engineers, Atlanta, Ga., 138, 1986.

CARHART, H. W., and THOMPSON, J. K., "Removal of Contaminants from Submarine Atmospheres", *ACS Symp. Ser.,* No. 17, 1 (1975).

CURRIE, K. L., MCDONALD, E. C., CHUNG, L. T. K., and HIGGS, A. R., *Am. Ind. Hyg. Assoc. J.,* **51**(1), 23 (1990).

DAISEY, J. M., and HODGSON, A., *Atmos. Environ.,* **23**(9), 1885 (1989).

DOE (U. S. Department of Energy and U. S. Department of Housing and Urban Development), "Energy Performance Standards for New Buildings", Federal Reg. 41, 68218-68220, 1979.

EIAN, G. L., "Sorbent Material for Reducing Formaldehyde Emission", U. S. Patent 4,443,354,1984.

ELKINS, H. B., *The Chemistry of Industrial Toxicology,* 2nd ed., Wiley, New York, 1959.

EPA (Environmental Protection Agency), "Draft Tier I Documents: Summary Review of Health Effects Associated with Exposure to Styrene Vapor", 1986.

EPA "A Summary of Formaldelyde Exposure in Residential Settings", Interim Final Report, Prepared by Verser, Inc., EPA Contract No. 68–02–3968, July 1985.

FENSKE, R. A., and STERNBACH, T., *Bull. Environ. Contam. Toxicol.,* **39,** 903 (1987).

FISK, W. J., SPENCER, R. K., GRIMSRUD, D. T., OFFERMANN, F. J., PEDERSEN, B., and SEXTRO, R., *Indoor Air Quality Control Techniques: Radon, Formaldelyde, Combustion Products,* Pollution Technology Review No. 144, Noyes Data Corporation, N. J., 1987.

GARRY, V. F., *Minn. Med.,* **63,** 107 (1980).

GEBEFUEGI, I. L., and KORTE, F., "Source of Organics in the Air of an Office Building", Proceedings 5th International Conference on Indoor Air Quality and Climate, Toronto, Canada, **2,** 701, 1990.

GIRMAN, J. R., HODGSON, A. T., NEWTON, A. S., and WINKES, A. W., *Environ. Int.* **12**(1-4), 317, 1986.

GODISH, T., and ROUCH J., "Efficacy of Residential Formaldehyde Control: Source Removal", Proceedings 3rd International Conference on Indoor Air Quality and Climate, Stockholm, Sweden, **2,** 127, 1984.

GODISH, T., and ROUCH, J., *Am. Ind. Hyg. Assoc. J.,* **47,** 792 (1986).

GODISH, T., *Indoor Air Pollution Control,* Lewis Publishers, Chelsea, Mich., 1989.

GOLDSMITH, R. L., McNULTY, K. J., FREEDLAND, G. M., TURK, A., and NWANKWO, J., "Contaminant Removal from Enclosed Atmospheres by Regenerable Adsorbents", NASA-CR-137626, Walden Research Corp., Cambridge, Mass., Accession No. N75-17969, 1974.

HAWTHORNE, A. R., GAMMAGE, R. B., DUDNEY, C. S., WOMACK, D., MORRIS, S., WESTLY, R. and GUPTA, K., "Results of a Forty Home Indoor Air Quality Monitoring Study", Report No. ORNL-7458, Oak Ridge National Laboratory, Oak Ridge, Tenn., 1985.

HAWTHORNE, A. R., and MATTHEWS, T. G., "Formaldehyde: An Important Indoor Pollutant", Indoor Air Quality Seminar, Implication for Electric Utility Conservation Program, Electric Power Research Institute, Palo Alto, Calif., 1985.

HEDGE, A., "Questionnaire Design Guidelines for Investigation of Sick Buildings", Proceedings 5th International Conference on Indoor Air Quality and Climate, Toronto, Canada, **1,** 605, 1990.

HENSCHLER, D., *Am. Ind. Hyg. Assoc. J.,* **51**(10), 523 (1990).

HEW (U. S. Department of Health, Education, and Welfare) "Smoking and Health—A report of the Surgeon General", DHEW Publication No. (PHS) 79-50066, Washington, D. C., 1979.

HINES, A. L., GHOSH, T. K., LOYALKA, S. K., and WARDER, R. C., Jr., "Investigation of Co-Sorption of Gases and Vapors as a Means to Enhance Indoor Air Quality", Report No. GRI-90/0194, Gas Research Institute, Chicago, Ill., NTIS Document No. PB91-178806, 1991.

HOEGG, V. R., "Cigarette Smoke in Closed Spaces", *Environ. Health Persp.,* **2,** 117 (1972).

IMMERMAN, F. W., and SCHAUM, J. L., "Nonoccupational Pesticide Exposure Study (NOPES)", Report No. EPA/600/3-90-003, Research Triangle Institute, Research Triangle Park, N. C., NTIS Document No. PB90-152224, 1990.

JEWELL, R. A., "Reduction of Formaldehyde Levels in Mobile Homes", Symposium on Wood Adhesives-Research, Applications, and Needs, Madison, Wis., 1980.

JEWELL, R. A., "Reducing Formaldehyde Levels in Mobile Homes Using 29% Aqueous Ammonia Treatment on Heat Exchangers", Weyerhaeuser Corp, Tacoma, Wash., 1984.

KERNS, W. D., DONOFRIO, D. J., and PAVKOV, K. L., in *Formaldehyde Toxicity,* (ed. J. E. Gibson), Hemisphere, New York, 1983.

KLAUSS, A. K., TULL, R. H., ROOTS, L. M., and PFAFFLIN, J. R., *ASHRAE J.,* **12**(6), 51 (1970).

KNIGHT, C. V., HUMPHREYS, M. P., and PINNIX, J. C., "Indoor Air Quality Related to Wood Heaters", Proceedings IAQ '86: Managing Indoor Air for Health and Energy Conservation, American Society of Heating, Refrigerating and Air-Conditioning Engineers, Atlanta, Ga., 1986.

KRAUSE, C., "Occurrence of Volatile Organic Compounds in the Air of 500 Homes in the Federal Republic of Germany", Proceedings 4th International Conference on Indoor Air Quality and Climate, West Berlin, West Germany, **1,** 102, 1987.

LAWLESS, P. A., MICHAELS, L. D., and WHITE, J., "Demonstration of EPA's Database of Indoor Air Pollutant Sources (DIAPS)", Proceedings 5th International Conference on Indoor Air Quality and Climate, Toronto, Canada, **3,** 673, 1990.

LEVIN, L., and PURDOM, P. W., *Am. J. Pub. Health,* **73**(6), 251 (1983).

MAIN, D. M., HOGAN, T. J., *J. Occup. Med.,* **25**(12), 896, 1983.

MATTHEWS, T. G., and HAWTHORNE, A. R., DAFFRON, C. R., and COREY, M. D., "Formaldehyde Release from Pressed-Products", 17th International Washington State University, Particleboard/Composite Materials Symposium, Pullman, Wash., 1983.

MATTHEWS, T. G., "Preliminary Evaluation of Formaldehyde Mitigation Studies in Unoccupied Research Homes", "Indoor Air Quality in Cold Climates: Hazards and Abatement Measures", (ed. D. S. Walkinshaw), Air Pollution Control Association Specialty Conference Proceedings, Pittsburgh, Pa., 1985.

MIKSCH, R. R., ANTHON, D. W., FANNING, L. Z., HOLLOWELL, C. D., REVZAN, K., and GLANVILLE, J., *Anal. Chem.,* **53,** 2118 (1981).

MIKSCH, R. R., HOLLOWELL, C. D., and SCHMIDT, H. E., *Environ. Int.,* **8,** 129 (1982).

MOKLER, B. V., WONG, B. A., and SNOW, M. J., *Am. Ind. Hyg. Assoc. J.,* **40,** 330 (1979a).

MOKLER, B. V., WONG, B. A., and SNOW, M. J., *Am. Ind. Hyg. Assoc. J.,* **40,** 339 (1979b).

MØLHAVE, L., "Indoor Air Pollution Due to Building Materials", Proceedings 1st International Indoor Climate Symposium, Copenhagen, Denmark, 89, 1979.

MØLHAVE, L., and MOLLER, J., "The Atmospheric Environment in Modern Danish Dwellings-Measurements in 39 Flats", Proceedings 1st International Indoor Climate Symposium, Copenhagen, Denmark, 171, 1979.

MØLHAVE, L., *Environ. Int.,* **8,** 117 (1982).

MØLHAVE, L., BACH, B., and PEDERSON, O. F., "Human Reactions During Exposures to Low Concentrations of Organic Gases and Vapors Known as Normal Indoor Air Pollutants", Proceedings 3rd International Conference on Indoor Air Quality and Climate, Stockholm, Sweden, **3,** 431, 1984.

MURATZKY, R., "Formaldehyde Injuries in Prefabricated Houses: Causes, Prevention, and Reduction", Proceedings 4th International Conference on Indoor Air Quality and Climate, West Berlin, West Germany, **2,** 690, 1987.

NAGDA, N., KOONTZ, M., LUMBY, D., ALBRECHT, R., and RIZZUTO, J., "Impact of Increased Ventilation Rates on Office Buildings Air Quality", Proceedings 5th International Conference on Indoor Air Quality and Climate, Toronto, Canada, **4,** 281, 1990.

NAS (National Academy of Science), *Particulate Polycyclic Organic Matter,* Committee on Biologic Effects of Atmospheric Pollutants, National Academy Press., Washington, D. C., 1972.

NAS, *Indoor Pollutants,* Committee on Indoor Pollutants, National Academy Press, Washington, D. C., 1981.

NOVOSEL, D., MCFADDEN, D. H., and RELWANI, S. M., "Dessicant Air Conditioner to Control IAQ in Residences", Proceedings IAQ 88, Engineering Solution to Indoor Air Problems, American Society of Heating, Refrigerating, and Air-Conditioning Engineers, Atlanta, Ga., 148, 1988.

OLSEN, J. H., and DOSSING, M., *Am. Ind. Hyg. Assoc. J.,* 43, 366 (1982).

OLSEN, J. H., JENSEN, S. P., HINK, M., FAURBO, K., BREUM, N. O., and JENSEN, O. M., *Int. J. Cancer,* **34**(5), 639, 1984.

OZKAYNAK, H., RYAN, P. B., WALLACE, L. A., NELSON, W. C., and BEHAR, J. V., "Sources and Emission Rates of Organic Chemical Vapors in Homes and Buildings", Proceedings 4th International Conference on Indoor Air Quality and Climate, West Berlin, West Germany, **1,** 3, 1987.

PEDERSEN, B. and FISK, N. J., "The Control of Formaldehyde in Indoor Air by Air Washing", Report No. LBL-17381, Lawrence Berkeley Laboratory, Berkeley, Calif., 1984.

PICKRELL, J. A., *Environ. Sci. Technol.,* **17,** 753, (1983).

RELWANI, S. M., MOSCHANDREAS, D. J., and BILLICK, I. H., "Indoor Air Quality Control Capabilities of Desiccant Materials", Proceedings 4th International Conference on the Indoor Air Quality and Climate, West Berlin, West Germany, **1,** 236, 1987.

RELWANI, S. M., MOSCHANDREAS, D. J., and NOVOSEL, D., "Indoor Air Quality Control Capabilities of the Humidity Pump: A Field Experiment", Proceedings 5th International Conference on the Indoor Air Quality and Climate, Toronto, Canada, **3,** 225, 1990.

SAMET, J. M., MARBURY, M. C., and SPENGLER, J. D., *Am. Rev. Respir. Dis.,* 136, 1486 (1987).

SAMET, J. M., MARBURY, M. C., and SPENGLER, J. D., *Am. Rev. Respir. Dis.,* 137, 141 (1988).

SCHENKER, M. B., WEISS, S. T., and MURAWSKI, B. W., *Environ. Int.,* **8,** 359 (1982).

SEITZ, T., "NIOSH Indoor Air Investigation", in *The Practitioners Approach to Indoor Air Quality Investigations,* (eds. D. W. Weekes and R. B. Gammage) American Industrial Hygiene Association, Akron, O., 1990.

SHELDON, L., ZELON, H., SICKLES, J., EATON, C., HARTWELL, T., and WALLACE, L., "Indoor Air Quality in Public Buildings: Volume II", NTIS Document No. PB 89-102 511, 1989.

SHIELDS, H. C., and WESCHLER, C. J., "The Effects of Ventilation on the Indoor Concentrations of Vapor Phase Organic Compounds", Proceedings IAQ '89: The Human Equation: Health and Comfort, American Society of Heating, Refrigerating and Air-Conditioning Engineers, Atlanta, Ga., 137, 1989.

STAYNER, L. T., "A Retrospective Study of Workers Exposed to Formaldehyde in the Garment Industry", National Institute of Occupational Health and Safety, Washington, D. C., 1986.

STERLING, T., WEINKMAN, J., and STERLING, D., "Exposure of Homemakers to Toxic Contaminants: I. Differences in Chronic Conditions Between Homemakers and Employed Persons", Proceedings 5th International Conference on Indoor Air Quality and Climate, Toronto, Canada, **1,** 471, 1990.

TEICHMAN, K. Y., and WOODS, J. E., "Ventilation and Indoor Air Quality: Regulations, Codes, and Voluntary Consensus Standards", Proceedings IAQ '87: Practical Control of Indoor Air Problems, American Society of Heating, Refrigerating and Air-Conditioning Engineers, Atlanta, Ga., **3,** 1987.

THUN, M. J., LAKAT, M. F., and ALTMAN, R., *Environ, Res.,* **29**(2), 320, 1982.

TICHENOR, B. A., "Organic Emission Measurements Via Small Chamber Testing", 4th International Conference on Indoor Air Quality and Climate, West Berlin, West Germany, **1,** 8, 1987.

TUCKER, W. G., "Buildings with Low-Emitting Materials and Products: Where Do We Stand", Proceedings 5th International Conference on the Indoor Air Quality and Climate, Toronto, Canada, **3,** 251, 1990.

VAUGHAN, T. L., STRADER, C., DAVIS, S., and DALING, J. R., *Int. J. Cancer,* **38,** 677 (1986).

WAKEHAM, H., "Recent Trends in Tobacco Smoke Research", in The Chemistry of Tobacco Smoke, (ed. I. Schmeltz) Plenum Press, New York (1972).

WALLACE, L. A., PELLIZZARI, E. D., HARTWELL, T. D., SPARACINO, C., WHITMORE, R., SHELDON, L., ZELON, H., and PERRITT, R., *Environ. Res.,* **43,** 290 (1987).

WALLACE, L. A., "The Total Exposure Assessment Methodology (TEAM) Study", Report No. EPA/600/6-87/002, 1987.

WANG, T. C., *ASHRAE Trans.,* **81**(Part 1), 32 (1975).

WEISEL, C. P., LIOY, P. J., and JO, W. K., "Exposure to Volatile Organic Compounds Resulting from Showering with Chlorinated Water", Proceedings 5th International Conference on Indoor Air Quality and Climate, Toronto, Canada, **2,** 495, 1990.

4

Inorganic Gaseous Pollutants

4.1 INTRODUCTION

Inorganic gaseous pollutants that are recognized as major contributors to indoor air pollution problems include carbon dioxide (CO_2), carbon monoxide (CO), sulfur dioxide (SO_2), nitrous oxide (NO), and nitrogen dioxide (NO_2). Combustion appliances are the major source of these gases, along with other organic pollutants and respirable particulate matter. Tobacco smoke is another combustion product that can contribute to increased indoor concentration levels of CO, NH_3, and CO_2.

The emission patterns and concentrations of these indoor pollutants from combustion appliances depend on several factors, such as fuel type, combustion efficiency, design of the appliance, ventilation system, operating conditions, maintenance, and frequency of use. Depending on the type of fuel and its application, combustion appliances can be divided into three major categories: unvented space heaters; wood-burning stoves, furnaces, and fireplaces; and gas cooking ranges and ovens. Although space heating of a home can be accomplished by using an externally vented heating system, such as a central furnace or heater, a significant portion of the homes in the United States and Europe use unvented gas and kerosene heaters. The combustion products from these appliances are introduced directly indoors. In unvented spaces, pollutant concentrations tend to increase sharply within the first hour of furnace operation and then reach a steady-state level that is maintained until the appliance is turned off. Concentration profiles of CO_2, CO, NO, and NO_2 in a 27-m^3

environmental chamber produced from the operation of an unvented convective kerosene heater are shown in Figure 4.1 (Traynor et al., 1982). In a similar study, Girman et al. (1982) investigated an unvented gas space heater with no mechanical ventilation in the chamber, but with a 0.5 air exchange rate per hour. They found that the peak concentrations of these pollutants were attained within 10 min (Figure 4.2). Once the heater was turned off, concentrations dropped to their original background levels within an hour. Similar results were also observed with gas stoves.

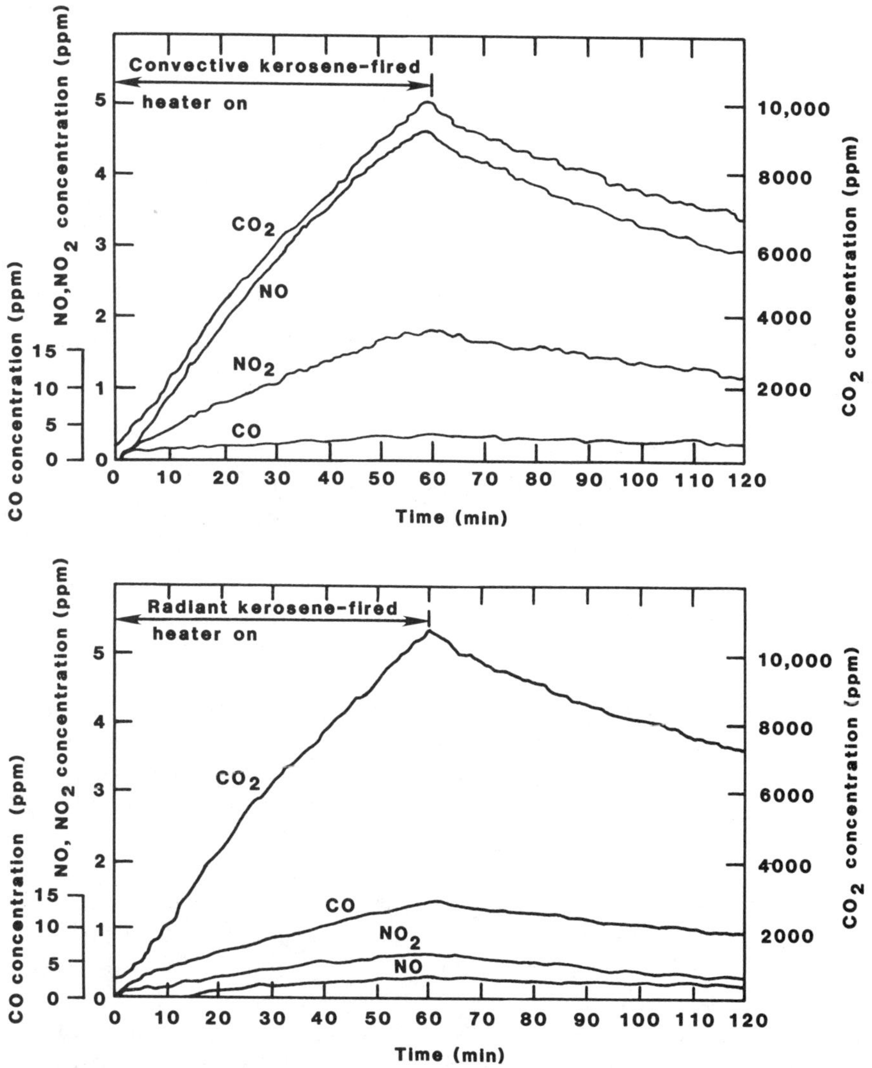

Figure 4.1 *Concentrations of combustion products measured during the operation of kerosene heaters. (Source: Traynor et al., 1982.)*

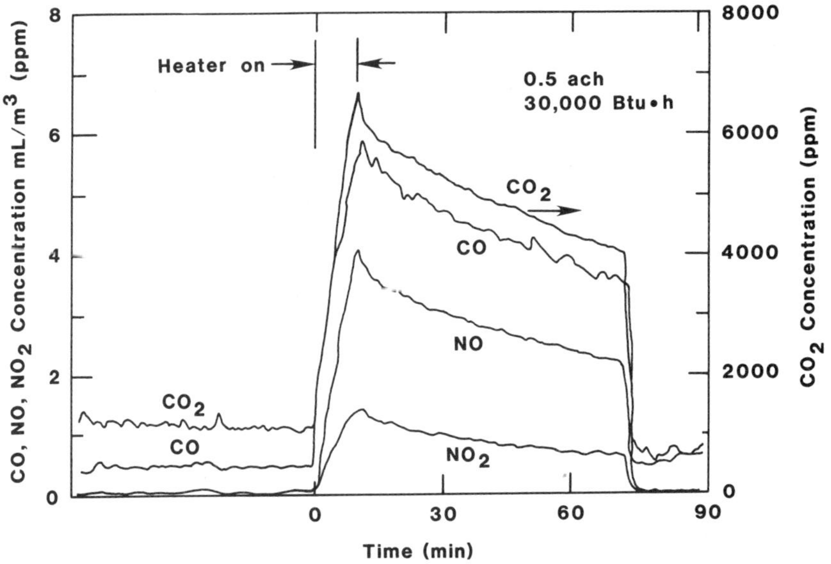

Figure 4.2 *Concentration of combustion products during the operation of an unvented gas space heater. (*Source: *Girman et al., 1982.)*

Wood-burning appliances generally emit pollutants intermittently, although the types of pollutants generated are highly variable and depend on the type of wood used. Along with the production of inorganic gases, the burning of wood also produces a significant quantity of particulates and a number of organic compounds, including formaldehyde (HCHO). The emissions produced by burning wood can be two to three orders of magnitude greater than those produced by other heating appliances.

Approximately one-half of all homes in the United States use natural gas in cooking ranges and stoves. Although most of these homes are equipped with range hoods, a significant portion of the combustion products remains indoors. Yocom et al. (1974) monitored the indoor and outdoor concentrations of CO when unvented gas stoves were in use in homes, and they concluded that gas stoves definitely contribute to indoor air pollution. Inorganic gaseous pollutants produced by the combustion of natural gas in range-top burners and ovens include NO, NO_2, CO, CO_2, and SO_2. Coté and his co-workers (1974) measured NO_2, NO, and CO concentrations in the kitchen, living room, and bedroom of a 6-year-old home that was equipped with a gas-fired stove and a central heating system. Their findings, which are shown in Table 4.1, illustrate that the indoor to outdoor ratio of these pollutants can be well over unity.

Other inorganic gases found indoors include ammonia, chlorine, hydrogen sulfide, and ozone. Ammonia and chlorine are strong reducing agents and are major components of most cleaning fluids. In most cases, outdoor air is the primary source of ozone, unless there is an indoor source such as an air cleaner that uses a negative ion generator. Ozone is a strong oxidizing agent with a half-life of less than 30 min; it decays rapidly by reacting with indoor surfaces. As much as 120 ppb of ozone has been measured indoors, but the typical

TABLE 4.1 Average NO_2, NO, and CO Concentrations in a House

Season and Data Category	Sampling Location					Average Stove Use, min/day		
	Kitchen above stove	*Kitchen 1 m from stove*	*Living room*	*Bedroom*	*Outdoor*	*Oven*	*Burners*	*Total*
Spring 1973								
NO_2 concentration, $\mu\gamma/\mu^3$	114	—	75	58	51	0	11	11
NO concentration, $\mu g/m^3$	53	—	42	34	21	0	11	11
CO concentration, $\mu\gamma/\mu^3$	4310	—	3210	2680	2230	14	23	37
Fall 1973								
NO_2 concentration, $\mu g/m^3$	180	140	—	70	32	25	49	74
NO concentration, $\mu g/m^3$	111	101	—	64	20	25	49	74
CO concentration, $\mu g/m^3$	7130	6620	—	5500	2500	70	45	115

[a] Average concentrations are based on daily averages for those days in which 12 valid 2-h averages were obtained.

—, no data reported.

Source: Cote et al., 1974.

concentration range is from 0 to 20 ppb. Ozone, sulfur dioxide, and hydrogen sulfide are highly reactive species and have been found to be responsible for the decoloration of paintings and the destruction of other artworks in museums and art galleries. The hourly sulfur dioxide concentration found indoors is generally less than 20 ppb.

4.2 HEALTH EFFECTS AND STANDARDS

Occupants of homes and buildings are regularly exposed to low level concentrations of inorganic gaseous pollutants, with periodic exposure to higher concentrations. Most combustion products are colorless, tasteless, and odorless at the concentration levels at which they are present indoors. In some instances, such as for natural gas, methyl mercaptan is added deliberately to the mixture to increase its odor so that a leak can be easily detected.

Although the indoor to outdoor ratio of carbon dioxide is greater than 1 for more than 99% of all residences, exposure to a very high CO_2 concentration level would be necessary (above 30,000 ppm) to cause a serious health problem (NASA, 1973). As a result, CO_2 is generally not considered to be an air pollutant by most researchers. The typical concentration of CO_2 indoors is in the range of 700 to 2000 ppm, whereas the outdoor concentration is typically 400 ppm. However, peak indoor concentrations can often exceed 3000 ppm because of the operation of an unvented combustion appliance. Respiration is affected when the CO_2 concentration exceeds 15,000 ppm, and exposure to concentrations above 30, 000 ppm can cause headaches, dizziness, and nausea. Because the long-term exposure to a CO_2 concentration above 5000 ppm can increase the incidence of illness, this level is used as threshold limit value (TLV) for submarine crews.

Carbon monoxide and nitrogen dioxide have been studied more frequently than other combustion products because of their health hazards. Carbon monoxide combines with hemoglobin and myoglobin of the blood to form carboxyhemoglobin and CO-myoglobin. However, the physiologic effect of CO-myoglobin formation has not been established. Carboxyhemoglobin reduces the transport of oxygen to tissues. The brain, myocardium, and exercising muscle tissues are most affected because of their higher oxygen demand (Samet et al., 1987). A carboxyhemoglobin level of 3% to 5% may adversely affect the vigilance ability of a person. Hand to eye coordination can slow down at a carboxyhemoglobin level of 6% to 10%. After 15 hours of exposure to 15 ppm of CO or 1 hour of exposure to 60 ppm of CO, the carboxyhemoglobin level can increase to 2.5%. The carboxyhemoglobin levels due to exposure to CO at varying concentrations and exposure times are shown in Figure 4.3 as a percent of the TLV set by the EPA. The calculations were performed by using the Coburn model (EPA, 1979), by considering the exposure of a male of average physical characteristics undergoing moderate exercise (breathing rate 10 L/min) under standard atmospheric conditions. Although carbon monoxide poisoning is a well established clinical fact, the health effects due to chronic exposure at the lower levels typically encountered indoors are still controversial. Acute illness and deaths have been reported as a result of CO poisoning while indoors. However, these incidences are generally attributed to a sudden high-level exposure that might result from the blocking of the infiltration of outdoor air during the winter or from a malfunctioning heater. In summary, health effects that result

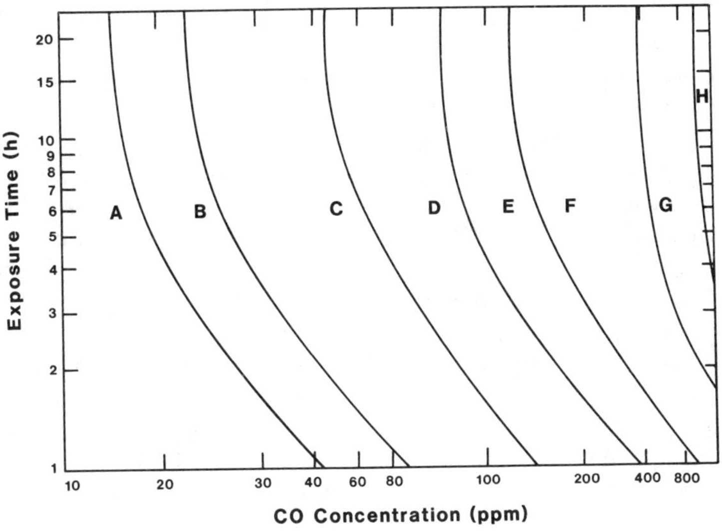

Key

Legends	% COHb Level of Threshold	Effects
A	0.3-0.7	Physiologic norm for non smokers
B	2.5-3.0	Cardiac function decrements in impaired individuals; blood flow alterations; and, after extended exposure, changes in red blood cell concentration
C	4.0-6.0[a]	Visual impairments, vigilance decrements, reduced maximal work capacity, norm for smokers
D	10.0-20.0	slight headache, lassitude, breathlessness from exertion, dilation of blood cells in the skin, abnormal vision, potential damage to fetuses
E	20.0-30.0	Severe headaches, nausea, abnormal manual dexterity
F	30.0-40.0	Weak muscles, nausea, vomiting, dimness of vision, severe headaches, irritability, and impaired judgment
G	50.0-60.0	Fainting, convulsions, coma
H	60.0-70.0	Coma, depressed cardiac activity and respiration, sometimes fatal

a, Smokers routinely have COHb levels of 3 to 8%.

Figure 4.3 represents combinations of carbon monoxide exposure concentration and exposure time required to produce the threshold levels of carboxyhemoglobin shown in the key in a male of average physical characteristics undergoing moderate exercise (breathing rate 10 L/min) under standard conditions of atmospheric pressure. Calculations are based on the "Coburn model" of carboxyhemoglobin concentration in humans. Thresholds are from a report by the EPA (1979). Only the lower limits of the threshold, representing the most sensitive individuals, are shown in the figure.

Figure 4.3 *Health effects of carbon monoxide.* (Source: *DOE, 1985.*)

from CO exposure include loss of alertness, impaired perception, learning disorders, sleep deprivation, drowsiness, confusion, and, at higher levels, coma and eventual death.

Several oxides of nitrogen (NO, NO_2, N_2O, and N_2O_3) can be present in combustion products, and all have the potential to affect human health. Although NO persists indoors for a longer time period, only NO_2 has been studied in sufficient detail to understand its effects. Like carbon monoxide, the health effects of oxides of nitrogen at present indoor levels are not properly understood. At high concentrations, NO_2 is known to cause lung damage (NAS, 1976). Both NO and NO_2 can interfere with the oxygen-transport properties of blood by forming methemoglobin. Most earlier studies focused on the exposure to NO_2 from outdoor air. However, with the recognition that the indoor concentrations of NO_2 are often higher than outdoor levels, numerous experimental and epidemiologic studies have been conducted with humans in a controlled atmosphere. Various health effects related to the exposure to NO_2 are summarized in Figure 4.4. These results show an increased incidence of acute respiratory infections, especially among infants and children, resulting from NO_2 and probably NO exposure. The exposure to NO_2 and NO during childhood can cause chronic lung disease later in life.

NO_2 and NO are highly reactive species that can easily combine with other indoor pollutants to produce more toxic substances. Both NO_2 and NO can react with amines, benzo(a)pyrene, and pyrelene to form carcinogenic nitrosoamines and mutagens such as hydroxybenzo(a)pyrene and 3-nitropyrelene.

Woodsmoke is a complex chemical mixture of gases and particles that has the potential for causing respiratory infections. Most of the data related to health effects from woodsmoke exposure have been obtained from less developed countries. These data suggest that smoke exposure can result in chronic lung diseases and acute respiratory illness. Although the concentrations of smoke particles, such as benzo(a)pyrene, are lower by two orders of magnitude in the U. S. residences that use wood-burning stoves (Samet et al., 1987), the sparse data on health effects related to the residential burning of wood do support those findings. Honicky et al. (1985) conducted a study of respiratory symptoms in 62 children in Michigan, 31 from homes with and 31 from homes without wood-burning appliances. A total of 84% of the children from homes that had woodstoves had at least one severe respiratory symptom, compared to only 3% of the children from homes with no woodstoves. However, in a similar study conducted in Massachusetts, Tuthill (1984) concluded that chronic respiratory diseases of children are not necessarily associated with the burning of wood. Various laboratory scale *in vitro* experiments are underway to determine the health effects from woodsmoke. A laboratory study by Alfheim and Ramdahl (1984) showed that woodsmoke is mutagenic in nature as determined by the Ames salmonella assay test.

Studies have been conducted to determine the indoor concentration levels at which sulfur dioxide, ozone, and hydrogen sulfide become a sensory irritant. Sulfur dioxide concentrations ranging from about 0.25 to 0.50 ppm can cause significant bronchoconstriction in exercising asthmatics. Although the short-term health effects of SO_2 at this level tend to be related to irritation, the long-term effects may cause changes in lung function, making them more susceptible to damage by other pollutants. Ozone generally causes irritation of the eyes and respiratory system. These effects depend on the ozone concentration and the

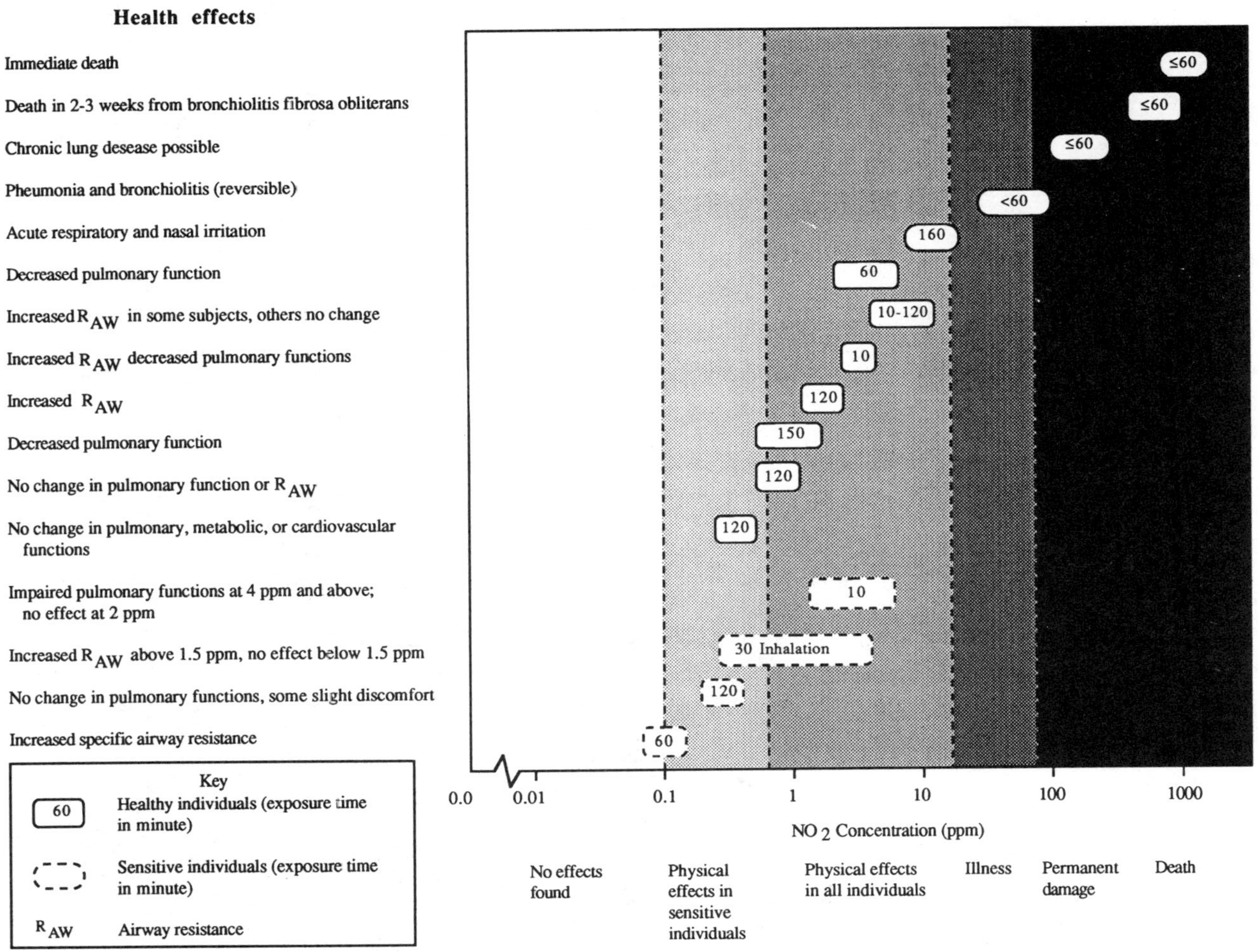

Figure 4.4 *Effects of short-term exposure (<3 h) to NO_2 in healthy and sensitive humans.* (Source: *DOE, 1985.*)

duration of exposure. Lippmann (1989) suggested that, at concentration levels below 120 $\mu g/m^3$, ozone may not produce a chronic effect on human health. However, the outdoor concentration, which is the main source of ozone indoors, frequently exceeds this level. In Europe the ozone concentration in rural areas is higher than 150 ppb; whereas, in urban areas its concentration level exceeds 180 ppb. More than one-half of the population in the United States is exposed to an outdoor ozone concentration that exceeds 120 ppb.

Toxic effects of chlorine gas on the respiratory tract are well documented clinically. Exposure to chlorine gas in a domestic environment generally occurs when two cleaning agents, one containing an acidic compound (most often hydrochloric acid) and another containing sodium hypochlorite, are mixed together with the intent of improving cleaning power. Although no long-term health effects from chlorine inhalation have been reported, short-term symptoms may include coughing, breathlessness, irritation of the upper airway, bronchospasms, nausea, and vomiting.

4.3 SOURCES AND INDOOR CONCENTRATIONS

The major combustion products are CO_2, CO, NO_2, NO, and SO_2, along with formaldehyde, hydrocarbons, and respirable particles. In addition to the emissions from combustion appliances, formaldehyde is produced by several other sources. The characteristics and other aspects of respirable particles are discussed in detail in Chapter 6. The U. S. Consumer Product Safety Commission developed a matrix that interrelates various factors that determine indoor pollutant concentrations from combustion sources. This diagram is shown in Figure 4.5.

Unvented Space Heaters

The increase of energy costs over the last two decades has significantly increased the use of unvented space heaters in the United States. Space heaters are used as a primary heating source for homes in warm climates, such as in the southern states, and as a secondary heat source during winter months in cold climates. For homes with central heating systems, it is a very common practice to lower the central thermostat setting during the nighttime. The required heating in bedrooms is then provided by a gas or kerosene heater. Barnes et al. (1990) estimated that from 15 to 17 million kerosene heaters have been sold in the United States, and one-third of those were purchased by residents of mobile homes.

Unvented space heaters can be classified into two categories. Those heaters that use natural gas as the fuel are called unvented gas space heaters, while those that use kerosene as a fuel are referred to as unvented kerosene space heaters. Kerosene heaters can be further classified into convective, radiant, convective/radiant, two stage, and wickless, depending on the burner design. Similarly, two types of gas space heaters, radiant and convective, are available.

Convective kerosene heaters are generally cylindrical in shape and operate at a relatively high combustion temperature. They produce a blue or white flame, depending on the burner design; a white flame burns hotter than a blue flame. The cool air drawn through

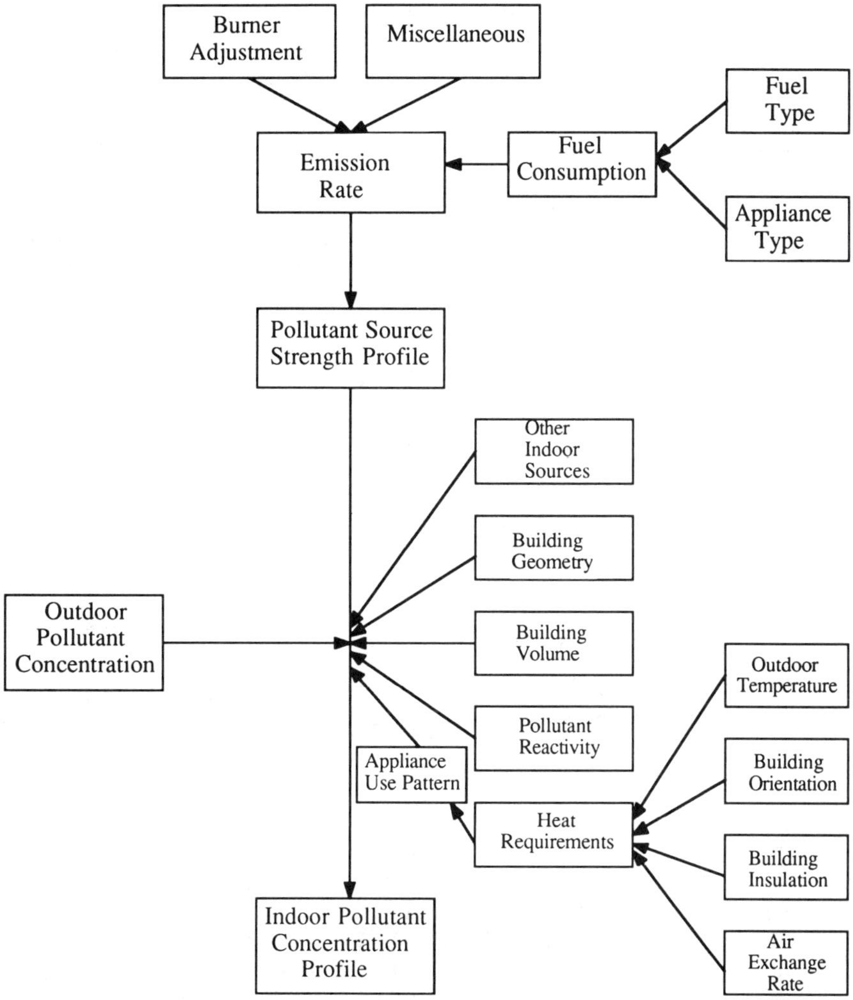

Figure 4.5 *Interrelated factors that determine indoor pollutant concentration from combustion appliances.* (Source: *CPSC, 1983.)*

slots at the base surrounding the burner is heated by the flame and then travels upward. The heat is distributed in the room through slots at the top of the heater, which creates the convective current. Most convective heaters are rated at about 6000 to 10,000 Btu/h by the manufacturer. Radiant or infrared heaters incorporate metal reflectors or perforated ceramic plates that become red hot and radiate heat from the burner. Most units have blue flames and have a rated heat output in the range of 12,000 to 20,000 Btu/h, according to the manufacturer. Radiant/convective heaters combine the design concept of both a radiant and convective heater. In a two-stage kerosene heater, a second chamber above the radiant element is used to increase the combustion efficiency by introducing secondary combustion air, thus preventing the emission of uncombusted hydrocarbons and CO. Wickless kerosene heaters contain a chamber in which the fuel and air are mixed prior to combustion, and a fan is used to disperse the heated air outward. Unvented gas space heaters generally produce a blue flame when well tuned and, depending on the fuel input ratings, can produce from 12,000 to 40,000 Btu/h.

The emission rates of pollutants from kerosene space heaters depend on several factors, including usage pattern, type of heater, fuel type, and the age of the heater. Most studies have been conducted in laboratories or environmental chambers under controlled conditions. Pollutant emission rates from these studies are summarized in Table 4.2. Leaderer (1982) studied emissions from portable convective and radiant kerosene heaters in a room similar in size to a modern bedroom using a normal ventilation rate (<1 air change per hour). He observed that the concentration of CO was in the range of 5 to 50 ppm, CO_2 varied from 1000 to 10,000 ppm, SO_2 exceeded 1 ppm, and NO_2 was in the range of 0.5 to 5 ppm. Yamanka et al. (1979) and Weston (1985) placed a hood over heating units to collect combustion products. A fan was used to draw combustion gases into the duct for sampling. Yamanka et al. (1979) studied 11 kerosene space heaters, of which 6 were the radiant type. They measured the NO_2 concentration only by placing a hood over the heating units to collect samples, and they found that the emissions from a convective heater operating with a blue flame are significantly higher (60 μg/kg of fuel burned) than emissions produced by radiant heaters (11 μg/kg). Weston (1985) tested three heaters (a new radiant heater, an identical old radiant heater, and a new white-flame convective heater) in an occupied residence, also by using a hood to collect samples. The kerosene used in the tests had a sulfur content of 0.02%. As expected, the emission rates of NO_2, CO, and SO_2 were higher with the old radiant heater than with the new unit. The Consumer Product Safety Commission (CPSC, 1983) conducted extensive studies with kerosene heaters in 130 homes. Results from these studies can be summarized as follows: CO was the major pollutant from heaters that burned with a blue flame. A heater operated with a blue flame produced almost five times more CO than heaters operated with a white flame. The NO_2 emission rates produced by blue and white flames and wickless heaters were significantly higher than other pollutants. Maladjusted wick heights can result in a two- to four-fold increase in CO concentration. As expected, the sulfur dioxide emission rate was dependent on the sulfur content of the fuel.

Ritchie and Oatman (1983) studied the effect of ventilation, heater age, and heater size, in relation to the room size, on pollutant levels for three convective kerosene heaters in a townhouse. As can be seen from Table 4.3, pollutant concentrations exceeded the ASHRAE recommended guidelines in several cases. Traynor et al. (1982) tested two radiant and two convective kerosene heaters (a new and an old unit) in a 27-m^3 environmental chamber. Later Apte and Traynor (1986) included two, two-stage kerosene heaters and several unvented gas space heaters in their study. All heaters emitted NO, NO_2, CO, CO_2, respirable particulates, and formaldehyde. Their findings on kerosene heaters are presented in Table 4.2, while the results of their study for gas space heaters are shown in Table 4.4. They used both natural gas and propane in the gas space heaters. The emission rate of CO_2 depended on the type of fuel used and the duration of burning. The average CO_2 emission

TABLE 4.2 Emission Factors of Kerosene Space Heaters

Investigators	Type	No. of Heaters	Pollutants (%g/kJ)[a]						
			CO	NO	NO_2	NO_x	HCHO	Particles	SO_2
Apte and Traynor (1986)	Convective	3	25.0	14.1	16.3	11.6	0.31	ND	—
	Radiant	4	64.0	1.3	4.7	1.9	0.29	0.49	—
	Two stage	2	9.2	4.3	2.2	2.7	0.20	1.7	—
Yamanaka et al. (1979)	Radiant	6	—	—	11.0	—	—	—	—
	Convective	5	—	—	60.0	—	—	—	—
Caceres et al. (1983)	Radiant	2	54–107	0.8–1.6	3.2–5.0	—	0.1–0.11	—	20.5–58
	Wick	1	145	4.6	4.4	—	0.06	—	—
Woodring et al. (1985)	Radiant	5	76–221	—	3.8–6.0	—	0.18–0.53	1.3–5.1	4.7–46.7
	Convective	3	123–213	—	4.1–20.7	—	0.02–0.45	3.9	3.9–33.2
Leaderer (1982)	Radiant								
	Normal flame	1	39	0.53	3.3	—	—	—	10.8
	High flame	1	36	0.4	3.8	—	—	—	14.2
	Low flame	1	39	3.0	—	—	—	—	9.0
	Convective								
	Normal flame	1	14.9	7.8	9.7	—	—	—	10.6
	High flame	1	10.9	17.7	7.5	—	—	—	16.3
	Low flame	1	6.6	2.7	4.4	—	—	—	4.5

[a] Emission rates log normally distributed; geometric means are given.

—, no data reported.

ND, not determined.

HCHO, formaldehyde.

TABLE 4.3 Average and Maximum Pollutant Concentrations Measured in a Residence During the Use of Three Convective Kerosene Heaters Operated for 4-h Periods

Nature of Study	*Particulate Matter*[a], *mg/m*^{3} *(avg)*	*Sulfur dioxide, ppm (max)*	*Nitric oxide, ppm (max)*	*Nitrogen dioxide, ppm (max)*	*Carbon monoxide, ppm (max)*	*Carbon dioxide, % avg (max)*
11,500 Btu new heater smaller than required[b], without ventilation (0.49 ach)[c,d], 65°F start, 81°F end[e]	168	0.141(0.16)	0.28(0.34)	0.05(0.06)	3.1(4)	0.32(0.41)
With ventilation[f], (1.1 ach), 64°F start, 76°F end	117	0.112(0.14)	0.14(0.2)	0.05(0.06)	1.8(2)	0.15(0.2)
18,000 Btu new heater larger than required, without ventilation (0.81 ach), 63°F start, 82°F end	149	0.183(0.225)	0.43(0.50)	0.04(0.06)	4.5(5)	0.37(0.45)
11,000 Btu 5 year-old heater smaller than required, without ventilation (0.65 ach), 62°F start, 72°F end	349	0.102(0.12)	0.21(0.3)	0.11(0.13)	6(7)	0.23(0.25)
Central, no heater, without ventilation, 62°F start, 65°F end	39	0.005(0.005)	0.02(0.020)	none detected	2(2)	0.05(0.05)
Recommended indoor concentration by ASHRAE	260 (24-h avg)	0.140 (24-h avg)	0.41 (24-h avg)	0.05 (annual avg)	9.0 (8-h avg)	0.25 (continuous)

[a] Formaldehyde was not detected (lower detection level = 0.1 ppm) during any of the tests.
[b] A heater rated at 13,800 Btu/h would be required based on the floor area (490 ft^2) of the space to be heated.
[c] CO_2 decay was used to determine ach (air changes per hour).
[d] O_2 decreased by 1% from 21%.
[e] Indoor temperatures at start and end of 4-h tests.
[f] Ventilation provided by opening a window; 44 in^2/11,000 Btu/h heater output.
(max) - Maximum concentration recorded during a test.
Source: Ritchie and Oatman, 1983.

TABLE 4.4 Emission Factors for Unvented Gas Heaters

				Pollutants (μg/kJ)[a]					
Investigator	*Type*	*Fuel*	*No. of Heaters*	*CO*	*NO*	NO_2	NO_x	*HCHO*	*Particles*
Apte and Traynor (1986)	Gas convective heaters	Natural gas	9	33	17.5	13.8	12.4	1.2	0.48
		Propane	3	16	26.6	13.0	16.4	1.0	0.27
	Gas infrared heaters	Natural gas	4	47	0.4	4.6	1.6	0.68	0.27
		Propane	1	45	0.1	5.5	1.7	1.0	0.13
Yamanaka et al., (1979)	Bath furnace gas burner	LPG	2	—	—	45.0	—	—	—
	Bath furnace gas burner	Town gas	3	—	—	30.0	—	—	—
	Gas stove	Natural gas	1	—	—	2.0	—	—	—
Caceres et al., (1983)	Radiant type gas stove	LPG	5	64–127	0.46–1.1	1.2–4.4	—	1.3–1.6	—
		LPG	1	26	—	—	—	2.47	—
		Town gas	1	161	3.0	14.6	—	2.37	—

[a] Emission rates log normally distributed; geometric means are given.

—, no data reported.

rates were 49,800 ± 2400 mg/kg from a natural gas heater, 60,500 ± 1200 mg/kg from a propane heater, and about 70,000 mg/kg from a kerosene heater. Coté et al. (1974) also measured emission rates of CO, NO, and NO_2 from unvented gas space heaters under various operating conditions. All three pollutant concentrations were found to be higher when the heater burned at a high-flame setting with a heat output of 24,500 Btu/h, compared to a low-flame setting with a heat output of 11,100 Btu/h. The American Gas Association compared 17 different gas-fired room heaters to determine their pollutant emission rates; five of these were unvented. Differences in emission levels of vented and unvented heaters were statistically insignificant. Although they analyzed the flue gas for CO, NO, NO_2, CH_4, C_2H_4, and total aldehydes when these heaters were operated at various conditions, only concentrations of NO, NO_2, CO, and formaldehyde were reported. These are presented in Table 4.5.

In summary, convective kerosene heaters produced higher concentrations of nitrogen oxides and lower CO levels. Older units had higher emission rates for CO, NO_2, and particulates. The adjustment of the flame appeared to have a greater influence on the CO emission rate than on the emission of other pollutants. Sulfur dioxide emission rates depended only on the sulfur content of the fuel and were independent of the heater type, age, and flame setting. Emission rates from unvented gas-fired space heaters were different for different brands of heaters and changed with the air/fuel ratio and with flame adjustment.

Wood-burning Stoves, Fireplaces, and Furnaces

Rising energy costs are prompting an increasing number of homeowners to use wood as a source of energy for residential heating. As noted by the National Research Council (NAS, 1981), this trend is particularly strong in the northeastern part of the United States, which has traditionally used oil as the heating fuel. An estimate by the Office of Technology Assessment (GPO, 1984) indicated that more than 11 million wood stoves are in existence in the United States and are used primarily during the cold months of the year.

Wood-burning appliances are potential sources of pollutants for both indoor and outdoor air. An increase in the total suspended particles in the ambient air due to residential wood burning was observed in various communities during the heating season. Combustion products from woodstoves and fireplaces, which include NO, NO_2, CO, CO_2, SO_2, respira-

TABLE 4.5 Emission Rates from Vented and Unvented Gas-fired Room Heaters[a]

Operating Conditions	*Concentration*	*NO (mg/h)*	*NO_2 (mg/h)*	*CO (mg/h)*	*HCHO (mg/h)*
Vented	average	1562	187	493	29
	range	317–4578	<3–1225	12–5004	0.2–300
Unvented	average	865	143	122	4.1
	range	183–2009	26–331	23–315	4.4

[a]Average emission rate based on 17 heaters of each type.

Source: Thrasher and DeWerth, 1979.

ble suspended particles, benzo(a)pyrene, and HCHO, are generally vented to the outdoor atmosphere. However, numerous circumstances can cause combustion products to be vented indoors. These include insufficient stack height, cracks and leaks in stove pipes, downdrafts, log rollover in fireplaces, and a negative air pressure indoors.

Several manufacturers produce a wide variety of wood-burning appliances, with efficiencies that vary from a minus 10% to as high as 70% (DOE, 1983). The negative efficiency indicates that the fireplace draft actually removed more heated air from the room than was supplied by the appliance. Efficiencies of various wood-burning units are shown in Table 4.6. Although wood appliances are a potential source of pollutants, very few studies have been conducted that fully characterize combustion products from the burning of woods. Duncan et al. (1980) measured emission rates of various chemicals, particles, and metals from a wood-fired stove. The emission of sulfur compounds ranged from 0.5 to 1.5 lb per cord of wood burned; whereas, from 0.7 to 2.6 lb of NO_x and from 300 to 1200 lb of CO were produced per cord of wood. Emission rates of pollutants are also strongly dependent on the type of wood and the appliance used to burn it. Table 4.7 provides a summary of the emission rates for several types of wood. Although the burning of wood emits the above pollutants and a number of polynuclear aromatic hydrocarbons (PAH) and benzo(a)pyrene, the contribution of these gases to indoor air pollution is rather controversial. Moschandreas et al. (1981) observed elevated levels of CO, but not of NO, NO_2, and SO_2. Godish and Ritchie (1985) obtained similar results, but found no difference in NO_2 levels between houses that used wood-burning stoves and houses that used electrical appliances. However, their results were contradicted by Traynor et al. (1985) and Knight et al. (1986). Traynor et al. field tested wood-burning stoves and furnaces in three houses and found that CO concentrations ranged from 0.4 to 2.8 ppm in nonairtight stoves. They also observed elevated levels of NO, NO_2, SO_2, and five polycyclic aromatic hydrocarbons. Knight et al. measured the concentration of CO, NO, and NO_2 during the operation of different types of airtight and nonairtight woodstoves. They also reported elevated levels for these pollutants. Their results are shown in Table 4.8.

The pollutant emission rates from woodstoves depend on whether the stove is clean

TABLE 4.6 Efficiencies of Various Wood-burning Appliances

Appliance Type	*Estimated Efficiency (%)*
Masonry	−10 to +10
Manufactured fireplace	−10 to +10
Manufactured fireplace with circulation and outside air	10 to 35
Free standing fireplace	−10 to +20
Fireplace stove	20 to 40
Radiant stove	50 to 70
Circulator stove	40 to 55
Fireplace insert	35 to 50
Furnaces	50 to 60

Source: DOE, 1983.

TABLE 4.7 Emission Rates from Different Types of Wood

Appliance Type[a]	Wood Type[b]	Burning Rate, kg/h	Emission Factor, g/kg						
			CO	Particulates	Condensable Organics	Volatile Organics	Total Hydrocarbons	NO_x	POM[c]
WS	sro	0.82–2.58	138–196	0.8–22.2	ND	ND	ND	ND	ND
WS	sro	3.9–7.7	55–148	0.5–2.3	ND	ND	ND	ND	ND
WS	sro	0.82–2.58	ND	ND	3.0–8.4	ND	ND	ND	0.004–0.04
WS	sro	3.9–7.7	ND	ND	0.8–2.1	ND	ND	ND	0.008–0.03
WS	o, gp, fb	2.6–14.5	33–400	ND	ND	ND	2–112	0.1–4.9	0.011–0.05
WS	o, p	0.93–3.79	ND	1.3–24	ND	ND	ND	ND	ND
WS	o, p	1.2–5.3	4–147	7.6–45	ND	ND	0.25–16.5	1.6–7.3	ND
FP	o	12.1–21.2	62–120	17–64	ND	ND	7.6–4.85	1.0–2.3	ND
WS	ssm	1.15–4.46	87–184	ND	ND	ND	20–43	ND	ND
FP	ssm	2.64	177	ND	ND	ND	42	ND	ND
WS	so, go	6.0–8.4	91–370	1.8–6.3	1.3–12.0	0.3–3.0	ND	0.2–0.8	0.19–0.37
FP	sp, gp	9.6–11.4	15–30	1.8–2.9	5.4–9.1	19	ND	1.4–2.4	0.025–0.036
WS	df	1.5–4.5	ND	2.6–42.5	7.0–32.1	ND	ND	ND	ND
WF	df	4.7–16.4	ND	0.8–4.2	1.8–14.9	ND	ND	ND	ND

[a] WS, wood stove; WF, wood furnace; FP, fireplace.

[b] sro, seasoned red oak; o, oak; gp, green pine; fb, fir brand; p, pine; ssm, seasoned sugar maple; so, seasoned oak; go, green oak; sp, seasoned pine; df, Douglas fir.

[c] Polycyclic organic material.

ND, not determined.

Source: DOE, 1983.

TABLE 4.8 Emission Rates from Airtight and Nonairtight Wood Heaters

		Average Emission Rate of Pollutants (mg/h)				
Investigators	*Type*	*CO*	*NO*	*NO_2*	*TSP*	*RSP*
Knight et al. (1986)	Airtight					
	Radiant heater	55–182[a]	0.4–2.6[a]	0.9[a]	4.1–7.5[a]	2.3–3.6[a]
		70–375[b]	0.6–0.8[b]	0.3–0.7[b]	4.2–42.9[b]	7.5[b]
	Conventional	69[a]	3.2[a]	2.4[a]	4.3[a]	3.4[a]
		95[b]	0.7[b]	0.2[b]	5.7[b]	2.3[b]
	Nonairtight					
	Box type, radiant heater	210–416[a]	2–4[a]	2.1–3.6[a]	11.9–12.7[a]	5.5–6.9[a]
	Franklin type freestanding fireplace heater	530[b]	9.4[a]	3.1[a]	20.6[a]	11.1[a]
Imhoff (1984)	Clean woodstove	86	1.2	1.3	2.6	-
	Dirty woodstove	560	3.9	7.0	100.0	-

[a], 12-h tests.

[b], 24-h tests.

—, no data reported; TSP, total suspended particles; RSP, respirable suspended particles.

or dirty, its design, installation, maintenance, and operation practices. Emission rates of CO and NO_2 from a clean stove can be about 86 and 1.3 mg/h, respectively, whereas these values can jump to 560 and 7.0 mg/h, respectively, for a dirty stove. In summary, the burning of wood can elevate the levels of CO, NO, NO_2, SO_2, formaldehyde, and particulates indoors, but emissions vary considerably, depending on the type of wood burned.

Gas Stoves and Ovens

Gas stoves and ovens are typically used in residences for only a relatively short period of time each day, but on almost every day throughout the year. These stoves are significant sources of CO, CO_2, NO, NO_2, and aldehydes indoors. When cooking with a gas stove, the hourly average CO concentration in most homes ranges from 2 to 6 ppm; in some conventional homes it may exceed 12 ppm. The use of gas stoves may increase the CO levels by as much as 25 to 50 ppm. Traynor et al. (1982) found that CO levels increased to about 20 to 25 ppm within the first 30 minutes of operation of a gas stove in the kitchen of a test house with no occupants. The nitrogen oxide levels rose to 1.2 ppm in the first 45 minutes. The average NO_2 level can increase by 25 ppb over the background level during the normal use of gas stoves; peak levels in the kitchen may reach 200 to 400 ppb when cooking. Using a computer model, Sexton et al. (1983) predicted that more than 25% of the occupants in homes with gas stoves would be exposed to over 50 ppb annually of NO_2, which is the national annual average.

A number of experimental studies have been conducted to measure emission rates of pollutants from gas burners and ovens. Generally, the production of NO_2 occurs at the start-up of the burner, but NO formation occurs throughout the combustion process. The emission of these two pollutants, therefore, can be reduced substantially by modifying the burner design and by using more primary air. Coté et al. (1974) tested a new gas stove that had pressed steel burners and an old stove that had cast iron burners, under various operating conditions. Concentrations of NO, CO, and NO_2 were measured, but a direct comparison between emission rates was not possible because of differences in the heat input. However, CO emissions from ovens operated at steady-state conditions were higher from the new stove than from the old one. Their results are shown in Table 4.9. A more detailed study of emissions from gas stoves was carried out at the Lawrence Berkeley Laboratory. Emissions of CO, CO_2, NO, NO_2, SO_2, HCN, HCHO, and particulates were measured from both a top burner and an oven and are presented in Table 4.10. The oven was operated at 350°F for one hour, while the burners had water-filled cooking pots on the top. Coté et al. (1974) and Hollowell et al. (1976) also studied pollutant emission rates from top burners under similar conditions, but they found no appreciable differences in emission rates, even when pans made of different materials were placed on the burners. From experiments with 18 different natural-gas-fired ranges, the American Gas Association found that the cooking utensil material and size had a significant effect on the nitrogen oxide emission rate. Lower emissions of nitrogen oxides were reported when larger pots were on top of the burners.

In summary, the air exchange rate, number of heaters, frequency of use, and air space volume all have an impact on the quality of the air in those homes that have gas heaters and stoves. However, the design of the range-top burner had the greatest influence on emission rates of pollutants, particularly on those of oxides of nitrogen. The fuel composition had minimal effect on CO, NO, and NO_2 emissions, but the emission of particulates was higher from the top burner than from the oven.

TABLE 4.9 Emissions from Gas Stoves

Appliance Type	*Mode of Operation*	*Input Heat Rate, kcal/h*	*Pollutant Emission Rates, mg/h*		
			NO	*NO_2*	*CO*
Older gas stove with cast-iron burners	Pilot lights	150	6.8	8.2	62.9
	1 Burner, high flame	2700	250	140	1031
	3 Burner, high flame	6780	793	494	3220
	Oven steady state	2200	201	161	1166
New gas stove with pressed-steel burners	Pilot lights	100	0.5	1.9	84.2
	1 Burner, high flame	3500	455	277	1795
	3 Burner, high flame	10,200	1408	669	3213
	Oven steady state	2200	171	111	3564

Source: Coté et al., 1974.

TABLE 4.10 Emission Rates from Gas Range and Oven with Cooking Pots

		Emission Rate, mg/h[a]						
Appliance Type	*Fuel Consumption*	*NO*	*NO_2*	*CO*	*CO_2*	*SO_2*	*HCN*	*HCHO*
Gas-fired oven operated at 350°F for 1 h	7970 Btu/h	56	85	1898	358–680	0.92	0.126	23
Gas-fired burner operated for 16 min	8730 Btu/h/burner	89	136	1840	416–944	1.47	4.6–3.8	16

[a]Average emission rates.

Source: Hollowell et al., 1976.

Other Combustion Sources

Several other sources can elevate indoor levels of combustion products significantly. These sources include tobacco smoke, water heaters, washers, dryers, attached garages, and hobbies involving combustion processes, such as soldering and welding. Although tobacco smoke is the primary indoor source of respirable particles, it also can contribute significant amounts of combustion products to indoor air pollution, depending on the number of cigarettes being smoked by the occupants. According to Weber (1984), smokers in the workplace can elevate indoor concentration levels of CO, NO_2, and particulates by 30% to 70%. The amounts of the various chemicals released from a cigarette during smoking are listed in Table 3.10. Indoor air pollution that results from cigarette smoke is also discussed in more detail in Chapters 3 and 6.

The design of an attached garage can strongly affect indoor air quality. In a number of instances, the elevated concentration level of CO indoors was attributed to automotive emissions from the garage (NAS, 1981). Higher indoor concentrations of nitrogen oxides have been observed in homes that have gas water heaters. If the flue collar on the top of the water heater is not designed properly to vent combustion products to the outside, nitrogen oxides and other combustion products will be emitted indoors.

4.4 SAMPLING AND MEASUREMENT

The measurement protocol depends largely on the objectives. Personal monitors are available only for CO, NO_2, and SO_2 and can work in either the passive or active mode. The two most frequently used personal CO monitors are known as COED-I and COED-II (CO exposure dosimeter). These were originally developed for the EPA by the General Electric Company. The COED-II was designed for use in a moving automobile. The COED-I is capable of recording 1600 data points over a 24-hour exposure period with an automatic data logging system. The COED-I monitor works in the active mode by employing a small pump installed within the monitor that continuously draws CO-laden air into an electrochemical cell which contains deionized water. The CO reacts with the water and changes its electrical properties in a manner that is proportional to the CO concentration in the air. The concentration of CO is read directly from the LCD display in parts per million. The measurement can be affected by temperature fluctuations and the presence of other chemical pollutants, such as SO_2 and NO_2. The interference from NO_2 is generally taken care of by using a filter coated with an oxidizing agent (potassium permanganate impregnated on activated alumina). With this filter, the only other pollutants that interfere with the response from the monitor are H_2, C_2H_2, and C_2H_4 in the concentration range of 100 ppm.

Several types of NO_2 personal monitors have been developed using various working principles. These monitors have been tested and certified by the National Institute of Standards and Technology (NIST) and are recommended by the EPA. The monitor that is most frequently used for monitoring personal exposure to NO_2 is known as the Palmes tube. This monitor works in the passive mode and is based on the principle of molecular diffusion. The sampler consists of three stainless steel screens coated with triethanolamine and is fitted to one end of an acrylic tube. One end of the tube is capped permanently, while the other end has a removable cap. The tube is exposed to air by removing the cap. The NO_2 accumulated on the triethanolamine-coated wire screens is collected by washing with a special reagent, known as Griess-Saltzman solution, which consists of 1 part water, 1 part sulfanilamide, and 0.1 part N-1-naphthyl ethylene-diaminedihydrochloride. The concentration of NO_2 is determined from the intensity of the resultant color by using a spectrophotometer set at a wavelength of 540 nm. On the basis of Fick's law of diffusion, the Palmes tube collects approximately 93% ± 3% of the theoretical amount. A NIST study estimated that a minimum of 0.05 mg must be collected on the wire screens to provide an accurate measurement. The concentration level at which NO_2 is present indoors will require approximately 24 hours of sampling time. The interference from nitrogen oxide (NO) is insignificant. The NIST also developed an active sampler for NO_2 by impregnating triethanolamine on a 0.8-mm Nuclepore filter. A higher sampling rate can be used, thus reducing the measurement time by one to eight hours.

West and Reiszner (1980) used a dimethylsilicone membrane that was highly selective for NO_2 to separate NO_2 from air. The diffusing NO_2 that passed through the membrane was then collected on a polyester film coated with triethanolamine. This sampler was found to be about 10 times more sensitive than the Palmes tube. Interscan Corporation developed another passive sampler on the basis of an electrochemical reaction of NO_2. In the sampler, the NO_2 diffuses into an electrochemical cell that contains deionized water and changes its conductivity. This produces an electrical signal that is proportional to the NO_2 concentration. However, the presence of NH_3, CO_2, methyl and ethyl mercaptans, and SO_2 in air can interfere with the measurement. A Japanese company, Toyo Roshi, is marketing a NO_2 badge for collecting NO_2 from the air. A filter treated with triethanolamine is used to adsorb the NO_2 that diffuses through a series of hydrophobic fiber filters. The filters are used to reduce the interference from other gases and to reduce the effect of wind velocity and relative humidity. However, this badge is not certified by NIST nor is it recommended by the EPA.

Sulfur dioxide is not a high-priority indoor pollutant because its source is rather limited and its indoor concentration is rarely very high. As a consequence, very few personal monitors have been developed for SO_2. A personal passive SO_2 analyzer is available from the Interscan Corporation, which works on the principle of an electrochemical reaction of SO_2 with deionized water. The SO_2 diffuses into the cell that contains the water and changes the conductivity of the water, which is calibrated against the SO_2 concentration. The accuracy of the measurement can be affected by the presence of NH_3, H_2S, and methyl and ethyl mercaptans if they are present at the same level as SO_2.

The emission rates from combustion appliances can be measured either in an environmental chamber or by placing a hood over the appliance. The environmental chamber described in Chapter 3 can be used, with some modification, to study emissions from combustion appliances. The effectiveness of various range hoods as a means of removing air pollutants also can be tested in the same chamber. Although the instruments used for measuring inorganic gases are typically located outside the chamber, the emissions of respirable particles and formaldehyde need to be measured with the equipment placed inside. When studying appliances, such as unvented gas space heaters, a wall composed of solar panels is used to help remove radiant heat. In chamber tests, special care must be taken to ensure a proper air exchange rate to prevent O_2 deficiency in the chamber during combustion. Generally, an air exchange rate of 0.5 per hour (ach) is used in most of the studies. Girman et al. (1983) have described the design of an environmental chamber at the Lawrence Berkeley Laboratory and a Mobile Atmospheric Research Laboratory for studying the emissions from combustion appliances.

The in-situ measurements of combustion products can be conducted by placing a canopy-type hood over the appliance. Such an approach can provide real-life operating conditions and more realistic data. The sampling probes are inserted into the ventilation duct to collect samples. Moschandreas et al. (1987) used a specially designed quartz dome placed directly over the burner of a gas range to collect samples. The quartz dome prevented the dispersion of products and channeled them to a Teflon bag for later analysis in a nearby laboratory. The water vapor was removed from the sample by using an ice trap. The authors used this method to measure emissions from a number of gas stoves in homes in the Chicago metropolitan area.

Either portable or stationary instruments can be used to measure concentrations of various inorganic gases in a chamber or in homes. A number of portable instruments are available commercially for the measurement of most inorganic gases. The most commonly used method for measuring CO and CO_2 is by nondispersive infrared (NDIR) photometry. The NDIR principle is the basis for one of the EPA reference methods for measuring CO. This measurement technique involves determining the difference in the infrared energy absorption over all wavelengths passed by the optical system between the gas sample containing the pollutant of interest and a reference path. The difference in the infrared energy absorbed is proportional to the concentration of the compound in the sample gas. The quantitative absorption of energy by CO or CO_2 in the sample cell is measured by a suitable detector. The interference may arise from gases that absorb infrared radiation in bands that are close to those of CO or CO_2. Carbon monoxide has its absorption peak at 2165 cm^{-1}. At this wavelength the major interfering gases are water vapor and CO_2. Some other gases, including methane and ethane, may also interfere. However, the interference can be reduced or eliminated by using a dual-beam arrangement that incorporates an interference cell in line with the sample cell. The interference cell contains the interfering gases in a concentration sufficient to block the radiation from the overlapping portion of the absorption spectrum. If a continuous flow through the reference cell is used, the CO from the sample gas is removed by using a CO converter. Therefore, the CO_2 and water vapor concentrations in the reference cell are essentially the same as that in the sample cell, and they cancel out the interference.

The gas filter correlation (GFC) method is certified by the EPA for measuring CO concentrations. The basic principle is the same as the NDIR method, but a single source for the infrared is used. The radiation is passed through a spinning gas filter wheel that contains a CO gas cell and a N_2 gas cell. The alternating beam of radiation is passed through the cells to a solid-state detector. The difference in the absorption between the sample cell and the N_2 cell is proportional to the CO concentration in the gas sample. These two methods are generally very reliable and are sensitive to the indoor CO concentration level.

The indoor concentrations of nitric oxide and nitrogen oxide are determined from the rapid gas phase chemiluminescent reaction of nitric oxide with excess ozone. The reaction occurs in a light free chamber, producing NO_2 in a highly excited energy state. The NO_2 decays quickly to a stable energy state, emitting photons (or light) in a broad frequency band having a peak at about 1200 nm. The intensity of the emitted light is proportional to the nitric oxide concentration and is measured by a photomultiplier tube. The nitric oxide is determined directly in a sample stream. To measure NO_2 in a sample gas, the NO_2 is first converted to NO in a converter, and then the concentration of NO is determined by the above procedure. However, if both NO and NO_2 are present, this method provides the total nitrogen oxide concentration in air. Two types of converters are available for conversion of NO_2 to NO: thermal converters and catalytic converters. Thermal converters employ a very high temperature, 600° to 800°C, to decompose NO_2 to NO and O_2. This converter has the disadvantage that NH_3 can react with O_2 to produce NO at this temperature. Catalytic converters employ a much lower temperature. The NO_2 conversion takes place in the temperature range of 200° to 400°C in the presence of a molybdenum or carbon catalyst.

The individual concentration of NO and NO_2 in the gas mixture can be determined by employing the same technique, after some modification of the instrument. Two basic instrument designs are available commercially: the cyclic mode and a dual-mode arrangement. The cyclic-mode instrument consists of a single reaction chamber and a detector. The sample gas is alternately cycled through the reaction chamber for the NO determination or through the converter and the reaction chamber for the total NO_x determination. The NO_2 concentration is determined by subtracting the NO concentration from the NO_x concentration. The dual-mode instrument has two reaction chambers but a single detector. Generally, the sample air stream is split into two streams; one stream is sent to one of the reaction chambers directly, while the other stream is passed first through the converter and then to the reaction chamber. The single detector is exposed alternatively to the chemiluminescent output from the reactor. This procedure for measuring NO and NO_2 is the EPA reference method and is accurate to within ±5%.

The ozone concentration in the gas phase is also measured by the chemiluminescent method. However, in the case of ozone, the reacting gas is ethylene. The lower limit of detection of ozone using this method is 0.003 to 0.004 ppm or 6 to 8 μg of ozone/m^3 of air. A small amount of unreacted gas mixture may be emitted from this instrument during its use. Therefore, proper ventilation is required. Most of the pollutants normally present indoors do not interfere with the measurement, except for water vapor, which is found to interfere in the very low measurement range. A number of instruments have been designed on the basis of the ultraviolet (UV) photometric measurement method for determining ozone concentration. The amount of light absorbed at a wavelength of 254 nm can be correlated to the gas phase ozone concentration. To reduce the interference from other gases, an ozone-free stream is passed through a reference cell, and its absorbance is used for the baseline correction. Commercial analyzers that employ UV detection are capable of measurements in the range of 0 to 1 ppm, with the lower limit of detection being 1 ppb. Atmospheric gases generally do not interfere with the measurement, but some fine particles may do so by scattering the light beam. Most commercial units use a Teflon filter to remove fine particles from the gas stream.

Two methods are available for measuring the SO_2 concentration in ambient air; one is based on the colorimetric method, and the other one is based on flame photometric detection (FPD). In the colorimetric method, SO_2 from air is absorbed into a liquid solution of potassium or sodium tetrachloromercurate by flowing the sample air stream through an absorber containing the solution. The SO_2 reacts with the solution to form a stable chemical complex, mono-chlorosulfonatomercurate. This complex is next treated with a number of other solutions to eliminate interference from other oxidants that may have been absorbed simultaneously in the liquid during sampling. An acid-bleached pararosaniline solution containing phosphoric acid to control pH is added to the complex to form an intensely colored pararosaniline methylsulfonic acid. The intensity of the color, which is proportional to the SO_2 concentration, is measured in a spectrophotometer at a wavelength of 548 nm. This technique is used by the EPA as the basis for the determination of SO_2 in ambient air.

The SO_2 in a flowing gas stream can be burned in a hydrogen flame. The increase in light emissions from the flame (or flame chemiluminescence) during combustion can be detected by a suitable detector. This method is called flame photometric detection. Although a number of other sulfur compounds and hydrocarbons are also combustible in the flame, optical filters are available for selecting the wavelength range for the compound of interest. The intensity of the emitted light, which is proportional to the gas phase concentration of the component, is sensed by a photomultiplier tube mounted on an electrometer that is used for measuring the current output of the tube. The detection limit is about 2 ppb, but any sulfur compound, such as H_2S and methyl and ethyl mercaptan, can interfere with the measurement. Therefore, this instrument provides the total concentration of sulfur compounds in the gas phase, expressed as the equivalent of the SO_2 concentration. The sulfur emission has a band structure in the range of 300 to 425 nm, with the peak band being at a wavelength of about 394 nm. A number of other nonsulfur compounds, particularly hydrocarbons, may appear in the same band. The CH band appears near the 400-nm wavelength. The response of the FPD can be also affected by a high concentration of CO_2 in the sample stream. Often a gas chromatograph coupled with a flame photometric detector is used for measuring individual sulfur compounds in the gas mixture. Various gas chromatographic columns are available for separating the sulfur compounds.

The colorimetric method can also be employed for other inorganic gases, using the same basic methodology described earlier. A list of the reagents to be used for specific pollutants is provided in Table 4.11, along with the measurement ranges.

Sulfur dioxide and other inorganic pollutants, including Cl_2, NH_3, and H_2S, can be collected from air in liquid solutions. Either a special liquid medium or deionized water is used to react with or dissolve the pollutant for subsequent analysis. However, care must be taken in selecting the medium. After collecting the liquid samples, they are analyzed by either titration or using a colorimetric method. Often a unique liquid collecting medium is not available for a specific pollutant. For example, water can absorb most inorganic pollutants because of their high solubilities in water. Therefore, if both CO_2 and SO_2 are present in the air sample, only the total concentration of these pollutants can be determined.

The indoor concentrations of inorganic gases are measured either by using a portable monitor or by collecting the air samples in a Teflon bag or other plastic bag for subsequent analysis with a gas chromatograph. Most of these inorganic gases are called permanent gases due to their extremely low boiling points. These chemicals remain in their gaseous state at room temperature and even at temperatures far below the ambient temperature. Often the chemical characteristics of these gases are very similar, and it is difficult to quantify each gas when they are present at low concentrations. A gas chromatograph coupled with a thermal conductivity detector (TCD) and a flame photoionization detector in series can be used to determine the individual concentration of most inorganic gases in a mixture. The TCD is capable of detecting and quantifying CO, CO_2, and H_2O, while FPD can be used to measure the concentrations of SO_2 and H_2S. By properly choosing a chromatographic column, the interference from each gas present can be minimized.

4.5 SPECIFIC CONTROL STRATEGIES

Typical processes used to remove inorganic gases from indoor air include (1) source control, (2) increased ventilation, and (3) air cleaning. Although any one of these methods can be used to reduce indoor pollutant levels, an effective removal method might include all three control strategies.

TABLE 4.11 Reagents for Colorimetric Determination of Selected Inorganic Gases

Compound	*Reagent*	*Range (ppm)*
Ammonia	Modified Berthelot	0–5
Chlorine	DPD procedure	0–5
Ethylene oxide	Periodate/PRA	0–25
Hydrazine	*p*-DMAB	0–20
Hydrochloric acid	Thixyanate method	0–100
Hydrogen cyanide	Chloramine T/Pyr-Bard	0–1
Hydrogen sulfide	Methylene blue	0–10

Source Control

The complete elimination of combustion appliances from homes as a means of reducing combustion products is unrealistic. However, emission rates can be reduced by using different types of appliances. From the preceding discussion, it can be concluded that wood-burning appliances are one of the worst emitters of pollutants. Therefore, by eliminating wood burning indoors, the indoor pollution can be reduced substantially. Replacement of wood-burning furnaces and woodstoves by oil- or gas-fired furnaces can reduce the emission rates of most pollutants by a factor of 1000 (Table 4.12). A comparison of emission rates for other types of combustion appliances is shown in Table 4.13. One choice may be to replace a woodstove with an electrical appliance, which does not emit combustion products. However, the replacement of a gas stove or kerosene heater by an equivalent electrical heater might not be a practical solution for various cities, such as Los Angeles, where more than 90% of all heating and cooking appliances are operated with natural gas.

Modifications in the design of gas stoves and kerosene heaters and some other simple strategies can reduce the emission rate of combustion products significantly. As reported by Fisk (1986), a recently developed two-stage kerosene heater can reduce CO emission rates by 30% to 60% and NO_2 emission rates by 10% to 50%, compared to a single-stage heater. Shukla et al. (1985) demonstrated that the modification of the burner of a gas stove can lower the emission of CO, NO, and NO_2. Lower emissions of NO_2 were also reported when gas pilots were replaced by electronic pilots. DeWerth and Sterbik (1983) evaluated a number of modified burners and suggested two modifications that will reduce the emission of NO_2 by 25% and other nitrogen oxides by 63%. Other preventive measures include proper operation and regular maintenance of the burner.

Increased Ventilation

Local and mechanical ventilation are most effective in reducing indoor concentrations of inorganic gases. However, the contaminants must be removed through a separate exhaust system to vent them directly outside. The most common form of local ventilation is the hood, which is installed over the cooking range or other combustion appliance. Care must be taken when selecting an exhaust fan or a range hood for local ventilation, however. A high-powered fan can reduce the natural draft in a furnace vent and can cause combustion products from the furnace to be drawn into the living area. Various ductless cooking ranges have recently become available. Instead of having a duct to vent pollutants outside, these ranges are equipped with a carbon filter to adsorb pollutants from the air. The user of these range hoods should be aware of several factors. The filters are often designed poorly, and pollutant adsorption efficiency is very low. Also, the adsorption capacity of carbons for pollutants like CO, CO_2, NO, and NO_2 is very low. Since carbon filters do have a large adsorption capacity for oil vapor and other organic pollutants, they can remove organic pollutants effectively during cooking if properly designed. But the adsorbed pollutants can slowly desorb from the filters into the living space when the filter is not in use. This phenomenon is described later in more detail. Additionally, these filters should be replaced periodically to maintain effectiveness.

A limited number of studies have been conducted to evaluate the change in pollutant

TABLE 4.12 Comparison of Emissions from Residential Fireplaces, Wood Stoves, and Oil- and Gas-fired Furnaces

	Fireplace		*Stove*			
Pollutant	***g/kg***	***μg/J***	***g/kg***	***μg/J***	***Oil Furnace μg/J (10^{-3})***	***Gas Furnace μg/J (10^{-3})***
CO	13–220	0.8–14	10–420	0.6–30	15–30	8.4
Particulate	1.8–64	0.1–5	1.3–45	0.1–4	8–30	9–25
Total hydrocarbons	2.7–49	0.2–4	2.0–110	0.1–8	0.2–3	0.02–0.2
NO_x	0.85–7.7	0.05–0.5	0.2–7.4	0.01–0.5	37–85	28–45
SO_2	—	—	0.16–0.24	9×10^{-3}–10×10^{-3}	22–220	0.26
POM	0.02–0.04	0.001–0.003	0.01–0.4	10×10^{-3}–30×10^{-3}	10–30	—

Source: DOE, 1983.

TABLE 4.13 Pollutant Emission Rates from Combustion Appliances and Tobacco Smoke

Source	Appliance Type	Emission Rate (mg/h)							
		NO	*NO_2*	*CO*	*CO_2*	*SO_2*	*Particles*	*BaP*	*HCHO*
Kerosene space heater	Radiant	0.54–11	16–38	281–542	4×10^5–6×10^5	31–109	0.13–0.16	—	0.66–4.60
	Convective	62–195	33–530	35–635	4×10^5–6×10^5	37–94	<0.03–0.034	—	0.79–6.19
Gas space heater		80–4578	3–1225	12–5004	5×10^5–2.35×10^6	—	0.21–3.23	—	0.2–300
Wood heater		1.2–3.9	1.3–7.0	70–375	—	—	2.6	1.4×10^{-5} 3.5×10^{3}	—
Gas appliance	Range (1 h)	9.5–455	18–430	191–2700	4×10^5	1.29–1.66	1.9–30	—	8–23
	Oven	30–581	67–270	195–3564	5.5×10^5 3×10^5 4×10^5	0.67–1.09	0.118–0.126	—	20–28
Tobacco smoke	Sidestream plus mainstream (mg/cigarette)	—	0.79	51.6–100	443	—	30–40	2.0×10^{-5}	—
	Sidestream (mg/cigarette)	—	0.065	0.5–105	143	—	10.8	1.7×10^{-4}	1

Source: DOE, 1985.

BaP, benzo(a)pyrene

concentration due to increased ventilation. Hollowell and co-workers (1976) measured the concentrations of CO and NO_2 that resulted from the emissions of a gas oven in a 27 m^3-test chamber. Tests were conducted for one hour, and the oven of the gas stove was maintained at 350°F. The peak concentration of CO dropped from approximately 50 to 4 ppm when the air exchange rate was increased from 0.25 ach (air change per hour) to 7.0 ach. The NO_2 concentration also decreased from its peak value of about 1.5 to 0.1 ppm in the same tests. Traynor et al. (1981a, 1981b) studied the effects of infiltration, mechanical ventilation, and local ventilation on CO, CO_2, NO, and NO_2 concentrations in a 260-m^3 unoccupied research house. The emissions entering the house were from a gas oven. An air-to-air heat exchanger was used for mechanical ventilation, and a range hood over the stove was used for local ventilation. Their results show that the initial peak concentrations of combustion pollutants are relatively independent of the air exchange rate. Mechanical ventilation increased the air exchange rate to about twice the ventilation rate of the entire house. This reduced the nitrogen oxide concentration in the kitchen to about one-sixth of the concentration without the mechanical ventilation. A reduction of 60% to 87% in the concentrations of CO, CO_2, and oxides of nitrogen was noted in the living room when the range hood was used along with the gas stove. Their results for NO_2 are summarized in Figure 4.6, which suggest that local ventilation is more effective in reducing the level of combustion products than ventilation of the whole house.

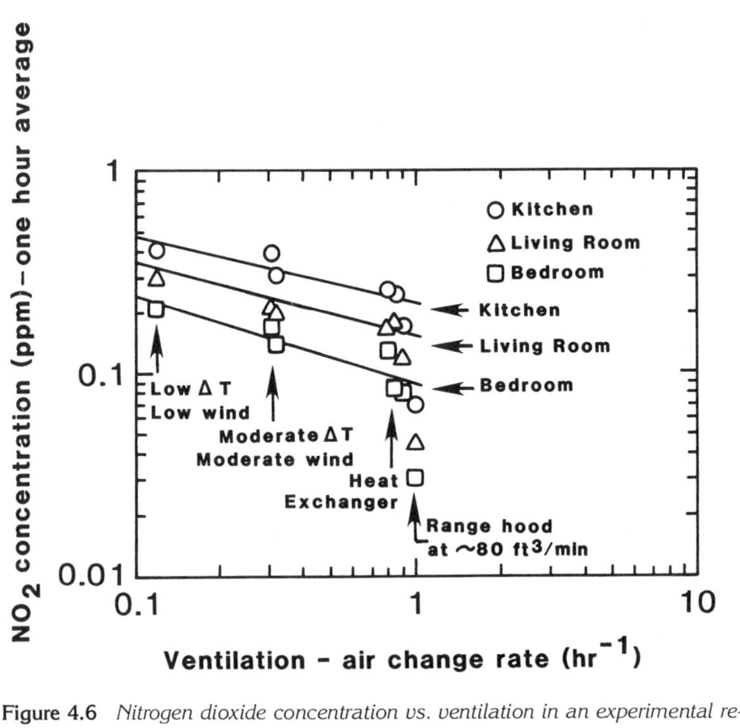

Figure 4.6 *Nitrogen dioxide concentration vs. ventilation in an experimental research house.* [Source: *Traynor et al., (1981a; 1981b).*]

Nagda et al. (1987) came to the same conclusion on the basis of their simulation study with CO. In their study, they assumed that infiltration was constant at 0.3 ach, and a mechanical ventilation system and a range fan installed over the stove each added an additional 0.2 ach. They assumed that each arrangement was used separately. Although both the mechanical ventilation and range fan had the same total air exchange rate (0.5 ach) during the period of operation, the range fan reduced the peak CO concentration from 10 to 3 mg/m^3. Nagda et al. pointed out that the fan was needed only when the range was in operation, whereas mechanical ventilation was used continuously. This increased the total energy consumption.

The CO and NO_2 concentrations resulting from the emissions from a gas stove were measured by Goto and Tamura (1984) in the kitchen of an energy-efficient house. An increase in the air infiltration rate from 0.1 to 0.9 ach reduced the concentration levels by only 20%. However, a 50% reduction in the peak concentration was achieved when the range hood was operated at 22 L/s. This reduction was increased to almost 100% when the rate was increased to 94 L/s.

Offermann et al. (1982) studied the effect of air exchange rates on indoor air quality in nine occupied houses in Rochester, New York, by measuring the concentrations of radon, formaldehyde, and NO_2, along with relative humidity. The air exchange rates varied from 0.22 to 0.50 ach, but were increased from 22% to 114% by mechanical ventilation. At low concentration levels, the outdoor concentrations of pollutants play a significant role. This is evident from the measurement of NO_2 concentrations. The NO_2 levels in five houses increased with an increase in the ventilation rate, but in all cases the outdoor NO_2 concentration was higher than the value indoors.

Air Cleaning

Absorption methods. Absorption processes can be highly effective in removing inorganic gaseous pollutants from an air stream, but the method receives little attention because of the size of the equipment. Liquid-desiccant-based air-conditioning systems are being employed more frequently in various indoor environments, such as hospitals, supermarkets, and commercial buildings. Because many inorganic gases are highly soluble in water, a properly designed liquid-desiccant system can improve indoor air quality along with dehumidification or air conditioning. Very little actual field data are available to evaluate the full capability of absorption in removing inorganic gaseous pollutants. A more in-depth discussion of this method is given in Chapter 9.

In submarines, an absorber that uses a monoethanolamine (MEA) solution is employed to maintain CO_2 levels between 0.8% and 1.2% (Carhart and Thompson, 1975). The spent MEA solution is regenerated by heating the loaded solution to release the CO_2, which is pumped overboard. Although the MEA solution can degrade to form ammonia, both ammonia and MEA vapors are oxidized to less harmful chemicals in a catalytic burner.

Moschandreas and Relwani (1990) evaluated the performance of a Kathabar humidity pump (a liquid-desiccant based, gas-fired dehumidification system) in three buildings in two cities. The indoor environment was assessed by measuring the indoor air quality, comfort, and ventilation parameters with and without the operation of the humidity pump. The removal rates of CO_2 and CO, along with VOCs, particulate matter, microbial contaminants, and nicotine, were measured in 10 locations inside each building. When the outdoor concentrations were high, changes in the CO_2 and CO concentrations were not statistically significant. However, the operation of the pump increased the indoor lithium chloride level, but by an insignificant amount.

Adsorption methods. Considerable effort has been directed toward the removal of moisture from indoor air by using solid adsorbents (desiccants). A number of porous solid materials are available commercially that have a very high affinity and capacity for water vapor. These adsorbents include silica gel, molecular sieve, and activated alumina. Recently, a number of other materials, including activated carbon, polymers, and manganese oxides, have been investigated. Water vapor or moisture is not considered to be an indoor pollutant in the general sense because it usually does not affect human health directly. However, excess water vapor indoors can condense on cold surfaces, such as ducts, driers, windows, and a variety of building materials, and it can promote the growth of various microorganisms, including molds, algae, and fungi, which do pose health hazards. Water vapor in indoor air can be a by-product of metabolic activity, combustion, evaporation from clothes, dish washing, cooking, and bathroom functions. A decrease in ventilation during the winter months tends to increase the relative humidity indoors. Water vapor can also absorb a number of gaseous pollutants, such as chlorine and ammonia, and increase their corrosive effect. Deterioration of furnishings and artworks has also been attributed to excess moisture in indoor air. Therefore, the control of humidity indoors can be an effective method of enhancing indoor air quality. As described later in Chapter 8, the concentration of airborne microorganisms can be reduced substantially by maintaining the humidity range between 40% to 60%. An increase in the relative humidity also can increase formaldehyde emissions from particle board. The ASHRAE Standard 62-1989 ASHRAE, 1989 now recommends that the indoor relative humidity be maintained between 30% and 60%. The previous standard recommended a range of 20% to 80%. Humidity control can therefore be a part of indoor air quality enhancement strategies.

All three adsorbents, silica gel, molecular sieves, and activated alumina, have comparable water adsorption capacity. The maximum amount of water vapor that can be adsorbed on these materials at a certain relative humidity is shown in Figure 4.7. Molecular sieves generally have a very high capacity in the low humidity range and approach saturation rather quickly. On the other hand, the adsorption capacities of silica gel and activated alumina increase gradually with an increase of the relative humidity of the air, providing a favorable working range for dehumidification applications. However, several other factors must be considered, including how easily or conveniently the material can be regenerated for repeated use. The water vapor adsorption characteristics of these adsorbents are discussed in Chapter 10.

Although CO_2 is not a pollutant of great concern, it should be noted that the indoor concentration of CO_2 is the highest of all pollutants. If an adsorbent has a high affinity for CO_2, its capacity for other pollutants may be reduced significantly. As a result, the adsorption capacity of an adsorbent for CO_2 is of great importance when designing a system for indoor applications. In some enclosed spaces, such as in a submarine or in the cabin of a space shuttle, the metabolic activity of the occupants is a source of water and CO_2.

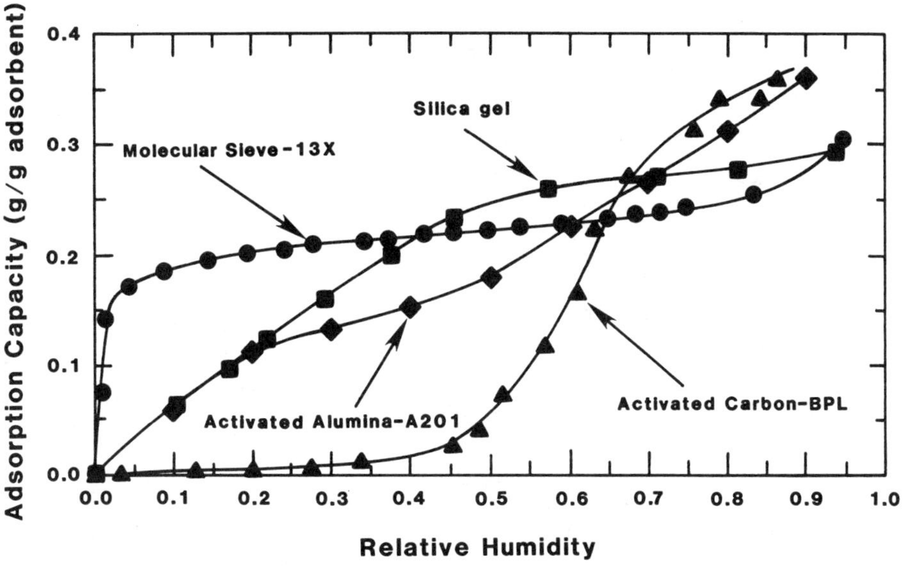

Figure 4.7 *Equilibrium adsorption isotherms of water vapor on various adsorbents at 298 K. (*Source: *Yeh, 1991.)*

Since the 1960s, the National Aeronautics and Space Administration (NASA) has sponsored several research projects for development of a regenerative CO_2 removal system in the presence of water vapor. In one such project, Wright et al. (1973) screened several molecular sieves, 3A, 4A, 5A, and 13X, and silica gel for simultaneous adsorption of CO_2 and H_2O. No single adsorbent was found to perform satisfactorily. A combination of silica gel and molecular sieve was found to provide the best results. The water vapor was first removed from the air by silica gel, which had a negligible capacity for CO_2, followed by a molecular sieve bed to remove CO_2. Molecular sieve also adsorbed a small amount of N_2 and O_2 from the air. In a space shuttle, the regeneration is usually carried out under a vacuum, and the adsorbed H_2O, CO_2, N_2, and O_2 are removed permanently from the cabin. Care must be taken not to remove too much oxygen when co-adsorbing two or more pollutants.

In a separate study, 5A molecular sieve that had preadsorbed water vapor was exposed to CO_2. A negligible amount of water vapor was desorbed during CO_2 adsorption, suggesting a strong interaction between water vapor and the 5A surface. Even when a mixture of CO_2 and H_2O was passed through a fixed bed of 5A molecular sieve, H_2O displaced the already adsorbed CO_2. As a consequence, CO_2 left the bed rather quickly and appeared at the bed outlet at a concentration higher than its value at the inlet. Our own measurements also showed that 13X can adsorb almost 100 times more CO_2 than either silica gel or activated carbon. The amount of CO_2 adsorbed on 13X at various gas phase concentrations is shown in Figure 4.8. Although our data taken on Davison-13X agree very well with the literature data on Linde-13X, the adsorption capacity was 15% lower than that reported by the manufacturer.

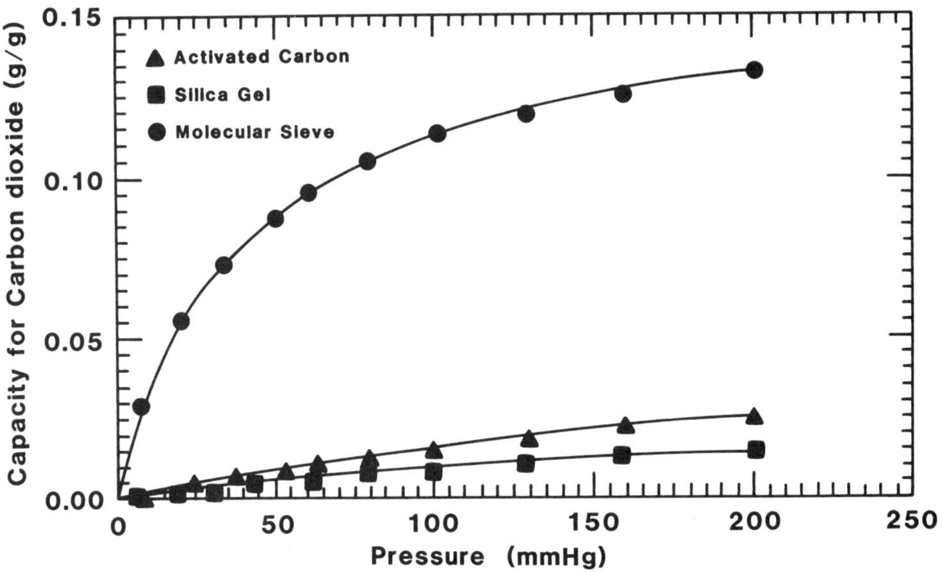

Figure 4.8 *Comparison of adsorption capacities for carbon dioxide on various adsorbents at 298 K.* (Source: *Yeh, 1991.)*

Most of the adsorption studies for carbon monoxide were conducted at very low temperatures of 77.3 to 145 K. These studies showed that CO is a very weak adsorbate whose uptake is very low on most adsorbents. Alfani et al. (1982) observed that a natural zeolite, cabasite tufa, has a greater adsorption capacity for CO than either 5A or 13X molecular sieve. The capacity for the tufa was almost five times greater than that of 5A and two times greater than that of 13X at 20°C. However, the actual amount adsorbed was extremely small, only 0.005 mole of CO per gram of tufa at 20°C when the gas phase partial pressure of CO was 0.01 atm.

Molecular sieves are usually considered to be good adsorbents for sulfur compounds. Although the molecular diameters of SO_2 or H_2S (4.12 and 3.4 Å, respectively) are higher than the average pore diameter of 3A, 4A, and 5A molecular sieves, significant amounts of SO_2 and H_2S can be adsorbed by these adsorbents. Because of the relatively large pore diameter of 13X molecular sieve, sulfur dioxide molecules can rather easily diffuse into the 13X. It is not surprising that 13X has higher capacities for both SO_2 and H_2S than either 4A or 5A. The naturally occurring zeolite H-mordenite has also been found to be a good adsorbent for removing H_2S from a gas stream (Talu and Zwiebel, 1987), and its retention capability is good. Neither CO_2 nor propane displaced the adsorbed H_2S from the H-mordenite surface when present together in a gas mixture.

It is interesting that silica gel was found to be a better adsorbent for ammonia than was molecular sieves. Kuo et al. (1985) used Davison silica gel grade 59 in a packed column to remove ammonia from a helium gas stream. The lowest concentration of ammonia in their study was 997 ppm. Approximately 0.06 kg of NH_3 was removed at room temperature (25°C) by 1 kg of silica gel. The removal capacity increased with an increase in the gas phase concentration of ammonia. Considerable effort has been made to develop new adsorbents and to modify existing ones in order to improve their adsorption capacities for various inorganic gases.

As described earlier, ozone is typically not produced indoors, but it does occur outdoors. The removal of ozone from outdoor air prior to circulating the air indoors can be important in some buildings, such as museums and art galleries. An activated carbon filter system has been in operation at the Huntington Art Gallery since the early 1950s. Shair (1981) evaluated the ozone removal efficiency of activated carbon beds from field tests. A carbon bed that contained 1000 kg of carbon was used to treat 14,000 ft^3/min of air. The initial ozone removal efficiency was 95% ± 5%, but declined to 50% after 3600 hours of operation. Encouraged by these results, a number of southern California museums and other museums throughout the United States have employed activated carbon filters to control ozone. Cass et al. (1988) studied the effectiveness of activated-carbon ozone-removal systems by comparing the ozone concentration inside of six museums, four of which were equipped with carbon filters and two that were not. The ozone concentration in those museums that had carbon filters was substantially lower than the ones without a carbon filter system. These findings are summarized in Table 4.14.

Parmer and Grosjean (1989) also tried to remove ozone, NO_2, SO_2, and H_2S by employing various types of solid adsorbents. These pollutants were targeted because of their relevance to museum air quality. Experiments were carried out both in the active and passive modes. In the passive mode, a tray containing a specific amount of adsorbent was exposed to pollutant-laden air in a 1-m^3 Plexiglas chamber. The objective was to simulate a museum display case. In the active-mode experiments, polluted air was drawn through a cartridge containing 200 g of the adsorbent at a flow rate of 100 cm^3/min. Twenty-three adsorbents were tested. The pollutant concentrations were in the range of 76 to 136 ppb for NO_2, 64 to 304 ppb for O_3, 125 to 182 ppb for SO_2, and 210 to 230 ppb for H_2S. Tests were carried out at room temperature and at a relative humidity of 55% ± 5%. The removal capabilities of the adsorbents are given in Table 4.15. Activated carbon appeared to perform the best, providing nearly 100% removal for all four inorganic gases. Activated alumina removed only SO_2 effectively. Although the removal efficiency of molecular sieve 13X was very high initially, it dropped rather quickly as the experiment progressed. Silica gel, which is often used in museums for humidity control, performed very well initially, but it failed to remove either NO_2, SO_2, or ozone after 4 to 5 h.

Daisey and Hodgson (1989) noted that activated carbon filters are best suited for removing NO_2. Four portable air cleaners that incorporated carbon filters were tested in a 20-m^3 chamber for 4 h at an air temperature of 23° ± 2°C and a relative humidity of 47% ± 9%. Panel and extended surface filters that contained the greatest amount of activated carbon (approximately 100 to 150 g) exhibited the highest cleaning rate for NO_2, about 50% for the panel filter and 40% for the extended surface filter.

Desiccant humidity control devices undergo adsorption-regeneration cycles alternately, with the adsorption cycle typically lasting from 10 to 20 minutes. Therefore, the pollutant removal capacity of silica gel or molecular sieve in the initial period is of great importance. Relwani et al. (1987) investigated the removal of NO_2, SO_2, CO, and total

TABLE 4.14 Removal Capabilities of Various Adsorbents for NO_2, SO_2, O_3, and H_2S

	Air Pollutant (% removed)			
SORBENT	*NO_2*	*SO_2*	*O_3*	*H_2S*
Activated carbon	100	100	100	100
Carbon powder on glass wool	—	—	100	—
Purafil	95	100	37[a]	100
	—	—	43[b]	—
Molecular sieve-13X	100–11(*1*)	—	100–24(*7*)	100
Silica gel	82–0(*2*)	100–36(*5*)	28	—
Tenax-TA	61–37(*3*)	100–10(*6*)	100	—
Chromosorb-102	100	100	100	—
Amberlite XAD-2	100	100	100–55(*8*)	—
Alumina (neutral)	30	100	—	—
Alumina (basic)	35	100	0	—
Colortech detector badge	—	—	—	100
Zinc	—	—	—	65
Zinc acetate	—	—	—	13
Lead	—	—	—	95.6
$KMnO_4$ (powder)	41.6–11(*4*)	—	—	—
C_{18}	2.50	10	0	—
C_{18} alkaline	17.10	100	0	—
KI on C_{18} alkaline	100	100	100	—
Phenoxazine on C_{18} alkaline	100	—	100	—
Triethanolamine on C_{18} alkaline	100	—	100	—
Guaiacol on C_{18} alkaline	100	—	—	—
$KMnO_4$ on C_{13} alkaline	100	—	45	—

1) 100% for first 3 h then decreased to 11% in 20 h (target time = 26.8 h).
2) Decreased from 82% initially to 0% after 24 h (target time = 30 h).
3) Decreased from 61% to 37% in 7 h (target time = 16 h).
4) 41.6% for first 3 h then decreased to 11% in 20 h (target time = 73 h).
5) Decreased from 100% (first 40 min) to 36% after 2 h 40 min (target time = 19 h).
6) Decreased from 100% (first 5 min) to 10% in one hour (target time = 0.3 h).
7) Decreased from 100% (for first 37 h) to 24% in 11 h (target time = 76 h).
8) Decreased rapidly from 100% first five min) to 55% after 25 min (target time = 25 min).
[a]Initial O_3 concentration = 274 ppb.
[b]Initial O_3 concentration = 140 ppb.
— no data reported.
Source: Parmer and Grosjean; 1989.

hydrocarbons by a commercially available rotary desiccant unit. Both silica gel and molecular sieve 5A were employed as the desiccant, and a total of 60 experimental runs were conducted with air mixtures in which the initial concentration levels of pollutants varied from 0.30 to 0.50 ppm for NO_2, 2 to 3 ppm for CO, 20 to 30 ppb for SO_2, and 8 to 12 ppm for total hydrocarbons. Molecular sieve 5A was the most effective in removing these pollutants, as shown by the test results in Table 4.16. However, a higher regeneration tempera-

TABLE 4.15 Effectiveness of Activated Carbon for Ozone Removal

Site	*Date*	*Indoor Maximum O_3 (ppm)*	*Outdoor Maximum O_3 (ppm)*	*Indoor O_3 as Percentage of Outdoor O_3*
Without an activated carbon air-filtration system:				
Scott	7/25/84	0.043	0.179	24%
	7/26/84	0.065	0.221	29%
Montgomery	7/30/84	0.060	0.150	40%
	7/31/84	0.067	0.171	39%
With an activated carbon air-filtration system:				
Southwest Museum Library	7/13/84	0.026	0.149	17%
	7/14/84	0.008	0.174	5%
	7/15/84	0.003	0.116	3%
Huntington Art Gallery	7/27/84	0.004	0.110	4%
	7/28/84	0.010	0.172	6%
Los Angeles County Museum of Art	7/1/85	0.010	0.165	6%
J. Paul Getty Museum	7/2/85	0.022	0.104	21%
	7/8/85	0.009	0.095	9%
	7/15/85	0.028	0.075	37%

Source: Cass et al., 1989.

TABLE 4.16 Removal of CO, NO_2, and SO_2 by Silica Gel and Molecular Sieve

		Silica Gel		*Molecular Sieve 5A*	
Pollutant	*Concentration Range (μg/m^3)*	*Percent Removal (%)*	*Absolute Reduction (μg/m^3)*	*Percent Removal (%)*	*Absolute Reduction (μg/m^3)*
CO	550–1000	21–36	115–279	45–61	187–427
	1000–2000	17–32	268–437	19–39	301–414
	2000–3000	10–18	411–1871	10–22	495–957
NO_2	50–100	16–23	13–22	22–36	16–21
	100–1000	15–25	35–160	20–38	173–404
	1000–2000	19–33	230–406	43	485–541
SO_2	50–110	10[a]	11a	73–82	29–69
	110–500	9–17	16–67	82–94	93–336
	500–1000	16–18	90–132	95	468[a]

[a]only one experiment.

Source: Relwani et al., 1987.

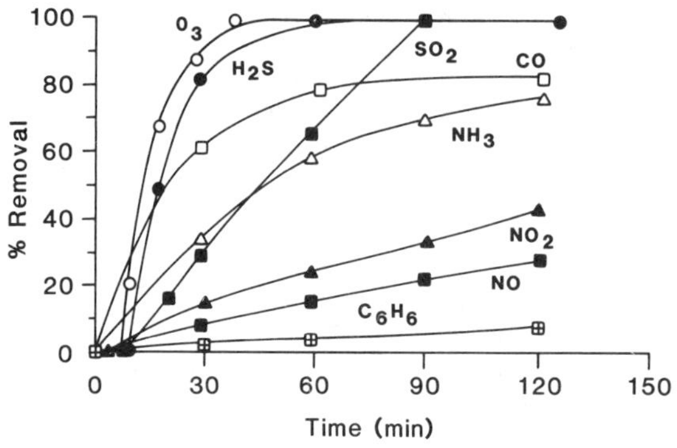

Figure 4.9 *Removal efficiency of low temperature catalyst. (Source: Collins, 1986.)*

ture was required for 5A than for silica gel. The final choice of the adsorbent will depend on the results of a detailed economic analysis.

Carbon monoxide, ozone, nitrogen oxide, and sulfur compounds are generally highly reactive molecules. Catalytic destruction is possible, but a high temperature may be required for the catalyst to be effective. As a result, research is underway to develop a catalyst that is effective at room temperature and is also insensitive to moisture. Teledyne Water Pik of Fort Collins, Colorado, tested a room-temperature catalyst (called LTC) to evaluate its effectiveness for removing CO, SO_2, H_2S, O_3, NH_3, NO, and NO_2 from indoor air in a 1008-ft^3 stainless-steel room (Collins, 1986). An air filtration unit containing 50% LTC catalyst distributed in activated carbon was designed to hold 230 g of the mixture. The performance tests were conducted by using an air stream that contained the individual pollutants flowing at a rate of 300 ft^3/min. The test results at the end of a 2-hour period are given in Figure 4.9. Pollutants such as H_2S may slowly poison the catalyst and reduce its activity. Although catalytic technology can be incorporated with a desiccant humidity control device, various factors need to be studied further, such as the catalytic activity in a mixture of pollutants, how frequently the catalytic bed would need to be replaced, and the effect of room temperature and humidity.

REFERENCES

ALFANI, F., GRECO JR., G., and IROIO, G., "Removal of Carbon Monoxide-Physical Adsorption on Natural and Synthetic Zeolites", Translation from *Inquinamento* (Italy), **20**, 51, May, 1978, Report No. NASA-TM-76934, NTIS Document No. N82-30765, 1982.

ALFHEIM, I., and RAMDAHL, T., *Environ. Mutagenesis*, **6**, 121 (1984).

APTE, M. G., and TRAYNOR, G. W., Proceedings IAQ '86: "Managing Indoor Air for Health and Energy Conservation", American Society of Heating, Refrigerating and Air-Conditioning Engineers, Atlanta, Ga., 405, 1986.

ASHRAE (American Society of Heating, Refrigerating and Air-Conditioning Engineers), "Ventilation for Acceptable Indoor Air Quality", ASHRAE Standard 62-1989, The Society, Atlanta, Ga., 1989.

BARNES, J., HOLLAND, P., and MIHLMESTER, P., "Characterization of Population and Usage of Unvented Kerosene Space Heaters", Report No. EPA/600/7-90/004, Applied Management Sciences, Oak Ridge, Tenn., NTIS Document No. PB90-155573, 1990.

CACERES T., SOTS, H., and LISSI, E., *Atmos. Environ.*, **17**(5), 1009 (1983).

CARHART, H. W., and THOMPSON, J. K., *ACS Symp. Ser.*, No. 17, 1, 1975.

CASS, G. R., DRUZIK, J. R., GROSJEAN, D., NAZAROFF, W. W., WHITMORE, P. M., and WITTMANN, C. L., "Protection of Works of Art from Photochemical Smog", Getty Conservation Institute, Marina del Rey, Calif., 1988.

COLLINS, M. F., Proceedings of the 79th Annual Meeting of the Air Pollution Control Association, Minneapolis, Minn., 153, 1986.

COTÉ, W. A., WADE, W. A. III, and YOCOM, J. E., "A Study of Indoor Air Quality", Report No. EPA/650/4-74-042, Research Corp. of New England, Wethersfield, Conn., NTIS Document No. PB-238 556/5, 1974.

CPSC (Consumer Product Safety Commission), *Kerosene Heaters*, Project Status Report and Staff Recommendation, Washington, D. C., 1983.

DAISEY, J. M., and HODGSON, A., *Atmos. Environ.*, **23**(9), 1885 (1989).

DEWERTH, D. W., and STERBIK, W. G., "Development of Advanced Residential Cooktop Burner with Low NOx Emissions", Report No. GRI-80-0177, Gas Research Institute, Chicago, Ill., NTIS Document No. PB83-228338, 1983.

DOE (U. S. Department of Energy), "Wood Combustion: State-of-Knowledge Survey of Environmental Health and Safety Aspects", NTIS Document No. DE83-005070, 1983.

DOE "Indoor Air Quality Environmental Information Handbook: Combustion Sources", Report No. DOE/EV/10450-1, Brookhaven National Laboratory, Upton, N. Y., NTIS Document No. DE85-006589, 1985.

DUNCAN, J. R., MORKIN, K. M., and SCHMIERBACH, M. P., Proceedings of the 73rd Annual Meeting of the Air Pollution Control Association, Montreal, Canada, Paper No. 80-7.2, 1980.

EPA (U. S. Environmental Protection Agency), "Air Quality Criteria for Carbon Monoxide", Report No. EPA-600/8-79-022, Informatics, Inc., Rockville, Md., NTIS Document No. PB81-244840, 1979.

FISK, W. J., Proceedings IAQ '86: "Managing Indoor Air for Health and Energy Conservation", American Society of Heating, Refrigeration, Air-Conditioning Engineers, Atlanta, Ga., 568, 1986.

GIRMAN, J. R., ALLEN, J. R., APTE, M. G., MARTIN, V. M., and TRAYNOR, G. W., "Pollutant Emission Rates from Unvented Gas Fixed Space Heaters: A Laboratory Study", Report No. LBL-14502, Lawrence Berkeley Laboratory, Berkeley, Calif., 1983.

GIRMAN, J. R., APTE, M. G., TRAYNOR, G. W., ALLEN, J. R., and HOLLOWELL, C. D., *Environ. Int.*, **8**, 213 (1982).

GODISH, T., and RITCHIE, I., Indoor Air Quality in Cold Climates: Hazards and Abatement Measures, (ed. D. S. Walkinshaw), Air Pollution Control Association Specialty Conference Proceedings, Pittsburgh, Pa., 261, 1985.

GOTO, Y., and TAMURA, G. T., Proceedings of the 77th Annual Meeting of the Air Pollution Control Association, 102, 1984.

GPO (Government Printing Office), "Wood Use: U. S. Competitiveness and Technology, Vol 11", Report No. OTA-M-226, Office of Technology Assessment, Washington, D. C., 1984.

HOLLOWELL, C. D., BUDNITZ, R. J., CASE G. D., and TRAYNOR, G. W., "Combustion Generated

Indoor Air Pollution", Report No. LBL-4416, Lawrence Berkeley Laboratory, Berkeley, Calif., 1976.

HONICKY, R. E., OSBORNE, J. S. III, and AKPOM, C. A., *Pediatrics*, **75**, 587 (1985).

IMHOFF, R. E., Proceedings of the 77th Annual Meeting of the Air Pollution Control Association, 134, 1984.

KNIGHT, C. V., HUMPHREYS, M. P., and PINNIX, J. C., Proceedings IAQ '86: "Managing Indoor Air for Health and Energy Conservation", American Society of Heating, Refrigeration, Air-Conditioning Engineers, Atlanta, Ga., 430, 1986.

KUO, S. L., PEDRAM, E. O., and HINES, A. L., *J. Chem. Eng. Data*, **30**, 330 (1985).

LEADERER, B. P., *Science*, **218**, 1113 (1982).

LIPPMANN, M., *J. Air Pollut. Control Assoc.*, **39**, 672 (1989).

MOSCHANDREAS, D. J., and RELWANI, S. M., "Impact of the Humidity Pump on Indoor Environments", Report No. GRI-90/0193, Gas Research Institute, Chicago, Ill., 1990.

MOSCHANDREAS, D. J., PELTON, D. J., and BERG, D. R., Proceedings of the 74th Annual Meeting of the Air Pollution Control Association, Philadelphia, Pa., Paper No. 81-22.2, 1981.

MOSCHANDREAS, D. J., RELWANI, S. M., BILLICK, I. H., and MACRISS, R. A., *Atmos. Environ.*, **21**, 285 (1987).

NAGDA, N. L., RECTOR, H. E., and KOONTZ, M. D., *Guidelines for Monitoring Indoor Air Quality*, Hemisphere, New York, 1987.

NASA (National Aeronautics and Space Administration), "Bioastronautics Data Book", Report No. NASA-SP 3006, Washington, D. C., 1973.

NAS (National Academy of Science), *Nitrogen Oxides*, Committee on Medical and Biological Effects of Environmental Pollutants, Sub-Committee on Nitrogen Oxides, National Research Council, National Academy Press, Washington, D. C., 1976.

NAS *Indoor Pollutants*, National Research Council, National Academy Press, Washington, D. C., 1981.

OFFERMANN, F. J., HOLLOWELL, C. D., NAZAROFF, W. W., and ROSEME, G. D., *Environ. Int.*, **8**, 435 (1982).

PARMAR, S. S., and GROSJEAN, D., "Removal of Air Pollutants from Museum Display Cases", Getty Conservation Institute, Marina del Rey, Calif., 1989.

RELWANI, S. M., MOSCHANDREAS, D. J., and BILLICK, I. H., Proceedings 4th International Conference on Indoor Air Quality and Climate, West Berlin, West Germany, **1**, 236, 1987.

RITCHIE, I. M., and OATMAN, L. A., *J. Air Pollut. Control Assoc.*, **33**, 879 (1983).

SAMET, J. M., MARBURY, M. C., and SPENGLER, J. D., *Am. Rev. Respir. Dis.*, **136**, 1486 (1987).

SEXTON, K., LETZ, R., and SPENGLER, J. D., *Environ. Res.*, **32**, 151 (1983).

SHAIR, F. H., *ASHRAE Trans.*, **87**(Part 1), 116 (1981).

SHUKLA, K. C., HURLEY, J. R., and GIRMANIS, M., "Development of an Efficient, Low-NOx Domestic Gas Range Cooktop, Phase II", Report No. GRI-85-0080, Gas Research Institute, Chicago, Ill., 1985.

TALU, O., and ZWIEBEL, I., *Reactive Polymers*, **5**, 81 (1987).

THASHER, W. H., and DEWERTH, D. W., "Evaluation of the Pollutant Emissions from Gas-fired Room Heaters", Research Report No. 1515, American Gas Association, Cleveland, O., 1979.

TRAYNOR, G. W., ANTHON, D. W., and HOLLOWELL, C. D., "Techniques for Determining Pollutant Emissions from a Gas-fired Range", Report No. LBL-9522, Lawrence Berkeley Laboratory, Berkeley, Calif., 1981a.

TRAYNOR, G. W., APTE, M. G, DILLWORTH, J. F., HOLLOWELL, C. D., and STERLING, E. M., "The Effects of Ventilation on Residential Air Pollution Due to Emissions from a Gas Fired Range", Report No. LBL-12563, Lawrence Berkeley Laboratory, Berkeley, Calif., 1981b.

TRAYNOR, G. W., APTE, M. G, DILLWORTH, J. F., HOLLOWELL, C. D., and STERLING, E. M., *Environ. Int.*, **8**, 447 (1982).

TRAYNOR, G. M., GIRMAN, J. R., APTE, M. G., DILLWORTH, J. F., and WHITE, P. D., *J. Air Pollut. Control Assoc.*, **35**, 231 (1985).

TUTHILL, R. W., *Am. J. Epidemiol.*, **120**, 952 (1984).

WEBER, A., Proceedings 3rd International Conference on Indoor Air Quality and Climate, Stockholm, Sweden, **2**, 297, 1984.

WEST, P. W., and REISZNER, K. D., "Personal Monitoring by Means of Gas Permeation", in Proceedings of the Symposium on the Development and Usage of Personal Monitors for Exposure and Health Effect Studies, Report No. EPA-600/9-79-032, Washington, D. C., NTIS Document No. PB80-143894, 1980.

WESTON, P., as cited in "Indoor Air Quality Environmental Information Handbook: Combustion Sources", Report No. DOE/EV/10450-1, Brookhaven National Laboratory, Upton, N. Y., NTIS Document No. DE85-006589, 1985.

WOODRING, J. L., DUFFY, T. L., and DAVIS, J. T., *Am. Ind. Hyg. Assoc. J.*, **46**, 350 (1985).

WRIGHT, R. M., RUDER, J. M., DUNN, V. B., and HWANG, K. C., "Development of Design Information for Molecular-Sieve Type Regenerative CO_2 - Removal System", Report No. NASA CR-2277, NTIS Document No. N73-27948, 1973.

YAMANKA, S., HIROSE, H., and TAKADA, S., *Atmos. Environ.*, **13**, 407 (1979).

YEH, R. L., "Adsorption of Water Vapor, Toluene, I, I, I-Trichloroethane, and Carbon Dioxide on Silica Gel, Molecular Sieve-13X, and Activated Carbon", M. S. Thesis, University of Missouri-Columbia, 1992.

YOCOM, J. W., COTÉ, W. A., and CLINK, W., "A Study of Indoor-Outdoor Air Pollution Relationship", Volume 1 and 2, Summary Report, National Air Pollution Control Administration, Washington, D. C., 1974.

5

Heavy Metals

5.1 INTRODUCTION

Indoor concentrations of trace heavy metals depend on their concentrations in the outdoor air and the surrounding soil and dust. Various industrial activities, such as mining, metal smelting, and the burning of coal in power plants, emit these metals or their compounds to the ambient air. They settle slowly and deposit in the surrounding soil or water. Although most ventilation systems in homes and buildings employ filters to remove dust particles from outdoor air before circulating it indoors, a significant amount of trace metals can be carried into homes by the soil attached to shoes or dust particles that attach to clothes. Therefore, the texture of the surrounding soil is very important in assessing indoor exposure to these heavy metals. The residences, schools, and commercial buildings of Hudson County, New Jersey, where waste slag from chromium processing plants was used for land reclamation, showed an unusually high level of chromium in indoor air (Coniglio et al., 1990). Indoor air samples from 22 locations, which included eight residences, one commercial building, and one school, had chromium levels in the range of 0.22 to 8.35 $\mu g/m^3$, while that in the ambient air was less than 0.008 $\mu g/m^3$. Some heavy metals, such as lead and mercury, do have indoor sources. Other than being components of particulate materials such as paint, these sources are rather limited. Foote (1972) found that the concentration of elemental gaseous mercury in several homes, offices, and laboratories in the Dallas area ranged from 5 to 5550 ng/m^3, which was substantially higher than the natural ambient background concentration of 3 ng/m^3. Measured lead concentrations in residences are generally low, often below 0.5 $\mu g/m^3$. However, concentrations can be four times higher in residences that have lead based wall paint. Automobile exhaust is another source of lead. About 90% of airborne lead is thought to be derived from automobile engines that use gasoline contains tetra alkyl lead. The concentration of lead in air is found to decrease as the concentration of tetra alkyl lead in gasoline is reduced. Therefore, it is not surprising that homes and buildings near a major thoroughfare have higher indoor lead concentrations. In addition, other heavy trace metals, such as arsenic, cadmium, and nickel, have been identified indoors and have been attributed to the burning of wood, to smoking, and to the use of some pesticides.

5.2 HEALTH EFFECTS AND STANDARDS

Health hazards associated with trace levels of heavy metals, especially lead, mercury, cadmium, and chromium, are well documented. The Agency for Toxic Substances and Disease Registry noted in a recent report (HHS, 1988) that about 17% of the small children in the United States have a blood-lead level above 15 $\mu g/dL$. Lead is a neurotoxin and can cause impaired metabolism, reading disorders, delay in early childhood development, and neurobehavioral problems. Young children who put their hands or toys in their mouth are more prone to lead poisoning from direct ingestion. Although the indoor lead concentration level may not be high enough to cause brain damage to adults, it can cause lethargy, headaches, and a loss of appetite.

Mercury is generally a sensory irritant. The lungs and skin can absorb mercury vapor rather quickly, which can cause skin burns, irritation in the mouth, rash, excessive perspiration, partial loss of hearing, and kidney damage by destroying cells in the tubular system. Hirschman et al. (1963) have noted that mercury poisoning may effect the sense of touch. Although Joselow (1973) has indicated that mercury vapor at high concentrations can be a health hazard, it is doubtful whether the indoor concentration level will ever become high enough to create a problem.

Cadmium, once inhaled, tends to deposit in the kidneys and liver. It can severely damage capillaries in the kidneys and will interact with nutrients in the liver (Elinder et al., 1976). The effects of other trace metals on humans are summarized in Table 5.1.

5.3 SOURCES AND INDOOR CONCENTRATIONS

The presence of a variety of trace metals found indoors may be the result of infiltration of dust from outdoor air or the result of its being carried inside on shoes and clothes. Indoor sources include old lead- and latex-based paints, domestic water supplies, smoke from the burning of wood, and tobacco smoke.

The white paint once used for painting the interior and exterior of houses contained lead carbonate and lead hydroxide as white pigment. In such paints, lead may constitute from 5% to 40% of the final dried solids. As the paint ages, it forms flakes that can become

TABLE 5.1 Toxicity of Trace Heavy Metals

Metal	*Route of Entry*	*Toxicity Effect*	*TWA by ACGIH (mg/m^3)*	*Carcinogen (suspected by NIOSH)*
Arsenic	Inhalation and ingestion	Irritation of respiratory system, liver and kidney damage; weakness, loss of appetite, nausea, and vomiting	0.20	Yes
Cadmium	Inhalation and ingestion	Lung, liver, and kidney damage; irritation of respiratory system	0.05	Yes
Chromium	Inhalation, ingestion, and absorption through skin	Lung damage and irritation of respiratory system	0.50	Yes
Mercury	Inhalation, ingestion, and absorption through skin	Irritation of respiratory system; lung, liver, and kidney damage	0.05 (vapor)	Yes
Lead	Inhalation and ingestion	Lung and liver damage, weakness, loss of appetite, nausea, and vomiting	0.15	No
Nickel	Inhalation	Lung, liver, and kidney damage	1.00	Yes

ACGIH, American Conference of Governmental Industrial Hygienists.
NIOSH, National Institute of Occupational Safety and Health.
TWA, time weighted average.

airborne because of peeling, oxidation, chipping, and abrasion from the opening and closing of doors and windows (Lemire, 1991). In the 1970s, lead-based paints were found to be responsible for a substantial proportion of all cases of severe lead poisoning found in children. This prompted the federal government in 1971 to limit the lead concentration in residential paints to 1%. Later, in 1977, the lead content was regulated to 0.06%, and a few years later it was totally banned from all paints. However, paint manufacturers had stopped using lead and mercury compounds several years earlier. The federal government has estimated that more than 57 million residences still contain lead-based paints. The office of Housing and Urban Development (HUD) estimates that the lead content of house dust, which contains paint chips, in approximately 20 million houses in the United States is above the recommended level. In these houses, the dust particles that were collected near windows had a higher lead content than dust that was collected from other parts of the house. Other common sources of lead are soldered cans, ceramic dishes, and cookware. Roberts et al. (1990) analyzed the dust and soil in rugs from 42 older homes in Washington state for lead content, and they found that the lead content of the dust increased dramatically as a result of walking on the carpet with shoes on. Their findings are given in Table 5.2. The lead concentration in outdoor air may vary from 0.0001 $\mu g/m^3$ in remote areas of the country up to about 5 $\mu g/m^3$ in cities that have heavy traffic and are highly industrialized. In closed parking garages, concentrations as high as 10 $\mu g/m^3$ have been reported. In most houses, the indoor to outdoor ratio rarely exceeds 0.6 to 0.8. However, the greatest

TABLE 5.2 Lead Content of House Dust

Activities	*Shoes off*	*Shoes on*	*Walk-off Mat*	*After Remodelling*[a]
No. of homes studied	5	32	6	9
Home age (years)	73	71	76	72
Total amount of dust, g/m^2	3.5	26	6.7	63
Lead content of dust, ppm	320	780	430	1320
Lead content of soil, ppm	860	1530	1350	2140

[a]Peeling of paint either from inside or outside of the home.
Source: Roberts et al., 1990.

threat, particularly to children, is the lead content of the dust. Lead in street dust near a lead smelter may be as high as 50,000 μg/g. Roels et al. (1980) analyzed dust samples obtained from school playgrounds, which were a distance of 1 and 2.5 km from a lead smelter, and in urban and rural areas. The lead concentrations of dust from the urban and rural playgrounds were in the range of 112 to 114 mg/kg, but they were 466 and 2560 mg/kg, respectively, in the school playgrounds that were 2.5 and 1 km from the smelter.

Another potential and dangerous source of lead indoors is the vapor that is generated during the removal of old paints by employing heat. The use of do-it-yourself heat guns by urban homeowners for the removal of old paint from buildings appears to be increasing. Hot air generated from the heat gun, open flames, or other strong heat sources is applied to the paint surface to soften it and facilitate its removal. This results in volatilization of the lead in the paint being removed. Although a very high temperature (above 500°C) is required to volatilize lead, some heat sources are capable of generating this kind of temperature, and often the users are not aware of the danger.

Foote (1972) noted that the concentration of mercury vapor found indoors depends primarily on the type of paint used, although broken thermometers and the mixing of a silver-mercury amalgam (such as in a dentist's chambers) can also be sources of mercury. Latex-based paints contain a mercury-based compound (diphenyl mercury dodecenyl succinate) to prevent fungus growth on the paint surface. This compound dissociates to elemental mercury over a period of time. High concentrations of mercury vapor were found in homes that had been painted with latex paint within the first few weeks, but the concentration decreased slowly over a period of time (see Table 5.3). About 20% to 25% of the mercury disappears from the air within three to four months after painting. Mercury vapor may be absorbed by other aerosol particles or condense on cold surfaces and create a problem at some later time.

A major contributor to the level of trace metals indoors is the particulate matter that enters a building in the form of dust and soil with the incursion of outdoor air. The fine particles, which have an aerodynamic cut diameter of 2.5 μm, generally contain trace metals such as lead, chromium, and silicon, along with sulfates, nitrates, ammonium salts, and organic compounds. Coarse particles, having a size distribution in the range of 3 to 10 μm, generally contain the same metals but in their oxide form. The "Particle Total Exposure Assessment Methodology" study sponsored by the EPA gives an analysis of the elemental

TABLE 5.3 Concentration of Mercury Vapor Indoors

Location	*Mercury Concentration (ng/m^3)*	*Comments*
House 1		21 months after painting with latex paint
Study	68.2	
Bedroom 1	66.5	
Living room	69	
Bedroom 2	139	
House 2		4 months after repainting with latex paint
Living room	164	
House 3		9 months after painting with latex paint
Bedroom	262	
House 4		New home, painted with latex paint 30 days before
Living room	1560	
Office building	203	New home, painted with latex paint 30 days before
Doctor's room	4950	Painted with latex paint 6 months before
Dentist's office	5550	Hg thermometer broken in the past
Dentist's office	1295	Mixing area for Hg-amalgam
Hospital laboratory	307	Inactive for previous 4 days
	930	Near the sink
Laboratory	592	Near the desk
	398	Office away from laboratory

Source: Foote, 1972.

components of the particulate matter collected from nine households and from the clothing of 18 individuals in the San Gabriel valley area of southern California (Ozkaynak et al., 1990). Of the 34 elements analyzed, silicon, calcium, sulfur, aluminum, and potassium were the most prevalent in terms of the total particulate mass. A greater number of trace elements were present in the 10-μm range samples than in the 2.5-μm range samples. The indoor to outdoor ratio for both sizes ranged from 0.6 to 1.6.

Other indoor sources of heavy metals are tobacco smoke, woodsmoke, and some pesticides. Cadmium, nickel, and arsenic are released from tobacco when burned. The average emissions of these metals from a smoked cigarette are 4.5×10^{-4} mg. Pesticides and fungicides are the primary contributors of arsenic and mercury indoors. DeAngelis et al. (1980) measured the emission rate of 29 trace metal elements from a nonbaffled wood-burning stove. Of those measured, zinc, silver, nickel, iron, calcium, and aluminum had the highest emission rates. Duncan et al. (1980) also found a similar result from tests on a wood-fired stove. Elemental emissions from these two studies are summarized in Table 5.4.

5.4 SAMPLING AND MEASUREMENT

Trace quantities of heavy metals are a part of respirable particulate matter, and their collection methodologies are described in detail in Chapter 6. The most common method of collecting particulate matter from air is to draw the air through a filter, typically at flow rates

TABLE 5.4 Trace Metal Emissions from Wood-burning Appliances

Element	*Nonbaffled Stove (g/kg)*	*Wood Stove (lb/cord)*
Aluminum	1.5×10^{-3}	1.3
Calcium	4.7×10^{-3}	10.2
Iron	8.1×10^{-3}	0.7
Magnesium	2.9×10^{-3}	2.0
Manganese	1.9×10^{-3}	1.6
Phosphorus	7.0×10^{-3}	1.0
Potassium	—	3.6
Silicon	2.7×10^{-3}	1.6
Sodium	3.0×10^{-3}	0.7
Titanium	1.0×10^{-3}	0.02
Lead	4.8×10^{-3}	—
Arsenic	1.3×10^{-3}	—
Chromium	9.0×10^{-3}	—

Source: DeAngelis et al., 1980; Duncan et al., 1980.

that range from 1000 to 1500 L/min, where particles with aerodynamic diameters less than the cut point of the inlet are collected by the filter. Several filters in series can be used for the size determination. The mass of the particles is determined from the weight difference of the filter paper taken before and after sampling. The concentration of suspended particulate matter in the designated size range is calculated by dividing the weight gain of the filter by the volume of air sampled. The identification and concentration of individual trace metals in the collected dust are determined by either atomic absorption spectrophotometry or x-ray fluorescence.

When using the atomic absorption spectrophotometry method, samples are collected on filters and treated with nitric acid to oxidize the organic matrix and to dissolve the metals present in the sample. Various types of filters can be used, including membranes, quartz and glass fibers, and cellulose filters, but the selection will depend on the specific application. Although some 68 elements can be detected by atomic absorption spectroscopy, it is essentially a destructive method. At least 1 to 2 mL of solution is necessary for each metal determination if a flame atomic absorption instrument is used. The sample solution is aspirated and sprayed as a fine aerosol into a mixing chamber where fuel and an oxidant gas are mixed and carried to the burner head. Combustion takes place in the burner head. A light of predetermined wavelength (based on the targeted metal) strikes a free ground-state atom. The atom absorbs the light by the process known as atomic absorption, with the amount absorbed being proportional to the concentration of the metal in the solution. A smaller sample size (10 to 100 μL) can be used in a graphite furnace atomic absorption instrument. Also, the sensitivity of the graphite furnace unit is 50 to 500 times greater than the flame instrument.

X-ray fluorescence (XRF) is more convenient than atomic absorption spectrophotometry for making an elemental analysis of dust samples. The XRF method does not require sample preparation and is nondestructive. This method also does not discriminate between the chemical state of the sample. After a suitable sampling time, the filter paper from the sampling unit can be directly mounted on the XRF instrument for analysis. The XRF spectrometry method consists of irradiating a solid sample with x-rays from a suitable source to excite the characteristic x-ray lines of the elements in the solid. Each line energy is characteristic of an element's atomic number, and the line intensity can be correlated with the concentration of the element in the solid. The x-ray source is generally directed toward a fixed area, and the measurement is done on a mass per unit area basis. If the volume of air to be sampled is known relative to the area of the filter, the mass of each element in the sample can be calculated. Therefore, the filter size is an important factor that should be taken into consideration when collecting samples. The exposure area in most XRF analyzers is approximately 2 cm in diameter. It is important that particulates deposit uniformly on the filter paper, otherwise the measurement may not be representative. Generally, a large-sized filter is used for air sampling so that it can be cut into several pieces, and the average of several readings can be obtained. Both the XRF and the atomic absorption spectrophotometry methods can be used to determine the indoor concentration of lead, cadmium, chromium, and arsenic.

A special sampling technique is required for determining mercury vapor in air. Mercury vapor is generally collected on cleaned silver wool by packing 1 to 2 g of it in a Pyrex tube that is 10 cm long with a 5 mm inside diameter. The air sample is drawn through the tube at a flow rate of 50 to 200 mL/min, depending on the anticipated concentration of mercury in the air. The collected mercury is vaporized from the wool by heating the tube to 400°C and is swept through the absorption cell of the atomic absorption spectrophotometer by a carrier gas. The response at 253.7 nm is measured to determine a quantitative analysis of the amount of mercury present. This method can be used to determine concentrations in the range of 0.02 to 500 $\mu g/m^3$. The recovery of mercury from the silver wool is at least 98% when its concentration in the air is in the range of 0.006 to 0.6 μg. A glass collection tube containing 30 mg of silvered Chromosorb P, which is commercially available and is recognized by the NIOSH, can also be used for collecting mercury vapor. Ambient concentrations of dimethyl mercury, SO_2, H_2S, and NO_2 do not seriously interfere with the measurement. However, at high concentrations, the interference may be significant. A second tube packed with Ascarite can be placed before the collection tube to reduce the interference.

5.5 SPECIFIC CONTROL STRATEGIES

Methods available for controlling trace metals found in indoor air include regular cleaning and the removal of sources such as paints. As mentioned earlier, a major entry route of soils into homes is with shoes. Therefore, removing one's shoes and the use of a long walk-off mat can be an effective control strategy (Roberts et al., 1990). From Table 5.2, it can be seen that a substantial reduction in lead content in home dust was noted when shoes were removed and a walk-off mat was used. The periodic vacuuming of a home can be also effective. However, regular vacuum cleaners are generally not very effective in removing fine dust, particularly from carpets or rugs. A vacuum cleaner with an agitator may pick up from two to six times more dust from a rug than a canister vacuum. Another measure that can be used to reduce soil track-in is simply to cover the exposed soil surrounding the home with grass.

The replacement of a wood-burning appliance by an equivalent gas or electrical appliance will reduce the concentration levels of trace metals indoors. Also, the elimination of smoking indoors will reduce the concentration levels of arsenic, nickel, and cadmium.

The removal of old lead-based paints from walls will, no doubt, reduce the indoor lead concentration. However, this work requires special equipment, trained personnel, and proper procedures. Untrained persons, including homeowners, should proceed with caution when attempting to remove paint from walls. Paint chips are heavy and can easily stick to exposed surfaces or settle in crevices or other openings; the removal of lead residue from floors is difficult. Homes should be tested properly after renovation to ensure that all residual metals have been removed.

The traditional methods of removing lead-based paints include scraping and sanding the wall up to a level of 4 to 5 ft above the floor. This will reduce the exposure to children by the hand-to-mouth route. An open flame or heat gun can be used to burn or soften the paint to facilitate the removal process, but this method has been found to be inadequate. When heating a painted surface, the lead vaporizes from the paint and creates an additional health hazard. The removal process itself generates a significant amount of particulates, which requires a thorough cleaning of floors and other surfaces after removal. Inskip and Atterbury (1986) noted that the deposition rate of particles on surfaces that were in close proximity to a wall from which paint was removed by using a heat gun with an open flame was 1 to 5 mg of lead/m^2/h. When the paint was removed by sanding, the deposition rate varied from 50 to 100 mg of lead per square meter per hour. A 10- to 100-fold increase over the preabatement levels in the lead content of house dust near the abated surface was also reported by Farfel and Chisolm (1990). Three percent of the airborne particles was in the respirable range, and 21% of the lead in the house dust was less than 44 μm in size. Therefore, an ordinary vacuum cleaner would not be effective in removing lead-containing particles from an indoor surface after the abatement process. These methods must still be considered as a temporary or partial solution, because the paint is removed only to a level of 4 to 5 ft above the floor.

Another problem with these methods is that the clean-up process was often left to the residents, and no systematic or standard procedure was available to residents for the disposal of the highly hazardous debris. Various states have modified their traditional practices by using trained personnel, and most city abatement programs now follow the guidelines set by the Center for Disease Control. The procedures include an extensive cleaning program, disposal of the debris, and protection of workers, occupants, and their belongings. The removal of paint by using an open flame or by sanding is no longer employed. In addition, the abated surface is usually repainted. The modified method, however, provides only a short-term improvement and has been found to be ineffective for long-term reductions. The lead content of house dust was found to be at the preabatement levels within 6 to 12 months following removal.

Mercury does not wet the surface of most materials that are commonly used in floors of houses or other buildings. Therefore, the containment of mercury and its subsequent removal are very difficult. Specially designed vacuum cleaners are available commercially for the removal of spilled mercury. In addition, mercury may be contained in one location by spreading powdered tin along with a catalyst, such as acetic acid or phosphoric acid, on the mercury to form an amalgam (Karpinski et al., 1975). The floor can be then scrubbed with water and vacuumed dry.

The removal of mercury vapor from air requires special techniques. One potential method is to use a packed bed of adsorbents. This method is most effective in laboratories and dentists' offices, where mercury vapor concentrations are often quite high. Various types of adsorbents are available for adsorbing mercury vapor. Most of these are impregnated with other chemicals, which will react with mercury and retain it on the adsorbent surface. The reaction product and the unreacted catalyst can be removed from the adsorbent by heat, and the regenerated adsorbent may be reimpregnated with the catalyst for reuse. Otani et al. (1988) compared the adsorption characteristics of activated carbon, activated alumina, and zeolite impregnated with various amounts of sulfur, in a packed bed under dynamic conditions. The bed was 1 mm in diameter and contained 0.1 g of the impregnated adsorbent. The concentration of mercury vapor in the air was 6.42 mg/m^3. Using air flow rates that varied from 162 to 234 cm^3/min and activated carbon impregnated with 13.1% sulfur, all the mercury could be removed from the air stream for a period of five days. The bed was maintained at a temperature of 309 K during the tests.

A gold-coated denuder can also be used for removing mercury vapor from air (Munthe et al., 1990). In a laboratory test, from 90% to 97% of the mercury vapor was removed from air and collected in the denuder during a 100-hour test period. The denuder was a 50-cm-long tube with a diameter of 0.4 cm that was coated with a gold solution. The air flow rate through the denuder was 1 L/min, and the mercury concentration in the air varied from 5 to 19 ng/m^3. The advantage of using the denuder is that it can be thermally regenerated for repeated use. Therefore, it appears that a better strategy for removing mercury vapor from buildings and dentists' offices may be to integrate either an adsorbent bed or a denuder with a ventilation system.

REFERENCES

CONIGLIO, W. A., FAGLIANO, J., GOLDOFT, M., UDASIN, I., and MILLER, S., Proceedings 5th International Conference on Indoor Air Quality and Climate, Toronto, Canada, **2**, 139, 1990.

DEANGELIS, D. G., RUFFIN, D. S., and REZNIK, R. B., "Preliminary Characterization of Emissions from Wood-Fired Residential Combustion Equipment", Monsanto Research Corp., Dayton, O., NTIS Document No. PB80-182066, 1980.

DUNCAN, J. R., MORKIN, K. M., and SCHMIERBACH, M. P., Proceedings of the 73rd Annual Meeting of the Air Pollution Control Association, Montreal, Canada, Paper No. 80-7.2, 1980.

ELINDER, C. G., KJELLSTROM, L., FRIBERG, B., LIND, B., and LINNMAN, L., *Arch. Environ. Health*, **31**, 292 (1976).

FARFEL, M. R., and CHISOLM, J., *Am. J. Public Health*, **80**(10), 1240 (1990).

FOOTE, R. S., *Science*, **177**, 513 (1972).

HHS (U. S. Department of Health and Human Services), The Nature and Extent of Lead Poisoning in Children in the United States, a report to U. S. Congress, 1988.

HIRSCHMAN, S. Z., FEINGOLD, M., and BOYLEN, G., *New England J. Med.*, **269**, 889 (1963).

INSKIP, M., and ATTERBURY, N., Proceedings of the International Conference on Heavy Metals in the Environment, Heidelberg, West Germany, **1**, 286, 1986.

JOSELOW, M. M., *Ann. Intern. Med.*, **78**, 449 (1973).

KARPINSKI, M. F., KARPINSKI, K. F., and KARPINSKI, C. M., U. S. Patent No. 3888268, 1975.

LEMIRE, W. A., *Gateway Engineer*, **71**(10), 7 (1991).

MUNTHE, J., SCHROEDER, W. H., XIAO, Z., and LINDQVIST, O., *Atmos. Environ.*, **24A**(8), 2271 (1990).

OTANI, Y., EML, H., KANAOKA, C., UCHIJIMA, I., and NISHINO, H., *Environ. Sci. Technol.*, **22**, 708 (1988).

OZKAYNAK, H., SPENGLER, J. D., LUDWIG, J. F., BUTLER, D. A., PELLIZZARI, E., CLAYTON, C. A., and WIENER, R. W., Proceedings 5th International Conference on Indoor Air Quality and Climate, Toronto, Canada, **2**, 571, 1990.

ROBERTS, J. W., CAMANN, D. E., and SPITTLER, T. M., Proceedings 5th International Conference on Indoor Air Quality and Climate, Toronto, Canada, **2**, 435, 1990.

ROELS, H. A., BUCHET, J-P., LAUWERYS, R. R., BRUAUX, P., CLAEYS-THOREAU, F., LAFONTAINE, A., and VERDUYN, G., *Environ. Res.*, **22**, 81 (1980).

6

Respirable Particulates

6.1 INTRODUCTION

Fine solid particles, aerosols, mist, smoke, dust, fibers, and fumes are collectively called particulates. The inhalation of airborne particulates can cause both beneficial and harmful effects. Asthmatics and sufferers of emphysema frequently utilize inhalers for relief of symptoms. Studies dealing with the targeting of pharmaceutical aerosols to the lung can provide insights into problems associated with indoor air pollution. Although all particles in the size range from about 0.0005 to 5000 μm (from the size of a molecule to the size of sand on the beach) are called particulates, only particles smaller than about 10 μm in size remain suspended in air for long periods of time. These airborne particulates can be detrimental to good health in two ways: (1) First, particulates themselves can penetrate into the respiratory system. (2) Second, they can be vehicles for the transport of other toxic elements into the human body.

Particulate matter is characterized by particle size, shape, phase, and chemical composition. These characteristics affect particulate deposition in the lungs and in air sampling and gas cleaning devices. Figure 6.1; indicates the relation of the size ranges of selected pollutant particles and the potential for respiratory tract damage. Generally, particles larger than 5 μm in diameter are removed in the upper respiratory tract (the nasal cavity, pharynx, and trachea, see Figure 6.2). Particles smaller than 5 μm in diameter can enter the lungs, and those in the size range from 0.5 to 5.0 μm can deposit in the bronchial region. Particles

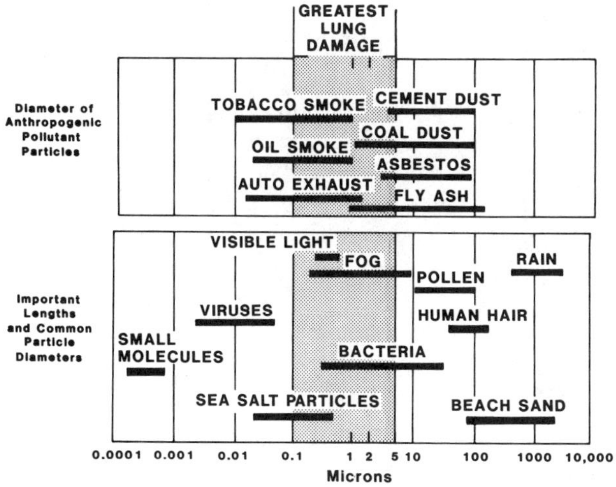

Figure 6.1 *Size range of selected particulate matter. (Adapted from* Aerosol Science: Theory and Practice, *M.M.M. Williams and S.K. Loyalka, Pergamon Press, Elmsford, N.Y., 1991.)*

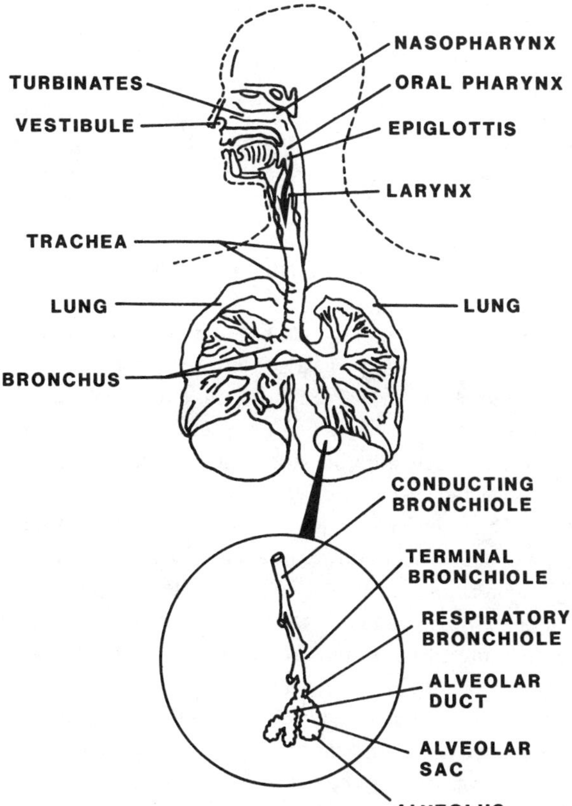

Figure 6.2 *Human respiratory tract. [Adapted with permission from* Aerosol Technology, *W.C. Hinds, copyright © 1982, John Wiley & Sons, Inc. as adapted from* Handbook of Air Pollution *USPHS 999-AP-44 (1968).]*

less than 0.5 μm in size can penetrate and remain in the alveolar region. Harmful substances, such as radon, alpha particles, and chemicals, can attach themselves to airborne particles and can reach the same regions of the lungs. Depending on the type, characteristics, and the toxicity of the chemicals attached to the particulates, they can cause a number of health problems, such as asbestosis (from asbestos fibers) and lung cancer (from tobacco smoke and attached radon daughters). The following sections describe particle deposition in the respiratory tract, health effects and standards, sources, sampling and measurement, and specific control strategies.

6.2 PARTICLE DEPOSITION IN THE RESPIRATORY TRACT

Deposition of particles in the human respiratory tract has been extensively documented since it was first described by Tyndall (1892). The term *deposition* refers to the mean probability of an inhaled (inspired) particle being deposited in the respiratory tract by collection on airway surfaces. The deposition of particles depends not only on the characteristics of the particle (for example, size, density, and shape) but also on the individual's breathing pattern (which determines the mean residence time of the particle in the respiratory tract), the mean volumetric flow rate, and on the morphology of the respiratory tract.

The airways of the human respiratory tract have a complex anatomical structure and a complex dynamical behavior. The shape and dimensions of these airways, particularly the lungs, change continuously during respiration. Modeling of the lung frequently involves the assumption that a constant air flow represents inspiration and that the lung geometry and volume are fixed. The deposition of inhaled particles in the human tracheobronchial tree is usually calculated with the assumption that airway bifurcations can be reasonably approximated by bent tubes. The air flow in the tracheobronchial depends on both the breathing cycle and the tidal volume of inspiratory air which, for a standard adult, can vary from 0.5 to 2.0 L.

Breathing systems are very complex, especially in the region of the tracheobronchial tree. Several different models are used for the lung airways; the most widely used for the calculation of particle deposition is Weibel's (1963) symmetrical lung model A. The model is based on the assumption that there are 16 generations of conducting airways in the tracheobronchial tree and 7 generations in the respiratory zone. The airways of the human lung are a repetition of bifurcated tubes wherein the number of airways multiply in a dichotomous pattern. Each bifurcation segment consists of a parent tube (the branch upstream of a bifurcation) and two daughter tubes (the branches downstream of a bifurcation). The leading edge of the inside wall of a daughter tube is called the carina. The branching angle of the lung bifurcation is assumed to be in the range of 30° to 60°. Flow through a bifurcation tube is a unique fluid mechanical feature of the human respiratory system. The air flow strongly influences the character of the local velocity distribution which, in turn, has a large impact on the local deposition and on the distribution of hot spots in the bifurcation.

Particle deposition has been studied both experimentally and theoretically for a wide range of particle sizes and various patterns of oral and nasal breathing, and there is a substantial body of literature. Total deposition usually refers to particle collection in the entire respiratory tract, and regional deposition refers to particle collection in a particular region of the tract. Predictions of particle deposition in the respiratory tract typically combine physical models of the anatomy of the mouth, nasal passages, and/or lungs with equations that describe the behavior of particles flowing in tubes of differing geometry. An extensive discussion of the theory of particle deposition and resuspension has been given by Williams and Loyalka (1991). The theoretical analyses to date have focused primarily on the motion of a single particle in isolation or near a surface and usually neglect agglomeration of particles or condensation on the particle surface. The major factor in these analyses is the particle size and its influence on the particle's sedimentation, impaction, and diffusion.

Sedimentation is the process whereby particles suspended in an air stream settle due to the effects of gravity. This process is relatively insignificant for particles less than 0.5 μm in size. *Inertial impaction* is the inability of relatively large mass particles flowing in an air stream to follow a change in flow direction. Consequently, larger particles will impact at a bend in the flow direction. In the lung, they will impact at or near an airway bifurcation. Larger particles have a greater chance of depositing by impaction. Inertial impaction is an important means for collecting and sizing large particles, and even small particles if the flow velocity is large. This process is relatively insignificant for particles smaller than 0.5 μm in size. *Diffusion* is the net transport of particles due to Brownian motion and is relatively important for particles smaller than 0.5 μm in size. The physical processes are not independent, but they are usually so treated, and their interactions are accounted for by superposition. In the following, the term *particle* will be restricted primarily to either a spherical particle or to an irregularly shaped particle described by an equivalent diameter, specified as an *aerodynamic diameter* or by a *volume equivalent diameter*. An aerodynamic diameter is the diameter of an equivalent sphere of unit density having the same settling velocity (due to gravity) as the particle in question. The term *fiber* will be used for nominally cylindrical or acicular (needle-shaped) objects. Detailed descriptions of these processes appear in a variety of sources (Hinds, 1982; NAS, 1986).

The following sections describe some aspects of particle deposition relevant to the human respiratory tract. For the purposes of this discussion, the studies have been grouped as either analytical or experimental. The latter investigations are divided into those that are directed at basic questions or phenomena, those involving replicas of the human airway passages, and studies with animals and humans.

Modeling of respiratory tract deposition initially took into consideration diffusion, sedimentation, and/or impaction of spherical particles in tubes and bifurcations. Refinements to the basic models include the effects of Reynolds number, turbulence levels, and the roughness of the surface. To understand the deposition of fibers, various models are used to represent these nonspherical particles. The representations include the equivalent diameter sphere model mentioned earlier, ellipsoids, cylinders, and straight chain aggregates of spheres. From studies with aggregates, it was found that deposition varied with the primary sphere diameter (d) and with the number of spheres in an aggregate. For $d = 0.001$ μm, diffusion was the predominant deposition mechanism. As the primary diameter of the spheres that comprise the aggregate increases, diffusional deposition decreases and interception deposition becomes increasingly significant, becoming about equal at sphere diameters of 0.01 μm. When $d = 1$ μm, the impaction deposition efficiency becomes significant in the upper airways, and some sedimentation occurs in the lower airways. From studies of sedimentation of ellipsoidal shaped fibers in a duct, it was found that for laminar flow, the deposition efficiency depends on the concentration profile and the fiber orientation at the entrance to the duct. When the fiber length is small compared to the duct radius, the deposition efficiency depends on the aspect ratio of the fiber, the duct aspect ratio, and a sedimentation parameter that represents the mean residence time of the fiber in the duct. For a fiber with a given volumetric diameter, the deposition efficiency decreases with increasing fiber aspect ratio due to the decrease in fiber mobility. The aspect ratio is defined as the ratio of the diameter to the length.

Turbulent fluctuations have been found to have a significant effect on the deposition process regardless of the particle diameter, while the effects of Brownian motion became negligible for particle diameters greater than 1 μm. Deposition of inhaled particles can be significantly enhanced if the particles carry electric charges. The deposition enhancement for spherical particles and the sedimentation of both conducting and nonconducting fibers have been considered. The deposition efficiency of the fibers is always lower than that of the spherical particles (for the same equivalent volumetric diameter), if the interception of particles at the wall due to finite particle size is neglected. Also, the incremental deposition could be larger for fibers because of their lower settling efficiency.

Experimental deposition studies are usually focused to obtain fundamental information about parameters for use with analytical models or to obtain relevant data using mechanical models of human airway passages, animal surrogates, or actual experiments with humans. There have been a number of studies using replica hollow casts of the human airways. Included in these investigations are both steady state and cyclic flow studies in the laminar and turbulent regimes. They have been focused primarily on deposition in the oral and nasal cavities.

When measured deposition efficiencies are compared with predictions of deposition by diffusion from a laminar flow stream, the measured deposition is about twice that predicted, except for tracheal deposition, which on average exceeds the predictions by a factor of nine. This is attributed to secondary swirling flows. There is less enhancement at higher inspiratory flow rates where the turbulence level increases (Kinsara, 1991). For both inspiratory and expiratory flows, with aerosol diameters between 0.005 and 0.2 μm, the deposition efficiency increases with decreasing flow rate and particle size, indicating that turbulent diffusion is the dominant deposition mechanism. Generally, diffusional deposition for nasal breathing is higher than that via the mouth and is weakly dependent on the flow rate. Studies indicate that the local flow in the nasal and oral passageways is predominantly turbulent. At fixed turbulence conditions, the deposition percentage increases with particle size for particles greater than 1 μm. With a fixed aerosol size, deposition increases with increasing fluid turbulence, but its contribution is less than with the larger-size aerosol.

Although the basic deposition mechanisms for cyclic flow conditions appear to be the same as with steady flows, the deposition efficiency has been found to be greater by about 25%, but with a broader deposition pattern. The increased deposition efficiency with cyclic flow may be related to enhanced inertial deposition during the peak flow range of a flow cycle. The effect of the cyclic flow is probably less likely to be a significant factor. For conditions in which compressibility effects cannot be ignored, deposition has been found to exceed values predicted by standard correlations that are based on the Reynolds and Stokes numbers.

There is a considerable body of literature on experimental studies of particulate inhalation and deposition in mammals, as well as studies that compare predictions or model studies with experiments. An ultimate goal of such experiments is to develop a reliable means to aid in the extrapolation of animal data to humans. Animals, such as rats, hamsters, dogs, baboons, and donkeys, have been exposed to particle sources, such as fibers, talc, rock dust, and cigarette smoke, in surrogate species studies. These studies included the effects of both short- and long-term exposures.

Animal studies have shown that the deposition pattern depends on both fiber diameter and length. Although inhaled fibers are effectively filtered by the nose and in the tracheobronchial region by impaction and interception, a significant fraction (2%–10%, depending on the fiber dimensions, lung volume, and tidal volume) of long fibers can still penetrate into the pulmonary region where they deposit.

A fraction of the inhaled particles are cleared from the respiratory tract during exhalation. This clearance depends on the location in the respiratory tract where the particles are deposited and on the particle morphology and solubility. A distinction is usually made between the short-term (fast) clearance in the upper airways (nasopharyngeal and tracheobronchial compartments) and the long-term or alveolar clearance. There are at least two problems with extrapolating animal pulmonary clearance data to humans: (1) Clearance rates may be species dependent. (2) The phenomenon of dust overloading the lungs, which is often used to facilitate measurements, may cause effects that interfere with clearance. The results of various studies support those models that assume that the dissolution rate of deposited materials depends on the material but not on the species, while the mechanical transport depends on the species but not the material.

Wehner (1986) has summarized a large number of studies dealing with the biological effects of inhaled aerosols in hamsters and rats. Laboratory animals were exposed to a variety of materials, including cobalt oxide, nickel oxide, chrysotile (asbestos), asbestos cement, cigarette smoke, talc, fly ash, volcanic ash, and quartz, in inhalation experiments. Exposure times ranged from 30 days to 2 years or throughout the lifetime of the animal. The exposure to chrysotile resulted in severe asbestosis in all animals, requiring discontinuation of the exposures, while animals exposed to asbestos dust developed slight pulmonary fibrosis. Particle clearances from the respiratory tract varied greatly.

An eight-part series of papers by Bailey et al. (1989) describes an extensive interspecies comparison of the lung clearance of inhaled monodispersed cobalt oxide particles. Clearance of deposited material was represented as a competition between mechanical clearance rate and blood translocation rates, that is, the dissociation of material from the particles by dissolution and subsequent absorption into the blood. Interpretation of the measurements in terms of this competition provided a quantitative description of the clearance kinetics, which permitted meaningful comparisons to be made of the lung clearance patterns for different particles and different species. Their results also supported the assumptions that mechanical transport and translocation to blood are independent, and that the mechanical transport rates are similar for different materials in the same species. In the absence of suitable human data, the best estimate of the clearance of a specific particulate material from the human lung is obtained by combining the translocation rate measured in animals with the particle mechanical clearance rate from the human lung.

A determination of the hazards to humans from exposure to materials in particle form requires an estimate of the rate at which the material is deposited in and cleared from the various regions of the respiratory tract. Usually it is impossible to obtain the pertinent data for hazardous materials directly from human observation. Measurements on humans, following occupational or accidental exposures to hazardous materials, are difficult to interpret because of uncertainties about the conditions of exposure. Consequently, the potential hazards of many materials have to be determined in the absence of human data. Nevertheless, substantial data exist on deposition and retention in the human respiratory tract.

There are many comparisons of experimental data with calculations for the total and regional deposition in the human lung. These comparisons make use of several lung structure models (equations that include deposition by sedimentation, diffusion, and impaction) and computational methods that take into consideration the existence of alveoli in different ways. It has been found that the calculated total deposition shows only small differences for the different lung structures, particle deposition equations, and computational methods. However, the bronchial deposition calculated using the available models is much larger than the corresponding experimental data, especially for particle sizes of less than 2 μm. The calculated pulmonary deposition is usually smaller than the experimental values. Pritchard et al. (1986) found that deposition in the upper respiratory tract of women can exceed that in men by a factor of two. Additional work is needed to gain a better understanding of these differences.

There has been a great deal of progress in the understanding of the deposition of particulates in the respiratory tract as a result of such studies; however, much remains to be done. The interested reader should consult the original papers and the associated references for details about particular studies.

6.3 HEALTH EFFECTS AND STANDARDS

Particles with very low solubility in the lungs, such as diesel soot, titanium dioxide, activated charcoal, and quartz, have been found to be carcinogenic in the lungs of rats. However, the significance of these results relative to human risk assessment is unclear. The heavy loading of toxic dusts may contribute to the findings, and there may also be a nonthreshold mechanism for carcinogenicity associated with the surface properties and small size of the particles.

Environmental tobacco smoke (ETS), which consists of sidestream smoke and exhaled mainstream smoke, is considered to be a major component of indoor air pollution. Numerous studies have shown that ETS can substantially increase the levels of particulate matter, mutagenicity, and polycyclic aromatic hydrocarbons in indoor air. The first report of the Surgeon General on smoking and health in 1964 asserted that cigarette smoking was a cause of lung cancer in men and was probably a cause in women. The 1979 Surgeon General's report on *Smoking and Health* (HEW, 1979) concluded that cigarette smoking was the major cause of lung cancer in both women and men. Subsequent studies (HHS, 1986; NAS, 1986) continued to support these conclusions, and additional concerns about the harmful effects of the low levels of exposure to ETS received by nonsmokers (involuntary or passive smoking) have emerged. These studies were based on the assumption that smoking is known to have a causal relationship to a variety of cancers other than lung cancer. It is a factor in cardiovascular disease and is the major cause of chronic obstructive lung disease. These findings have led to efforts by individuals, employee organizations, and employers to protect nonsmokers. The increasing numbers of laws and regulations enacted by local, state, and federal agencies reflect this concern. The banning of cigarette smoking on all domestic airline flights is a recent example of these efforts.

The number of studies relating to ETS that appear each year is voluminous. On the basis of 25 years of study, one could surmise that there would be unanimity about the deleterious effects of ETS on lung cancer, pulmonary function, and respiratory health. However, there remain differing (and sometimes vociferous) views regarding the severity of the effects of ETS on humans. Many of the objections relate to the fact that with ETS one is working near the bottom of a dose effect curve for a substance that is both poorly characterized and highly diluted. A lack of control or even knowledge of many of the experimental variables involved can lead to a great deal of variability in the results of the numerous studies. Consequently, the results of many ETS studies are open to criticism. There does not appear to be any controversy regarding the causal relationship between smoking and lung cancer, but rather as to the effects of ETS. Many of the critiques of studies that conclude that ETS has an adverse health effect on humans are summarized in a 1989 symposium supported by the tobacco industry (Ecobichon and Wu, 1989). The conclusions of these and other studies appear to be in conflict with the basis for the draft EPA regulations (IAR, 1991a) and will, no doubt, continue to be a matter of controversy.

There is experimental evidence that a variety of fibers with different chemical compositions that are sufficiently thin, long, and durable can induce tumors. The mechanism of fiber carcinogenesis appears to be substantially different from that caused by nonfibrous small particles. The carcinogenic potency of asbestos and man-made mineral fibers is believed to be especially significant. These cancers include lung cancer, mesothelioma (a cancer of the membrane lining the chest and abdomen), and cancers of the gastrointestinal tract. In addition, exposure to asbestos dust increases the risk of asbestosis, a fibrotic disease of the lung whereby imbedded dust fibers are surrounded by scar tissue. Lung cancer and mesothelioma are the diseases of greatest concern associated with indoor asbestos exposure. Recent data suggest that the fibers must remain in the respiratory tract for approximately one year, and those with a diameter less than 1 μm and a length greater than 5 to 10 μm are believed to be particularly dangerous. However, animal data suggest that very short fibers have significantly less carcinogenic activity than the longer fibers and may even be relatively inactive.

Asbestos standards have been reviewed by Rajhans and Bragg (1978). The literature related to asbestos in the air of public and commercial buildings is reported in a study commissioned by the EPA (HEI-AR, 1991). The potential for adverse health effects are described. Asbestos-containing materials within buildings that are in good repair are unlikely to expose most building occupants to airborne asbestos fiber concentrations higher than the levels found in the air outside such buildings. However, the situation is significantly different for custodial, maintenance, and renovation workers. Their added lifetime risk of cancer may be appreciably higher than the risk to other building occupants, because such workers may be intermittently exposed to higher levels of asbestos. In well-maintained buildings that have airborne levels of asbestos fibers similar to those found outside the buildings, removal or abatement actions, if done improperly, can cause an increase in fiber levels that may persist for some time.

The only voluntary consensus standard covering ventilation for indoor air quality is ASHRAE Standard 62-1989, which has also been approved by the American National Standards Institute (ANSI). The National Primary Ambient-Air Quality Standards for Outdoor Air (as set by the EPA) for total particulate matter are 75 mg/m^3 (0.075 μg/cm^3) for long-term exposure, as averaged over one year, and 260 mg/m^3 maximum for short-term exposure, as averaged over 24 hours (ASHRAE, 1989). This standard specifies annually averaged exposures of 55 mg/m^3 (long-time periods) and 150 mg/m^3 (short-time periods) for particle aerodynamic diameters less than 10 μm. These recommendations are based on a 24-hour average. This standard also applies indoors for the same exposure times. The ASHRAE standard specifies ventilation rates from 15 ft^3/min per person to 60 ft^3/min per person. The rate specified by the standard for office space is 20 ft^3/min per person. These rates are three times greater than those previously required and represent an effort to reduce the indoor air pollution produced by human activities, such as cooking, cleaning, and outgassing from synthetic materials.

6.4 SOURCES AND INDOOR CONCENTRATIONS

Respirable particulates are introduced indoors from outdoor air, tobacco smoke, wood combustion, and other sources. Unfiltered outdoor air is a major carrier of atmospheric dust, combustion particles from power plants, and automobile exhausts. Although outdoor particulates are introduced by both natural and human activities, natural sources such as volcanic eruptions, forest fires, and sea salts contribute almost 90% of the total suspended particulates. A number of sources of indoor particulates are listed in Table 6.1.

There have been a number of investigations of the indoor to outdoor ratio of respirable particles. While most studies have relied largely on filter sampling (time-averaged measurements) to determine the indoor particle level, Tu and Knutson (1988) obtained nearly simultaneous measurements of the temporal variations of the indoor-outdoor particle number concentrations and size distributions. A similar study by Clark et al. (1992) of an integrated circuit manufacturing cleanroom revealed a strong correlation between atmospheric particle concentrations and background particle concentrations inside the clean-

TABLE 6.1 Indoor Sources of Respirable Particles

Environment	*Sources*
Home	Tobacco smoking, outside air, unvented kerosene heaters, coal stoves, fireplaces, attached facilities, wood stoves, occupant activities, curtains, carpets, thermal and acoustic insulation
Workplace	Tobacco smoking, outside air, carpets, curtains, asbestos, ventilation systems, laser printers, photocopying machines, thermal and acoustic insulation, manufacturing processes
Transportation	Tobacco smoking, outside air, ventilation systems, automotive airbags

room. Although a broad range of ratios have been reported in various studies, it is apparent that respirable suspended particles are present in considerably higher concentrations indoors than they are in the outdoor air. A relatively airtight house in which there are human activities will exacerbate this difference. Indoor-outdoor ratios for respirable particles are summarized in Table 6.2.

Particle size distributions, both indoors and outdoors, are found to be similar, but submicron particles are present in a larger proportion indoors. Owen et al. (1990) recently completed a detailed study of particle concentrations and size distributions in offices under varying conditions of occupancy and outdoor-return air ratios. They found that the building particle concentrations, when unoccupied and using minimum outdoor air, were at least as low as those when occupied, and using maximum outdoor air. More recently Grot et al. (1991) reported on an extensive IAQ evaluation of a new office building. They measured carbon dioxide, carbon monoxide, respirable particulates in the 0.3 to 10 μm range, formaldehyde, radon, and VOCs. The indoor levels of respirable particles in six size ranges (0.3 to 0.5, 0.5 to 0.7, 0.7 to 1.0, 1 to 5, 5 to 10, and greater than 10 μm) were measured every 10 minutes on the fourth floor of a seven story building. The concentrations of particles in the 0.3 to 0.5 μm range remained fairly constant and showed little daily or hourly variation, while the next three ranges had much greater variations. The upper two size ranges were not considered to be respirable. After the lowest-level parking garages were cleaned with street sweeping machines, they also found excessively high particle levels in the 0.7 to 1.0 and 1 to 5 μm ranges, which required that all the filters in the air handlers be replaced. Respirable particulate concentrations (for particles less than 3 μm in size) were in the 10 to 15 million/m^3 range, which was comparable with their previous measurements for office buildings with no smoking. They were unable to compare the measured particle levels with standards (which are expressed in mg/m^3) due to a lack of information on the particle compositions. Studies have also shown that tobacco smoke is a major contributor to respirable particles indoors.

Environmental Tobacco Smoke

Environmental tobacco smoke (ETS) is a complex mixture of gases and particles that result from cigarettes, cigars, and pipes. However, our discussion here will focus on cigarette smoke. Cigarette smoke is difficult to characterize because it contains highly volatile materials that are susceptible to the influence of dilution air. Respirable particles, nicotine, poly-

TABLE 6.2 Indoor/Outdoor Air Ratios for Respirable Particles

	Study	*Indoor to Outdoor Air Ratio*
A:	Residences with smokers	4.4
	Residences without smokers	1.4
	Office buildings	1.1
B:	Residences with one smoker	1.7
	Residences with two or more smokers	3.3
	Residences without smokers	1.2

Source: Yocom, 1982.

cyclic aromatics, carbon monoxide, acrolein, nitrogen oxides, and a number of other chemicals are released in the atmosphere during the burning of a cigarette. Cigarette smoke has been found to be the largest single contributor to indoor air particle concentrations in an office environment. The emission rates of respirable particulates and chemicals depend on the burning characteristics of a cigarette. Smoke is characterized as environmental tobacco smoke (ETS), which consists of mainstream smoke (MS) and sidestream smoke (SS). The MS is the smoke exhaled after inhalation by a smoker, whereas the SS is the smoke emitted from the idle burning of a cigarette. Smoke composition is known to be very sensitive to the type of tobacco, its packing density, composition of the wrapping paper, puffing rate, and the person inhaling or exhaling the smoke. The ratio of the emissions of SS to MS depends on several factors, including the moisture content of the tobacco, air dilution from filter perforation, paper porosity, and puff volume. Furthermore, different sizing methods are used to measure different size-related properties of the particles.

Reported cigarette smoke particle sizes have varied over the range from 0.1 to 1.5 μm. Recently, Chen et al. (1990) undertook a detailed evaluation of the cigarette smoke produced by a Walton Smoke Machine. They obtained estimates of the total particulate matter, the aerodynamic size distributions, the geometric size distribution, number concentration, and coagulation coefficients with the same operating conditions and sampling techniques. They found that the total particulate matter averaged 3.37 mg per puff, the mass median aerodynamic diameter was calculated to be 0.45 μm at a dilution ratio of about 20, and the count median diameter was 0.22 μm. The effects of humidity in the room on particle size were minimal.

Particulates and other chemicals in the ETS have been studied extensively. Recently Eatough et al. (1989) have critically reviewed various studies of the chemical composition of fresh ETS, as well as the changes in the chemical composition and gas-particulate phase distribution of ETS that occur with time in the indoor environment. A great deal of effort is being expended to assess the suitability of various materials as markers for ETS. The National Research Council (NAS, 1986) recommended that any ETS marker chosen should be present in a consistent ratio to the ETS constituents of interest. Nelson et al. (1990) evaluated nicotine as an ETS marker and found that the ratio of nicotine to other ETS constituents, such as respirable particulates, was highly variable and depended on both the air exchange rate at the sampling location and the sampling time. The desorption of nicotine from clothing, cigarette butts, and interior surfaces led to measurable nicotine levels even in the absence of ETS. Eatough et al. (1990) investigated the use of cotinine (a metabolite of nicotine) as an ETS marker in commercial aircraft cabins. Cotinine urine concentrations were correlated relative to exposure to nicotine, but not with exposure to many other constituents of ETS.

Repace and Lowrey (1980, 1982) sampled the air in restaurants, bars, and cafeterias for aerosols and respirable particulates. Respirable particle levels were as high as 700 mg/m^3 in these locations, but varied with the intensity of smoking. A significant difference was reported in the respirable particle levels between the smoking and nonsmoking sections of two cafeterias and a sandwich shop. The concentrations of respirable particles in the nonsmoking sections of two cafeterias were found to be 43% and 60% lower than that in the smoking section. In the nonsmoking section of the sandwich shop, it was 50% lower than in the smoking section. Similar results were found in office buildings from a limited number of studies (HHS, 1986; NAS, 1986).

The contribution of respirable particles from tobacco smoke to indoor air pollution has been studied by Spengler et al. (1981), who monitored 80 homes over several years. A smoker of one pack of cigarettes daily contributes approximately 20 mg/m^3 in a 24-hour period to the indoor particulate concentration. This level encompasses about 50 to 75% of all American homes, which is the percentage of homes reported to have at least one smoker (NAS, 1981). When cigarettes are actually being smoked, the particulate concentration can range from 500 to 1000 mg/m^3 for a short period of time. Leaderer et al. (1990) investigated the impact of cigarette smoking on indoor aerosol mass and elemental concentrations over a 3-month period in a sample of nearly 400 homes. To investigate differences in aerosol mass and concentrations between homes with and without smokers, subsets of 75 homes without smokers and 141 homes with smokers were selected. They sampled for particles less than 2.5 μm in diameter and found that the average indoor mass concentration in the nonsmoking homes and the outdoor concentration are similar. Furthermore, homes with smokers present had mass concentrations that are approximately three times higher than homes with no smokers. Moschandreas et al. (1981) studied the indoor and outdoor concentrations of respirable particles in a number of locations. As can be seen from Figure 6.3, the indoor concentrations always exceed the outdoor concentrations whenever smoking oc-

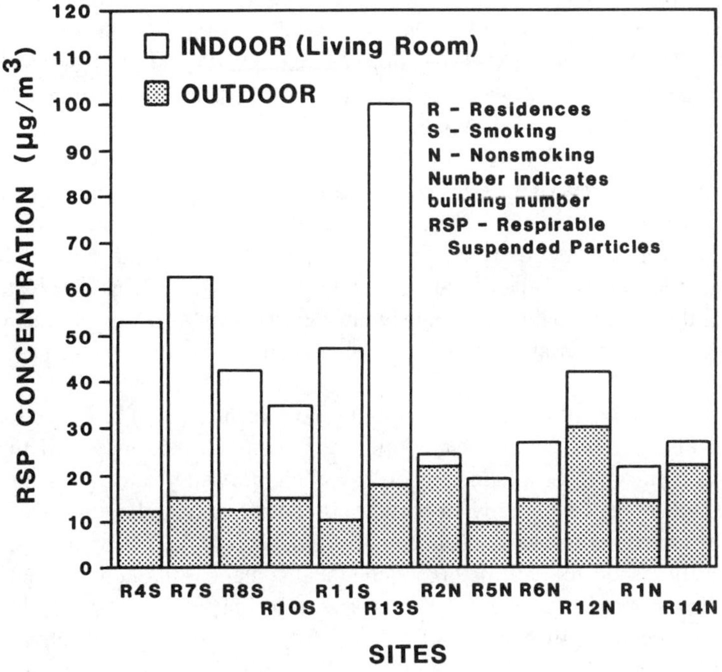

Figure 6.3 *Indoor/outdoor concentration of particles in various locations.* (Source: *Moschandreas et al., 1981.*)

cured. However, the indoor concentration level was dependent on the number of smokers and the ventilation rate of the home.

Asbestos and Other Fibers

Fiber concentrations indoors can be significantly greater than outdoor concentrations because of an increased number of sources and reduced dilution. Sources of indoor fibers include carpets, curtains, and insulation. The fiber lengths of greatest interest for respiratory toxicology are roughly between 5 and 200 μm. Most common fibers, such as asbestos, quartz, and fiber glass, are considered to be electrical insulators, whereas man-made carbon and silicon carbide fibers are considered to be electrically conductive. In practice, most airborne particles should be considered to be conductive, primarily because of the surface adsorption of water.

Assessment of exposure to asbestos is particularly difficult. Asbestos is a generic term applied to fibrous silicate minerals. Six common asbestos varieties are actinolite, amosite, anthophyllite, crocidolite, tremolite, and chrysotile. Chrysotile (white asbestos) accounts for more than 90% of all asbestos uses. Most airborne asbestos fibers are too small to be counted by optical microscopy. As a result, scanning (SEM) and transmission (TEM) electron microscopy are used to analyze samples collected on filters. Guillemin et al. (1989) carried out a detailed investigation of 12 different buildings with and without insulation materials that contained friable asbestos. They considered a variety of measurement methods, including a fibrous aerosol monitor (FAM), phase-contrast microscopy (PCM), SEM, and TEM. The simplified methods (PCM, FAM, and SEM) may be of use in some instances; however, they should always be backed up by the TEM method (see Section 6.5). Some of the data published on indoor concentrations in schools and public buildings as measured with TEM (see also Chapter 4 and Appendix A of HEI-AR, 1991, for an extensive review of asbestos concentration measurements in buildings) are summarized in Table 6.3. Guillemin et al. noted that in buildings where no known source of asbestos fibers exists, high levels of fibers from unsuspected sources can be found. The resuspension of settled fibers is one of the main sources of exposure. This observation was related primarily to the activities of people in the concerned area. During normal use, the air in buildings that contain asbestos has not been found to have fiber counts higher than those of outdoor air.

Indoor particulate concentrations have been found to be a function of foot traffic and ventilation rate. During the daytime when the foot traffic and ventilation rate are at a maximum, the concentration is higher than at night. Also, the total organic content (chemicals soluble in benzene) of indoor particles is considerably higher than that of outdoor particles. The attachment of toxic chemicals from tobacco smoke, aerosol cans, and cooking is primarily responsible for the higher chemical content. Although the relative contributions of different sources to indoor levels of respirable particles are not available in the literature, the particulate concentrations that result from different types of activities, such as smoking and burning of wood, are presented in Table 6.4.

There are many other sources of respirable particulates, such as natural-gas-fired ap-

TABLE 6.3 Indoor Asbestos Concentrations as Measured by TEM from Various Studies

Concentration Range	Building		Reference
	Number	Type	
ND–9.6	5	Schools	Chadwick et al. (1985)
< 2	3	Public	Chatfield (1985)
< 4–< 9[a]	7	Public	Chatfield (1986)
ND–0.3	19	Public	Pinchin (1983)
ND–1	43	Public	Burdett and Jaffrey (1986)
ND–8.6	10	Public	Guillemin (1989)

Concentration results expressed in fibers >5 μm/L of air.
ND, not detected.
[a]most of the values were below the limit of detection.
Source: Guillemin et al., 1989.

pliances, automotive airbags, dust from laser printers, and various industrial processes. For example, the activation of automotive airbag restraint systems can lead to exposure to respirable particulates, which is of particular concern for asthmatics. On the basis of estimates of the number of over 20 mph accidents and the number of asthmatics, it has been projected that by 1994 there will be about 100 accidents per one million miles driven annually that will result in significant exposure for asthmatics. However, total contributions from all such sources are usually small when compared to tobacco smoke. Industrial processes, such as welding and the laser cutting of steel components and concrete structures, are potential sources of particulates. Such processes may become increasingly commonplace in the dismantling of a variety of defense and nuclear power plant structures.

TABLE 6.4 Particulate Concentrations during Different Activities

Activity	Concentration
Tobacco smoking	>500 μg/m^3 in bars, meetings, waiting rooms 100–500 μg/m^3 in smoking section of planes 10–100 μg/m^3 in homes 500–1000 μg/m^3 during smoking in homes
Wood burning in fireplaces	1000 μg/m^3
Gas stove	0.6 mg/h/burner
Gas oven	0.1 mg/h
Gas-fired unvented space heaters	3.2 mg/h
Kerosene space heaters	
Radiant	0.49 mg/kJ of fuel energy
Two stage	1.70 mg/kJ of fuel energy

6.5 Sampling and Measurement

Aerosol particle concentrations in the home and workplace are usually measured by means of an aerosol sampler. Sampling can be either continuous or batchwise, although continuous sampling provides a more statistically consistent result. In continuous sampling, air is drawn through one of several particulate samplers, such as filters, impactors, and direct-reading samplers, at flow rates that in most cases are less than 2 L/min. A filter is used when sampling for total suspended particles. These samplers operate by aspirating an air sample through one or more openings in the solid casing of the filter holder and collecting the particles on a filter, which is subsequently examined. The physical presence of the sampler and the actual sampling process can disturb the flow and affect the particle motion. From an aerodynamic perspective, all samplers are blunt. Consequently, there have been a number of theoretical and experimental studies to ascertain how representative the collected aerosol sample is of the fluid being examined.

The selection of a filter depends on its collection efficiency, the pressure drop through the filter, background contamination, and compatibility with the analytical methods to be used for measurement. Cellulose, fiber, and membrane filters are suitable for collecting indoor particles, and they have reasonable efficiencies. The collection efficiency and pressure drop through a filter are of paramount concern when various particle removal methods are being compared (see Section 6.6). Filter efficiencies have been studied extensively, and several are summarized in Figure 6.4.

The size of a particle can be used to indicate its source, potential toxic effect, and residence time in the atmosphere. It can also help in the selection of suitable removal equipment. Density measurements of respirable particles are important because the dissolution rate of particles deposited in the lung is a function of particle density.

A cascade impactor, which is a multistage impaction device used to separate airborne particles into size classes on the basis of their aerodynamic diameter, is used to measure the size distribution of airborne particles. An aerosol is drawn through a series of progressively narrower jets, each of which is followed by an impaction surface that is usually placed perpendicular to the axis of the jet. The jet may issue from a rectangular or circular slit. Each jet and its associated impaction surface are termed a *stage*. After traversing the last stage, the stream is usually drawn through a final filter as it exits from the impactor. A monodisperse aerosol is never collected exclusively on a single stage, but is distributed among several stages. Losses to internal walls will also occur. Cascade impactors are used to determine the total suspended particles and the particle size distribution. However, the intervals between the cutoff diameters of subsequent stages are generally too large for an accurate determination of the particle size distribution of even fairly monodisperse aerosols (AIHA, 1986; Cohen, 1986; ACGIH, 1989). Impactors can be operated at flow rates ranging from about 0.1 to 4 ft^3/min, but the flow rate specified by the manufacturer should be followed for the best results.

Two general types of test aerosols (monodispersed and polydispersed) are typically used in particulate research to calibrate test instruments and to develop and test air-cleaning and air-sampling equipment. Monodispersed aerosols are used to calibrate particle size measuring instruments and to determine the effect of particle size on sampling devices.

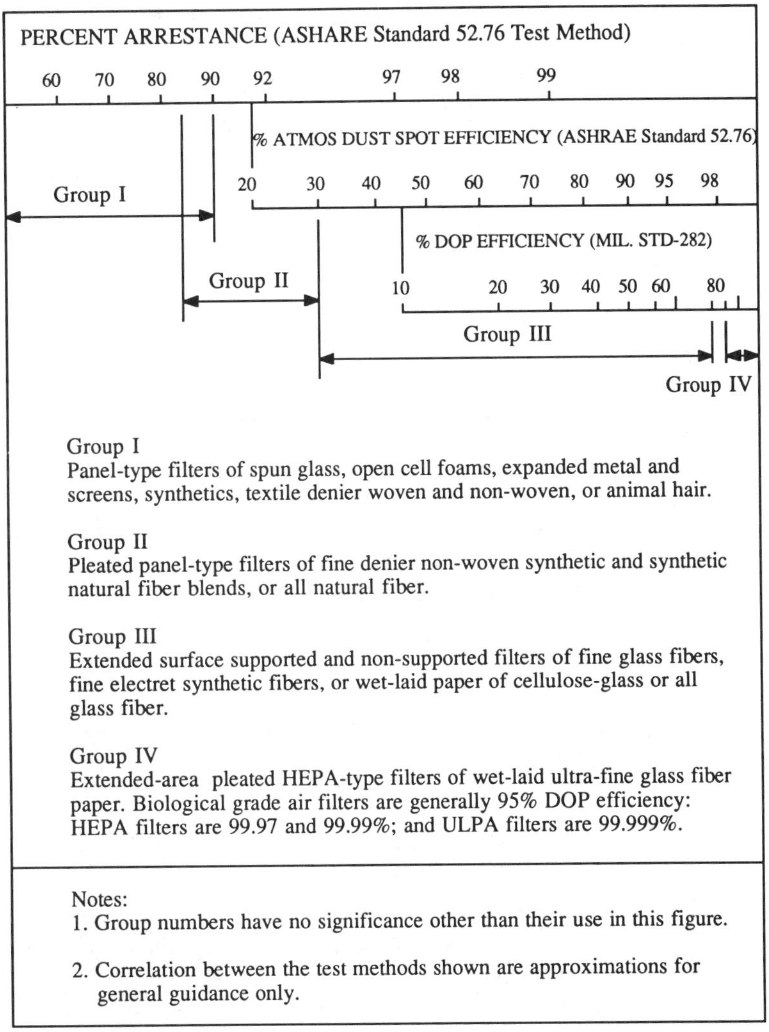

Figure 6.4 *Efficiency of various filters. (*Source: *ASHRAE Equipment Handbook, 1988.)*

Polydispersed aerosols can be used in the laboratory to simulate the conditions under which a sampler would be used.

Both liquid droplets and solid particles can be used for laboratory tests. Liquid droplets are generated by atomization of a liquid using a nebulizer or a vibrating orifice aerosol generator. Dioctyl phthalate (also called DOP, di-(2-ethylhexyl phthalate, di-sec octyl phthalate or DEHP) is frequently used in testing high-efficiency particulate aerosol filters and for calibrating instruments. There are standard test methods (cold-DOP and hot-DOP tests) that use DOP to produce a polydispersed aerosol to challenge the filter. The ratio of the challenge to the penetrating aerosol concentrations is measured using light scattering. Recently concerns about the toxicity of DOP, which is considered to be a carcinogen in laboratory animals, have lead to a search for replacement materials. Discussions of the types of equipment that are used to generate aerosols and their operating techniques are available in any standard textbook on aerosol technology (Hinds, 1982).

Kenny and Liden (1991) have developed a measurement system to test samplers for respirable dust using polydispersed aerosols. An especially interesting aspect of their findings was the daily variations in the results obtained using a properly calibrated aerodynamic particle sizer, which is usually considered to be the standard for such measurements.

There is no satisfactory sampling technique to measure the particulate matter and chemicals produced from environmental tobacco smoke. Although the MS smoke can be readily collected by a variety of techniques, the collection of representative SS smoke is a problem. The methods that are presently used to collect the MS smoke essentially duplicate the processes of puffing a cigarette. The burning end of a cigarette is held under conditions comparable to that of a cigarette being smoked. When compared to an actual smoking process, some changes in the MS smoke composition are expected. The collection of SS smoke is a difficult task because of the interference with the normal burning process of the cigarette and with the diffusion of SS smoke into the atmosphere. Several instruments have been developed to collect SS smoke, but they have met with limited success. The most frequently used instrument to generate SS smoke is the modified Neurath chamber (Brunnemann et al., 1977). Later Brunnemann et al. (1980) and Stehlik (1982) designed another instrument to collect SS smoke, but it did not perform as well as the modified Neurath chamber.

Fixed stations (area samplers) have traditionally been used to determine human exposure to air contaminants. More recently personal air samplers have been introduced. As compared with fixed station samplers, personal samplers should provide a more representative sample of what a person is inhaling. Such a sample is also more likely to be associated with actual health effects. Crouse and Oldaker (1990) compared area and personal sampling methods for determining concentrations of nicotine in ETS. They found that the concentrations of nicotine measured with the area sampling device were significantly greater than concentrations determined with personal samplers. Consequently, the results obtained by area sampling devices in ETS surveys provide a conservative estimate of personal exposure to ETS nicotine. It is important to note that the performance of a particular sampler used for area sampling may differ markedly when used as a personal sampler, because of localized turbulence and air currents associated with a person's movements.

Direct-reading particle samplers that are currently in use include electric aerosol analyzers, optical particle counters, and the integrating nephelometer. A list of instruments, analytical methods for particle analyses, and the particle size working range is provided in Table 6.5. The advantages and disadvantages of several analytical methods have been reported by Wadden and Scheff (1983) and the ACGIH (1989). Care must be taken, however, when sampling tobacco smoke using any of these methods. For example, it has been found that the piezoelectric balance underestimated the tobacco smoke concentration by 15% as compared to a low-volume filter sampler.

Other types of samplers suggested for the collection of fine particles include both

TABLE 6.5 Characteristics of Direct-reading Particle Monitors

Instrument	*Analytical Method*	*Type of Measurement*	*Effective Particle Size Range*	*Portable*
Aerosol photometer	Light scatter at a fixed angle	Indirect measure of aerosol mass or dust	<1.0–10.0 μm	Yes
Integrating nephelometer	Light scatter integrated over a wide angle (8°–170°)	Visual range; indirect measure of respirable particles	0.1–1.0 μm	No
Light scattering optical counters	Line scatter at a fixed angle	Particle number distribution	0.3–10.0 μm in up to 10 size ranges	Yes
Active scattering aerosol spectrometer	Light scatter of output of a He-Ne laser at a fixed angle	Particle number distribution	0.08–4.5 μm in up to 48 size ranges	No
Condensation nuclei counter	Condensation of nuclei by adiabatic expansion and optical detection	Condensation nuclei concentration	0.0025–1.0 μm	Yes
Electrical aerosol analyzer	Electrostatic particle mobility analyzer	Particle number distribution	0.0032–1.0 μm in 10 ranges	No
Respirable dust mass monitor	β attenuation	Cyclone followed by an indirect measure of mass	<10 μm	Yes
Aerosol mass monitor	Piezoelectric crystal	Particle mass concentration	0.01–10.0 μm	Yes

Source: Wadden and Scheff, 1983.

sonic and supersonic impingers. The dry collection efficiency of a small-hole, supersonic impinger is approximately 100% for a 0.5 μm DOP aerosol, but drops to 98% for a 0.3 μm aerosol and to 80% for aerosols of 0.1 μm and smaller. The dry collection efficiency for tobacco smoke at sonic velocity is about 90%, but can be increased to 99% by adding water to the collection flask.

The size distribution of fibrous particles is important in indoor air pollution control because long fibrous particles have been linked with cancer. Major experimental challenges usually associated with such measurements include the presence of agglomerates, the lack of a monodispersed aerosol for calibration purposes, and the relatively low concentration of such fibers, particularly when they are present in a complex that contains a large number of nonfibrous particles. A commercial instrument that detects fibers by the intensity of scattered He-Ne laser light is reported to detect fibers greater than 2 μm in diameter with lengths from 2 to 200 μm over the concentration range of 0.0001 to 30 fibers/cm^3 (ACGIH, 1989). Gentry et al. (1989) compared theoretical and experimental collection efficiencies of ultrafine asbestos fibers on Nuclepore filters. For filters with pore diameters of 1 μm or less, the efficiency was found by TEM analysis to be dependent on the fiber length. The efficiencies were 10% to 30% lower than would have been expected if there had been fiber rotation within the pores. For larger size filters (approximately 3.0 μm), they found an unusually high collection efficiency for the thinnest fibers. A comprehensive literature review of asbestos sampling and analytical methodologies has recently been completed (HEI-AR, 1991) and should be consulted for details.

Spengler et al. (1989) suggested that personal exposure to asbestos-containing materials tended to be higher than simultaneously measured area concentrations. However, Corn et al. (1991) observed no difference in the concentrations of asbestos as measured by simultaneous personal and area samplers. These conclusions differ from those found for ETS and underscore the difficulties associated with obtaining reliable determinations of the composition and concentrations of respirable particulates in indoor air.

6.6 SPECIFIC CONTROL STRATEGIES

Mechanical Filtration

Experimental studies of aerosol behavior form the basis of the development of many practical systems for dust measurement and air pollution control, such as filters. Generally, particle filtration occurs in two forms: fabric or cake filtration and fibrous filtration. Cake filtration is used for conditions that require cleanup of high particle concentrations and large quantities of gas flow. The most common cake filters are the fabric filters used in applications such as baghouses. The designs for such filters require consideration of the pressure drop through the filter and the dust mass and gas velocity. These characteristics are related empirically by constant parameters to the flow resistance of the fabric and the dust cake. Because the constants cannot be reliably predicted from the system's physical properties or from operating data, the design of filtration systems is usually quite conservative.

Fibrous filtration is generally used to clean gas streams that have relatively low particulate concentrations and low flow rates, and their particle removal efficiencies are very high, compared with fabric filter systems. There have been numerous attempts to model the collection characteristics of filters for over 30 years. The models are typically used to compute the flow behavior in an idealized filter medium for particles of a single size. The filtration of spherical particles is relatively well understood in clean filters that are comprised of regularly spaced fibers. It is difficult, if not impossible, to compare the model predictions with experimental data for filters that have complex structures. Some models have utilized geometries, such as an in-line array of parallel circular cylinders, which are more realistic than that of a single fiber. However, these models still fail to take into account the nonuniform, random orientation of different sized fibers that are found in commercial filters. Furthermore, a description of changes in filtration efficiency and pressure drop as particles accumulate within the filter is currently beyond the scope of available theories. Consequently, filter performance evaluation and selection will be based on empirical data for the foreseeable future.

Panel filters are the most frequently used mechanical filters for removing particles from air. A principal concern in evaluating filter performance is the pressure drop through the filter and the efficiency of removal. The collection efficiency of these filters depends on the particle size being filtered, and their ratings are usually based on ASHRAE Standard

52-76. Low-efficiency filters, including metal filters and polyester and glass pads, typically have ratings of less than 10%. Such filters will unload particulates onto fans, coils, and ductwork, where buildup can significantly reduce heat transfer efficiencies and increase energy requirements. Medium-efficiency filters have efficiencies in the 20% to 50% range. They are typically made from cotton and/or synthetic or microfine glass fibers and are frequently used as prefilters in multistage systems. High-efficiency filters have efficiency ratings in the 60% to 90% range. Their use is becoming more common in hospitals, office buildings, and in other applications where maximum protection is needed (IAR, 1991b). Other types of filters that have been employed include extended surface filters, high efficiency particle air (HEPA) filters, and ultra low particulate/penetration air (ULPA) filters. The HEPA and ULPA filters have high efficiencies for removing submicron particulates. HEPA filters have efficiencies that range from 99.99% to 99.9999% for 0.3 μm particles; whereas, ULPA filter efficiencies are around 99.9999%. An efficiency of 99.9999% means that one 0.3 μm particle in one million will penetrate the filter. Kozicki et al. (1991) discussed the use of such filters in cleanroom applications.

Several standard tests are suggested by ASHRAE for evaluating the performance of filters, including the weight arrestance (based on the weight of material it removes), DOP penetration (ability to remove specific particle sizes), dust spot efficiency (ability to reduce staining dust spots around diffusers), and the dust holding capacity test (ASHRAE, 1988). The efficiencies of various filters evaluated by these tests are listed in Table 6.6. It is essential to recognize that manufacturers ratings can be misleading. For example, a 91% arrestance efficiency filter may have a dust spot efficiency below 20% and a DOP particle size efficiency near zero. The efficiencies of various filters have been tested extensively in laboratory studies and by manufacturers, but there is minimal information on their efficiencies in occupied homes and buildings. In addition, most of the data refer to the initial efficiencies of fresh filters.

The removal and retention of particulates from air streams become increasingly difficult as their sizes approach the submicron range. Furthermore, methods are not readily available to assess the efficiency of such filters. Henkelmann et al. (1986) determined the retention efficiency of nine different types of industrial air filters that were specified as highly efficient for 0.1 μm particles. They used a radioactive tracer aerosol that had mean diameters ranging from 0.02 to 0.25 μm and determined the permeability of each type of filter. The duration of the aerosol exposure and the permeabilities measured are listed in Table 6.7. Independent measurements of the data reproducibility were made on four of the filters and showed that the permeability of an individual filter may vary by as much as a factor of two. This was attributed to differences within the filters, rather than to measurement errors. VanOsdell et al. (1990) investigated the performance of eight types of high-efficiency filters, including glass fiber filters, a composite fiber filter, and two membrane filters, using challenge aerosols with sizes in the range from 0.004 to 0.42 μm and filter face velocities ranging from 0.5 to 20 cm/s. The glass fiber filters were rated from 93% to 99.999% efficient using the standard 0.3 μm DOP test (see Section 6.5). The data were found to be in general agreement with conventional filter theory in that the filtration efficiency continues to increase as particle size is reduced. Based on a filter figure of merit (defined as $FM = -\log P/\Delta P$), the performances of the tested filters were remarkably sim-

TABLE 6.6 Filter Media and Their Effectiveness

Filter Media Type	*ASHRAE Weight Arrestance (%)*	*ASHRAE Atmospheric Dust Spot Efficiency (%)*	*MIL-STD 282 DOP Efficiency (%)*	*ASHRAE Dust-holding Capacity, grams per 1700 m^3/h cell*
Finer open cell foams and textile denier nonwovens	70–80	15–30	0	180–425
Thin, paperlike mats of glass fibers, cellulose	80–90	20–35	0	90–180
Mats of glass fiber multi-ply cellulose, wool felt	85–90	25–40	5–10	90–180
Mats of 5–10 μm fibers, 6–12 mm thickness	90–95	40–60	15–25	270–540
Mats of 3–5 μm fibers, 6–20 mm thickness	>95	60–80	35–40	180–450
Mats of 1–4 μm fibers, mixture of various fibers and asbestos	>95	80–90	50–55	180–360
Mats of 0.5–2 μm fibers (usually glass fibers)	NA	90–98	75–90	90–270
Wet laid papers of mostly glass and asbestos fibers, <1-μm diameter (HEPA filters)	NA	NA	95 95–99.999 99.97	500–1000
Membrane filters (membranes of cellulose acetate, nylon, etc., having holes 1-μm diameter or less)	NA	NA	~100	NA

NA, indicates that test method cannot be applied to this type of filter.
Source: ASHRAE Equipment Handbook, 1979.

ilar. Other than pressure drop and penetration, performance characteristics, such as particle loading, cost, and ease of cleaning, must be considered in selecting a filter for a specific application.

The performance of the above filters is not as encouraging for tobacco smoke. Offermann et al. (1985, 1992) examined the performance of various air cleaners for removing tobacco smoke and reported the results as the effective cleaning rate and system efficiency. The effective cleaning rate (ECR) is the difference in the particle decay rates observed with and without the air cleaner, multiplied by the indoor net air volume. The system efficiency is the effective cleaning rate divided by the air flow rate through the forced air system. The effective cleaning rate provides information on how fast the device cleans the air, whereas the system efficiency provides the overall removal efficiency for particles of a certain size. Their earlier study included a foam, two electrets, and an electret with corona charging. These four panel filters exhibited a very low cleaning rate for tobacco smoke, giving system efficiencies that ranged from 0% to about 40%. The largest efficiency was

TABLE 6.7 Permeabilities of Nine Particulate Filters

Filter No.	*Exposure Time (h)*	*Permeability ($\times 10^{-6}$)*	
1	1.6	0.088	(0.055), (0.042)[a]
2	1.6	0.191	(0.251), (0.235)[a]
3	1.0	0.984	(0.786), (0.863)[a]
4	1.4	1.06	(1.72), (1.56)[a]
5	1.5	1.29	
6	1.6	1.73	
7	1.5	7.89	
8	1.5	16.6	
9	1.5	72.9	

[a]Reproducibility checks.

Source: Henkelmann et al., 1986.

for a filter enhanced by corona charging. The term *electret* refers to a class of material that has been processed so as to impose a permanent, strong electrostatic charge. The two extended surface filters that Offermann et al. tested (a HEPA filter and an electret filter with a negative ion generator) had removal efficiencies of about 100% and 86% at cleaning rates of 180 and 57 ft^3/min, respectively. More recently they tested two panel filters and two extended surface filters in a three-room test space. Cigarette smoke was used as the test aerosol, and radon was injected into all three rooms. The panel filters had efficiencies of less than 5%, whereas the extended surface filters (a bag filter and a HEPA filter) had efficiencies of about 70%, with cleaning rates of about 800 and 570 ft^3/min, respectively. Based on both installation and operation cost estimates, Offermann et al. projected that properly selected air cleaners can reduce indoor concentrations of particles by 75% to 80% and cost between \$0.28 and \$0.56 per year per ft^3/min of cleaned air when operating 24 hours per day for 365 days each year. This should be contrasted with the costs associated with poorly performing air cleaners, which may reach \$4 to \$6 per ft^3/min of cleaned air.

Design procedures can be found in any of the several standard sources on aerosol or particulate control technology (see Hinds, 1982; ASHRAE, 1988; Flagan and Seinfeld, 1988; Licht, 1980).

Electronic Air Cleaning

The cited results suggest that filter effectiveness can be increased by using electrostatic technologies. Among the electronic air cleaners, the electrostatic precipitator (EP) is the most efficient and is widely used for removing fine particles from air. An efficiency greater than 99% can usually be obtained for particles less than 1 μm in size, but the removal efficiency is considerably lower for tobacco smoke. Kimmel (1987) evaluated the performance of electrostatic precipitators in seven restaurants and pubs in Finland and found that the particulate mass concentration was reduced by a factor of 20% to 50% in most of the restaurants for particles with aerodynamic diameters below 2 μm. The two EPs tested by Offermann et al. (1985) had an efficiency of about 58% with ECRs of about 120 ft^3/min. In their 1992 study, a flat plate electrostatic percipitator was found to have a system efficiency of about 70% at a cleaning rate of over 500 ft^3/min.

Offermann et al. also tested a negative ion generator for removing tobacco smoke. The ion generator failed to remove tobacco smoke (the effective cleaning rate was only 10 ft^3/min) from a 36-m^3 chamber unless circulating fans were turned on. While the cleaning rate increased to 30 ft^3/min when a higher voltage unit was used, the actual voltage was not specified. Olander et al. (1987) examined the effectiveness of 34 different air cleaners for removing tobacco smoke. These cleaners were of the following types: 12 electrostatic precipitators, 6 electret filters, 6 fiber filters, 7 ionizers, and 3 of undefined filtration materials. Some of these were also equipped with a second cleaning unit: nine with activated carbon beds, three with impregnated alumina beds, two with an ionizing lamp, and one with an electron generator. They used various combinations of removal systems to remove particulates together with gaseous pollutants, including an electret fiber filter along with an ionizer, an electrostatic precipitator in conjunction with an activated carbon bed, and a fibrous filter with an impregnated alumina bed. In addition to monitoring particulates in the size range of 0.01 to 7.5 μm, the cleaning rates of carbon monoxide, ammonia, formaldehyde, nitric oxide, nitrogen dioxide, hydrocarbons, hydrogen cyanide, and ozone were measured. The removal efficiencies for particulates ranged from 90% to 95%, but for gaseous constituents they were below 50%.

Sandberg and Mellin (1987) used an electrostatic field and an ion generator to remove tobacco smoke and mineral dust from a 3.8-m^3 test chamber. The reduction in mass concentration was about 60% when both the electric field and the ion generator were in operation. An increased ventilation rate had little effect on the removal efficiency. Similar results were obtained when only the electric field was in operation. Other equipment, such as gravitational settlers, cyclones, and scrubbers, have been used for removing particulates, but none of these systems have been employed to remove tobacco smoke, particularly from an indoor environment. Detailed descriptions of these devices and their advantages and disadvantages are available in the literature (see Licht, 1980; Theodore and Buonicore 1982; Calvert, 1984; Flagan and Seinfeld, 1988; ASHRAE, 1988).

Absorption

A number of wet collectors, including spray towers, cyclone scrubbers, impingement scrubbers, venturi scrubbers, and sieve-plate towers, have been used successfully in industry to remove particles from effluent gases (Calvert, 1984). Water is used primarily in the collectors and, accordingly, most of the experimental and design data are available for water only. Clausen et al. (1987) studied the washing of air to determine its impact on environmental tobacco smoke odor, but they did not examine the removal efficiency for particulates. Although washing did not change the intensity of the odor, it did alter the characteristics of the smoke such that it was considered to be more acceptable. Liquid desiccants, such as lithium chloride and triethylene glycol, which are commonly used for air dehumidification, can be also used to scrub tobacco smoke from air.

Adsorption

Although granular bed filters are not currently used for removing tobacco smoke, work related to the design, prediction of performance, pressure drop in the bed, and collection efficiency using standard test aerosols, such as DOP, oil mist, and mineral dust, is available in the literature. Olander et al. (1987) used an activated carbon and an impregnated alumina bed along with electronic air cleaners to remove gaseous constituents of tobacco smoke, but they did not provide a detailed description of the bed. Mann and Airah (1987) investigated the applicability of activated carbon adsorption to control gaseous contaminants from tobacco smoke and concluded that the addition of a carbon adsorption system to an HVAC unit would not introduce a cost disadvantage.

References

ACGIH (American Conference of Governmental and Industrial Hygienists) "Air Sampling Instruments for Evaluation of Atmospheric Contaminants", 7th ed., The Conference, Cincinnati, O., 1989.

AIHA (American Industrial Hygiene Association) "Cascade Impactor-Sampling and Data Analysis", The Association, Akron, O., 1986.

ASHRAE (American Society of Heating, Refrigerating and Air-Conditioning Engineers), *ASHRAE Handb. Prod. Dir.-Equipment*, The Society, Atlanta, Ga., 1979.

ASHRAE, *ASHRAE Handb.-Equipment*, The Society, Atlanta, Ga., 1988.

ASHRAE, "Ventilation for Acceptable Indoor Air Quality", ASHRAE Standard 62-1989, The Society, Atlanta, Ga., 1989.

BAILEY, M. R., KREYLING, W. G., ANDRE S., BATCHELOR, A., COLLIER, C. G., DROSSELMEYER, E., FERRON, G. A., FOSTER, P., HAIDER, B., HODGSON, A., MASSE, R., METVIER, H., MORGAN, A., MÜLLER, H-L., PATRICK, G., PEARMAN, J., PICKERING, S., RAMSDEN, D., STIRLING, C., and TALBOT, R. J., *J. Aerosol Sci.*, **20**(2), 169 (1989).

BRUNNEMANN, K. D., YU, L., and HOFFMANN, D., *Cancer Res.*, **37**, 3218 (1977).

BRUNNEMANN, K. D., FINK, W., and MOSER, F., *Oncology*, **37**, 217 (1980).

BURDETT, G. J., and JAFFREY, S. A., *Ann. Occup. Hyg.*, **30**, 185 (1986).

CALVERT, S., "Particle Control by Scrubbing", Chapter 10 of *Handbook of Air Pollution Technology* (eds. S. Calvert and H. M. England), Wiley, New York, 1984.

CHADWICK, D. A., BUCHAN R. M., and BEAULIEU, H. J., *Environ. Res.*, **36**, 1 (1985).

CHATFIELD, E. J., in 5th International Coll. Dust Measuring Technique and Strategy, Asbestos International Association, London, 1985, pp. 269-296; as cited in Guillemin et al. (1989).

CHATFIELD, E. J., in Asbestos Fibre Measurements in Building Atmospheres, Proceedings (ed. E. J. Chatfield) Ontario Research Foundation, Mississauga, Ontario, 1986, pp. 87-110; as cited in Guillemin et al. 1989.

CHEN, B. T., NANENYI, J., YEH, H. C., MAUDERLY, J. L., and CUDDIHY, R. G., *Aerosol Sci. Technol.*, **12**, 364 (1990).

CLARK, L. A., HASTIE, T., PSOTA-KELTY, L. A., SINCLAIR, J. D., and RAUCHUT, J., *Aerosol Sci. Technol.*, **16**, 43 (1992).

CLAUSEN, G. H., MOLLER, S. B., and FANGER, P. O., Proceedings 4th International Conference on Indoor Air Quality and Climate, West Berlin, West Germany, **1**, 47, 1987.

COHEN, B. S., "Introduction: The First 40 Years", in *Cascade Impactor-Sampling and Data Analysis*, American Industrial Hygiene Association, Akron, O., 1986.

CORN, M., CRUMP, K., FARRAR, D. B., LEE, R. J., and MCFEE, D. R., *Regul. Toxicol. Pharmacol.*, **13**, 99 (1991).

CROUSE, W. E., and OLDAKER, G. B. III, in *Measurement of Toxic and Related Air Pollutants*, Proceedings EPA/Air and Waste Management Association International Symposium, VIP-17, 562, 1990.

EATOUGH, D. J., HANSEN, L. D., and LEWIS, E. A., in *Environmental Tobacco Smoke*, Proceedings International Symposium at McGill University, Lexington Books, Lexington Mass., 1989.

EATOUGH, D. J., BENNER, C. L, CAKA, F. M., CRAWFORD, J., BRAITHWAITE, S., HANSEN, L. D., and LEWIS, E. A., in *Measurement of Toxic and Related Air Pollutants*, Proceedings EPA/Air and Waste Management Association International Symposium, VIP-17, 542, 1990.

ECOBICHON, D. J., and WU, J. M. (eds.), *Environmental Tobacco Smoke*, Proceedings International Symposium at McGill University, Lexington Books, Lexington Mass., 1989.

FLAGAN, R. C., and SEINFELD, J. H., *Fundamentals of Air Pollution Engineering*, Prentice-Hall, Englewood Cliffs, N. J., 1988.

GENTRY, J. W., SPURNY, K. R., and SCHÖRMANN, J., *Aerosol Sci. Technol.*, **11**, 184 (1989).

GROT, R. A., HODGSON, A. T., DAISEY, J. M., and PERSILY, A., *ASHRAE J.*, 16 (September 1991).

GUILLEMIN, M. P., MADELAIN, P., LITZISTORF, G., BUFFAT, P., and ISELIN, F., *Aerosol Sci. Technol.*, **11**, 231 (1989).

HEI-AR (Health Effects Institute-Asbestos Research), "Asbestos in Public and Commercial Buildings: A Literature Review and Synthesis of Current Knowledge", Cambridge, Mass., 1991.

HENKELMANN, R., BAUMGARTNER, F., KREBS, K, and PETZOLDT, O., *J. Aerosol Sci.*, **17**(6), 931 (1986).

HEW (Department of Health, Education, and Welfare) "Smoking and Health - A Report of the Surgeon General", DHEW Publication No. (PHS) 79-50066, Washington, D. C., 1979.

HHS (Department of Health and Human Services), "The Health Consequences of Involuntary Smoking. A Report of the Surgeon General", U. S. Government Printing Office, Washington, D. C., 1986.

HINDS, W. C., *Aerosol Technology*, Wiley, New York, 1982.

IAR, (Indoor Air Review), *Indoor Air Review*, 2 (June 1991a).

IAR, *Indoor Air Review*, 11 (July 1991b).

KENNY, L. C., and LIDEN, G., *J. Aerosol Sci.*, **22**(1), 91 (1991).

KIMMEL, J., Proceedings 4th International Conference on Indoor Air Quality and Climate, West Berlin, West Germany, **1**, 226, 1987.

KINSARA, A. A., "Deposition of the Radon Daughter Po-218 in a Lung Bifurcation" Ph. D. Dissertation, Nuclear Engineering, University of Missouri-Columbia (1991).

KOZICKI, M., HOENIG, S., and ROBINSON, P., *Cleanrooms: Facilities and Practices*, Van Nostrand Reinhold, New York, 1991.

LEADERER, B. P., KOUTRAKIS, P., BRIGGS, S. L. K., and RIZZUTO, J., in *Measurement of Toxic and Related Air Pollutants*, Proceedings EPA/Air and Waste Management Association International Symposium, VIP-17, 567, 1990.

LICHT, W., *Air Pollution Control Engineering: Basic Calculations for Particulate Collection*, Marcel Dekker, New York, 1980.

MANN, J. L., and AIRAH, J., *Aust. Ref. Air Cond. Heat.*, **41**(8), 21 (1987).

MOSCHANDREAS, J. D., ZABRANSKY, J., and PELTAS, D. J., "A Comparison of Indoor and Outdoor Air Quality", Electric Power Research Institute Report, EA-1733, Palo Alto, Calif. 1981.

NAS (National Academy of Science), *Indoor Pollutants*, National Research Council, National Academy Press, Washington, D. C., 1981.

NAS, *Environmental Tobacco Smoke: Measuring Exposures and Assessing Health Effects*, National Research Council, National Academy Press, Washington, D. C., 1986.

NELSON, P. R., HEAVNER, D. L., and OLDAKER III, G. B., in *Measurement of Toxic and Related Air Pollutants*, Proceedings EPA/Air and Waste Management Association International Symposium, VIP-17, 550, 1990.

OFFERMANN, F. J., SEXTRO, R. G., FISK, W. J., GRIMSRUD, D. T., NAZAROFF, W. W., NERO, A. V., REVZAN, K. L., and YATER, J., *Atmos. Environ.*, **19**(11), 1761 (1985).

OFFERMANN, F. J., LOISELLE, S. A., and SEXTRO, R. G., *ASHRAE J.*, 51 (July 1992).

OLANDER, L., JOHANSSON, J., and JOHANSSON, R., Proceedings 4th International Conference on Indoor Air Quality and Climate, West Berlin, West Germany, **2**, 39, 1987.

OWEN, M. K., ENSOR, D. S., HOVIS, L. S., TUCKER, W. G., and SPARKS, L. E., *Aerosol Sci. Technol.*, **13**, 486 (1990).

PINCHIN, D. J., "Asbestos in Buildings", in the report of the Royal Commission on Matters of Health and Safety Arising from the Use of Asbestos in Ontario, Queens' Printer for Ontario, Toronto, Ontario, Canada, 1983, p. 574; as cited in Guillemin et al., 1989.

PRITCHARD, J. N., JEFFERIES, S. J., and A. BLACK, *J. Aerosol Sci.*, **17**(3), 385 (1986).

RAJHANS, G. S., and BRAGG, G. M., *Engineering Aspects of Asbestos Dust Control*, Ann Arbor Science Publishers, Ann Arbor, Mich., 1978.

REPACE, J. L., and LOWREY, A. H., *Science*, **208**, 464 (1980).

REPACE, J. L., and LOWREY, A. H., *ASHRAE Trans.* **88**, Pt. 1, 895 (1982).

SANDBERG, M., and MELLIN, A., Proceedings 4th International Conference on Indoor Air Quality and Climate, West Berlin, West Germany, **1**, 231, 1987.

SPENGLER, J. D., DOCKERY, D. W., TURNER, W. A., WOLFSON, J. M., and FERRIS, B. G., Jr., *Atmos. Environ.*, **15**, 23 (1981).

SPENGLER, J. D., OZKAYNAK, H., MCCARTHY, J. F., and LEE, H., *Summary of Symposium on the Health Aspects of Exposure to Asbestos Buildings,* Harvard University Energy and Environmental Policy Center, Kennedy School Government, Cambridge, Mass., 1 (1989).

STEHLIK, G., *Ecotoxicol. Environ. Saf.*, **6**(6), 495 (1982).

THEODORE, L., and BUONICORE, A. J., *Air Pollution Control Equipment: Selection, Design, Operation, and Maintenance*, Prentice Hall, Englewood Cliffs, N. J., 1982.

TU, K. W., and KNUTSON, E. O., *Aerosol Sci. Technol.*, **9**, 71 (1988).

TYNDALL, J., *Essays on the Floating Matter of the Air, in Relation to Putrefaction and Infection*, D. Appleton, New York, 1892.

VANOSDELL, D. W., LIU, B. Y. H. , RUBOW, K. L., and PUI, D. Y. H., *Aerosol Sci. Technol.*, **12**, 911 (1990).

WADDEN, R. A., and SCHEFF, P. A., *Indoor Air Pollution: Characterization, Prediction and Control,* Wiley, New York, 1983.

WEHNER, A. P., *J. Aerosol Sci.*, **17**(3), 305 (1986).

WEIBEL, E. R., *Morphometry of the Human Lung*, Academic Press, New York, 1963.

WILLIAMS, M. M. R., and LOYALKA, S. K., *Aerosol Science: Theory and Practice*, Pergamon Press, Elmsford, N. Y., 1991.

YOCOM, J. E., *J. Air Pollut. Control Assoc.*, **32**(5), 500 (1982).

7

Bioaerosols

7.1 INTRODUCTION

Viable airborne particulates are an integral part of our ecological system, and they play an important role in its balance. These particulates include viruses, bacteria, fungi, amoebae, algae, mites, protozoa, pollen, and arthropods and are collectively called airborne biological contaminants, bioaerosols, or microorganisms. Various types of bioaerosols are present indoors and are responsible for numerous human diseases. Although medical treatment for most of these illnesses is available, rising medical costs and the loss of productivity are of growing concern to the general public, government health agencies, and employers.

Bioaerosols are present both indoors and outdoors. It is not usually possible to control their growth or concentration outdoors, but control indoors is achievable. Bioaerosols require a reservoir (for storage), an amplifier (for reproduction), and a means for dispersal (dissemination). All viruses and a few bacteria and fungi need a living host for survival (that is, storage and reproduction); these are called obligated parasites. Two common hosts are the human body and moist places, which also provide necessary nutrients for their growth. Although humans are a major source of indoor viruses and bacteria, most fungi identified indoors originate from spores present in outdoor sources (for example, the disturbance of dead vegetative matter can lead to suspension of fungi in air). Once these bioaerosols are introduced indoors, they can grow in moist areas, such as on damp carpet and walls, and in kitchens and bathrooms. As a result, large differences have not been found in the types, sources, and concentrations of most of these microorganism in indoor environments.

Since this book has been written for use by both the practitioner and the general public alike, it may be advantageous to summarize some of the basic terminology. According to *Webster's New International Dictionary,*

> *Virus:* any of a large group of submicroscopic infective agents that are regarded either as extremely simple microorganisms or as extremely complex molecules. They typically contain a protein coat surrounding an RNA or DNA core of genetic material but no semipermeable membrane and are capable of growth and multiplication only in living cells. Viruses cause various diseases in humans, lower animals, or plants. All viruses need a living host for growth and multiplication.
>
> *Bacteria:* any of a class (Schizomycetes) of microscopic plants having round, rodlike, spiral, or filamentous single-celled or noncellular bodies. They are often aggregated into colonies or motile by means of flagella, living in soil, water, organic matter, or the bodies of plants and animals, and being autotrophic, saprophytic, or parasitic in nutrition and important to humans because of their chemical and pathogenic effects. Some bacteria need a living host for survival.
>
> *Fungus:* any of a major group (Fungi) of saprophytic and parasitic lower plants that lack chlorophyll. These include molds, rusts, mildews, smuts, mushrooms, and yeasts. Some fungi need a living host for survival.
>
> *Mold:* (1) a superficial, often woolly growth produced on damp or decaying organic matter or living organisms, (2) a fungus (as of the order Mucorales) that produces mold.
>
> *Mildew:* a superficial, usually whitish, growth produced on organic matter or living plants by fungi (as of the families Erysiphaceae and Peronosporaceae).
>
> *Antigen:* a protein or carbohydrate substance (as a toxin or enzyme) capable of stimulating an immune response. Since we are interested in responses induced by indoor air pollutants, for our purposes an antigen carries the notion of an *allergen,* which is a substance that induces allergy.
>
> *Spore:* a primitive unicellular resistant or reproductive body produced by plants and some invertebrates. A spore is capable of developing into a new body, in some cases unlike the parent, either directly or after fusion with another spore.
>
> *RNA (ribonucleic acid):* any of various nucleic acids that contain ribose and uracil as structural components and are associated with the control of cellular chemical activities.
>
> *DNA (deoxyribonucleic acid):* any of various nucleic acids that are localized, especially in cell nuclei, are the molecular basis of heredity in many organisms, and are constructed of a double helix held together by hydrogen bonds between purine and pyrimidine bases that project inward from two chains containing alternate links of deoxyribose and phosphate.
>
> *Autotrophic:* needing only carbon dioxide or carbonates as a source of carbon and a

simple inorganic nitrogen compound for metabolic synthesis, or not requiring a specified exogenous factor for normal metabolism.

Saprophytic: obtaining food by absorbing dissolved organic material; especially obtaining nourishment osmotically from the products of organic breakdown and decay.

Parasite: an organism living in or on another organism for existence or support without making a useful or adequate return.

Cleaner indoor air, no doubt, will reduce the probability of illness among the general population. The development of better control strategies for bioaerosols requires a knowledge of the risks associated with their presence, their sources and sizes, accurate sampling and identification techniques, and information about removal methods.

7.2 HEALTH EFFECTS AND STANDARDS

The pathogenic properties of microorganisms are well understood and consequently have been documented. The role of *airborne* microorganisms in spreading diseases, however, was recognized more slowly. Also, the assessment of risk remains somewhat unexplored, because of the lack of comprehensive data at the bioaerosol levels found indoors. The synergism between microorganism-induced diseases and other causes that lower life expectancy remains mostly unquantifiable.

Following the account of Anderson (1973), Louis Pasteur in the 1860s conducted some of the first systematic and scientifically sound experiments to confirm the universal presence of bacteria in air. This prompted Joseph Lister in 1867 to use carbolic acid (phenol) sprays in surgical rooms to disinfect the air. The belief at that time, that infections were caused by contact rather than by bioaerosols, led to the replacement of antiseptic surgery by aseptic surgery, and the interest in studies of bioaerosols subsided. This perception was exacerbated by the experiments of Flugge (1897-1898) who exposed Petri dishes to the mists produced from sneezing or coughing, only to discover that the dishes did not show microorganisms upon culturing. He concluded that the infectious microorganisms did not travel far from a source via air. They either settled out, died, or evaporated near a source. Thus, the role of airborne microorganisms in spreading diseases could be disregarded. This view prevailed until Stillmen's experiments during the World War I period showed the possible role of airborne bacteria in spreading *pneumococcus*, *Klebsiella*, and *Streptococcus pyogenes* pneumonia. The revival of interest in the field was, however, due to the work of Wells, who beginning in 1933, published a series of papers that described the development of an atomizer which allowed aerosolization of bacteria from liquid suspensions, followed by their collection from air by a centrifuge sampler of his own design. Wells concluded that, while large droplets did not remain airborne for long, the sampler did collect a significant number of viable bacteria, and bacteria could remain airborne in a viable state. The centrifugal sampler was effective in its ability to sample a large volume of air. Thus, bacteria can be detected even when present in small numbers. By contrast, in Flugge's experiments the deposition on the Petri dishes would have occurred principally from

Brownian diffusion, which is not an efficient method for sampling bacteria from small volumes of air.

The subject matured during World War II, both with respect to the need for prevention of infectious diseases in army barracks and for use in army and civilian hospitals. The desires of some individuals to have the option of spreading diseases through germ warfare also created interest. Although a significant amount of information on the subject is known, much is not in the public domain.

In the 1940s and the 1950s, the existence of infectious bioaerosols in hospitals was monitored through the use of beagles. Apparently, the beagles were kept in cages through which the ventilation air was circulated. The beagles were euthanized and their lungs were dissected for determination of the bioaerosols. The situation changed with the development of the Andersen sampler (originally developed by Andersen for germ warfare experiments) and impingers.

It is now recognized that bioaerosols are introduced into the human body either through inhalation or deposition in wounds and can localize or migrate to other portions of the body. Complaints by building occupants of headaches, nausea, and other illnesses of a not readily identifiable nature or cause are termed the sick building syndrome. This phenomenon is often associated with the presence of bioaerosols. Infections caused by bioaerosols include Legionnaire's disease, humidifier fever, colds, and influenza. A study by the Medical Research Council and the Department of Health and Social Security (England) of the operating rooms in 19 hospitals showed that the bacteria that caused 80% to 90% of wound infections came from ambient air, and that cleaner air resulted in fewer cases of sepsis (Howorth and Hugh, 1987).

Since an individual inhales 10 to 20 m^3 of air daily, the respiratory tract provides a major entry and deposition site for bioaerosols. Once deposited, the microorganisms may either multiply locally or enter the bloodstream through absorption. The body has many defense mechanisms, but microorganisms can and do cause a number of diseases. Some of the diseases that can be caused by the inhalation of microorganisms were summarized by Atlas (1984) and are given below.

Viruses: the common cold (*Rhinovirus* and others), influenza (*Myxovirus influenzae*), measles (measles virus), German measles (rubella virus), mumps (*Paramyxovirus*), chicken pox and shingles (*Varicella-Zoster*), infectious mononucleosis (*Epstein-Barr*).

Bacteria: pneumonia (*Streptococcus pneumoniae*, *Staphylococcus aureus*, *Haemophilius influenzae*, *Klebsiella pneumoniae*), atypical pneumonia such as Legionnaire's disease (*Mycoplasma pneumoniae*, *Legionnella pneumophilia*) that requires special treatment, bronchitis (*Streptococcus* species, *Staphylococcus* species, *Haemophilius influenza*, *Mycoplasma pneumoniae*), rheumatic fever (*Streptococcus* species such as *Streptococcus pyogens*), Otitis media (*Streptococcus pneumoniae*, *Streptococcus pyogens*, *Haemophilius influenzae*), carditis (*Streptococcus* species, for example, *Streptococcus virdans*), diphtheria (*Carneybacterium diphtheriae*), whooping cough (*Bordetell pertussis*), tuberculosis (*Mycobacterium tuberculosis*), meningitis (*Neisseria meningitidis*), and Q-fever (*Coxiella burnett*).

Fungi: histoplasmosis (*Histoplasma capsulatum*), cocciodomycosis (*Cocciodioides immitis*), and blastomycosis (*Blastomyces dermatitis*).

Antigens: usually large compounds (greater than 10,000 atoms), but even some metals can act as antigens through their attachment to aerosol particles. Inhaled antigens induce immune responses from the human system and can lead to allergic diseases of hypersensitivity pneumonitis (HP), allergic asthma, allergic rhinitis, and allergic aspergillosis.

7.3 SOURCES

Most bacteria and viruses in indoor air originate from humans and pets, particularly from sneezing, coughing, dander, and saliva. Crowded conditions, during and following epidemics of respiratory tract infections, are often correlated with a rise of the pathogenic bacteria in air.

Various microorganisms, including most fungi, many bacteria, protozoa, algae, and green plants (pollen), are present in outdoor air and are introduced indoors by either natural ventilation or through the air intake of the ventilation system. To a lesser extent, water supplies can also be a source of bioaerosols. Although the concentration of microorganisms outdoors is rarely high enough to cause illness or epidemics, once these organisms infiltrate an indoor environment, many places can provide amplification sites. Such a site would have moisture, the correct pH, and the appropriate temperature for the growth of the organism. Among these factors, moisture is critical for the growth of all microorganisms. Moist surfaces on materials such as leather, wood, carpets, soaps, cloth fabrics, and some pastes and adhesives are examples of amplification sites. Other growth sites for microorganisms include bathrooms, damp or periodically flooded basements, and areas with water leaks. Evaporation pans of refrigerators, shower heads, and hot tubs can also aerosolize bacteria.

Humidifiers, air-conditioning systems, cooling towers, mechanical ventilation systems, air-distribution ducts, and areas of water damage are found to provide suitable breeding sites for a number of fungi and bacteria. Building ventilation systems that employ either humidifiers or chillers have been associated with building-related illnesses, such as infections from legionellosis, hypersensitivity pneumonitis, humidifier fever, asthma, and allergic rhinitis. Elevated levels of fungi have been observed in the immediate vicinity of wet surfaces. When furnishings and heating, ventilating, and air-conditioning system components are disturbed, similar results are obtained. Fungi that have been identified indoors include *Aspergillus*, *Penicillium*, *Cladosporium*, and *Aureobasidium*, with *Cladosporium* being predominant. Outbreaks of *Aspergillus* infections have been attributed to contaminated ventilation systems and to fireproofing materials.

Cold water humidifiers that are used in hospitals for patients who suffer from asthma, bronchitis, or rhinitis are also breeding grounds for organisms, including gram-positive organisms and common air molds (*Staphylococcus aureus, Staphylococcus epidermidis, Bacillus sp.*, and *Penicillium*). Similarly, furnace and console humidifiers have been identified as sources for thermophilic bacteria and mesophilic fungi (*Pseudomonas aureginosa, Serratia marcescens,* and bacillus species).

Thermophilic actinomycetes and varieties of fungi, bacteria, amoebae, and nematodes are often responsible for hypersensitivity pneumonitis and humidifier fever in office buildings and have been found to originate from water-damaged furnishings, air-handling systems, central and room humidifiers, and cool mist vaporizers.

Cooling towers used in central air-conditioning systems are of special concern because of the repeated identification of bacterial and fungal contamination in water. Two particular diseases, Legionnaire's disease and Pontiac (Q) fever, are documented as having started with contaminated cooling towers. Bacteria that cause Pontiac fever have been found in contaminated air-conditioning systems, whirlpool spas, steam turbine condensers, and industrial coolants. The bacteria responsible for Legionnaire's disease, *Legionella pneumophila,* has been identified in evaporative condensers, cooling towers, aerosolized tap water from respiratory devices, water supplies of hospitals, shower baths, shower heads, and plumbing systems.

Airborne antigens found indoors arise from pets, humans, insects, organisms, spores, and effluents or are attached to particles. Humidifiers, water spray cleaners, flushing of toilet bowls, showers, mites, cockroaches, and cats can and do lead to the production of airborne antigens. For example, vacuuming or making a bed can stir up dust and resuspend the mite antigens, which are usually attached to particles greater than 10 μm in diameter. A summary of the various indoor bioaerosols, their sizes, diseases that they can cause, and their sources is provided in Table 7.1.

7.4 SAMPLING AND MEASUREMENT

Bioaerosols can be collected by using the same principles as used for other particulates, but the variety and viability of the organisms complicate the collection as well as the counting. The viability of the bioaerosols and contamination free samplers must be ensured during collection and subsequent transfer to the laboratory for identification and counting. The collection methodology is different from that for other particulates in the sense that the microorganisms must be deposited in a medium (saline water or gel) which ensures their viability. The collected sample should be placed in either ice or Dry Ice (solid carbon dioxide) for transfer to the laboratory. Culturing is specific to the organism and is generally carried out by exposing the collected sample in an appropriate culture (nutrient or growth) medium, which could be agar gel and/or broth. The intermediate steps might involve dilution followed by incubation for a certain period of time, which may be from one to seven days, depending on the microorganism to be cultured. The number of colonies is then counted or, alternatively, a suspension is prepared and its opacity is compared to that of a known standard that is prepared under controlled conditions. The kind of microorganisms either must be known or must be determined from culturing. On-line, noninvasive sampling processes for bioaerosols are presently not available.

The objective of air sampling is typically to identify the presence, source, and types of bioaerosols and not to demonstrate that they are a direct cause of disease. Positive results

TABLE 7.1 Types, Sizes, Pathogenic Properties, and Common Sources of Selected Airborne Microbes

Types	*Sizes(μm)*	*Pathogenic Properties*	*Sources*
Bacteria			
Escherichia coli	1.1–1.5 × 2–6	Urinary tract infections, diarrhea, electrolyte imbalances, dysentary	Human body
Pseudomonas aeruginosa	0.5–0.7 × 1.5–3.0	Respiratory, gastrointestinal, and skin infections	Soil, water
Staphylococcus aureus	0.5–1.0 (diameter)	Bacteremia, food poisoning, toxic shock, syndrome, furuncles, carbuncles, impetigo, pneumonia, meningitis, endocarditis	Human body
Bacillus subtilis	0.7–0.8 × 2–3	Opportunistic pathogen	Soil, air
Klebsiella	0.3–1.0 × 0.6–6.0	Bacteremia, pneumonia, urinary tract infections	Soil, plants, water, etc.
Legionella pneumonphilia	0.3–0.9 × 2–20	Legionnaire's disease, Pontiac fever	Cooling towers, air conditioners
Staphylococcal pneumonia	0.5–1.5 (diameter)	Pneumonia	Human body
Streptococcal pneumonia	0.5–1.25 (diameter)	Pneumonia, meningitis, Otitis media, abcesses	Human body
Micrococcus luteus	0.9–1.8 (diameter)	Opportunistic pathogen	Soil, dust, human body, animals
Staphylococcus epidermis	0.5–1.5 (diameter)	Cardiovascular infections, wound infections, Ottis media, urinary tract infections	Human body

TABLE 7.1 (continued)

Types	*Sizes(μm)*	*Pathogenic Properties*	*Sources*
Bacillus circulans	0.6–1.0 × 2–5	Opportunistic pathogen	Soil
Bacillus anthrasis	1.0–1.2 × 3.0–5.0	Pulmonary anthrax	Air, soil
Fungi			
Thermoactinomyces vulgaris, Acanthamoli spp., Naegleria spp.	1–2 (diameter)	Humidifier fever, allergic reactions	Humidifers
Pencillium	4–6 (diameter)	Irritation to respiratory tract	Air, soil
Aspergillosis fungus	4–6 (diameter)	Aspergillosis and aspergilloma	Cooling towers, humidifiers
Coccidiodomycosis immitis	60–200 (diameter) vegetative cell	Lung infections	Air, soil
Cladosporium	8–15 (diameter)	Allergies	Moist surfaces, plants
Viruses			
Rhinovirus	0.02–0.03 (diameter)	Common cold	Human body
Influenza virus	0.08–0.120 (diameter)	Influenza	Human body
Lymphocytic choriomeningitis	0.2–0.3 (diameter)	Congenital infections, neurological infections	—
Coxsackie virus	0.02–0.03 (diameter)	Gastrointestinal infections, upper respiratory infections	—

from sampling are informative, but negative results are often inconclusive. Therefore, several points must be taken into consideration when selecting a sampler:

1. No one sampler is effective for collecting all types of microorganisms.
2. No sampling device provides for 100% recovery of bioaerosols.
3. The viability of bioaerosol samples should be maintained in the sampler for subsequent growth and identification; that is, the sample must be kept within a well-defined temperature range (frequently in an ice bath) and should be transported to a laboratory for analysis as soon as possible.
4. The efficiency of the sampler depends on the size of the particular organism.
5. The growth and survival of individual organisms depend on the humidity, temperature, and pH of the culture medium.
6. Proper selection of sampling and identification media is necessary for accurate sampling analysis. If a liquid medium is used, dilution will be necessary. However, if a solid agar is used, the collection plate can be analyzed directly. It is also important to set criteria for interpretation of the data. Although guidelines are often available, in their absence experimental data should be compared with a control experiment. For example, identification of only viable organisms results in an underestimation of the allergen load in the air. This is important since dead organisms and parts of organisms are often the cause of allergic reactions.
7. Samplers must be calibrated and used according to the manufacturer's flow rate specifications.
8. Since no one method is available for collection and identification or counting of all bioaerosols of interest, the selection of a method is guided by what is suspected to be present.

Because of these reasons, the sampling for bioaerosols should be carried out carefully and, generally, be undertaken after a thorough examination of the premises. After steps have been taken to identify bioaerosol reservoirs, amplification sites, dispersal mechanisms, and removal methods, only then should microorganism specific sampling be undertaken.

A wide range of samplers has been used, including diffusive samplers, Moulton air samplers, settling plates, dipstick samplers, filters, all-glass impingers (AGI), and impactors. The samplers most commonly used for collecting bioaerosols indoors include high-volume filters, high-volume electrostatic precipitators, all-glass impingers, and impactors. Principles of operation, sampling rates, and recommended sampling times for these samplers are presented in Table 7.2.

The collection efficiency of a sampler depends on the sizes and types of organisms to be collected. When the size of an organism is below 2 μm, the efficiency of most samplers drops dramatically. Because most bioaerosols of interest are in the 1 to 10 μm range, a filter is often used prior to the primary sampler to remove particles larger than 10 μm preferentially. The size distribution of microorganisms often depends on the size of the particles that carry it, such as dust, body fluids, dead cells, and other particulates. Organisms associated

TABLE 7.2 Recommended Samplers

Sampler Type	*Principle of Operation*	*Sampling Rate (L/min)*	*Recommended Sampling Time*
Slit impactor	Impaction on rotating or stationary plate	30–700, continuous	1–60 min or 7 days
Sieve impactor			
a. Single–stage impactor	Impaction on agar "rodac" plate	90 or 185	0.5 or 0.3 min
b. Single–stage impactor	Impaction on agar 100–mm plates	28	1 min
c. Two–stage impactor	Impaction on agar	28	1–5 min
Cassette filter	Filtration	1–2	15–60 min or 8 h
High volume filtration	Filtration	140–1400	5 min–24 h
High volume electrostatic	Electrostatic collection	up to 1000	Variable
All–glass impinger	Impingement into liquid	12.5	30 min
Centrifugal impactor	Impaction on agar	4077	0.5 min

Source: ACGIH, 1989.

with many human diseases are generally carried by particles in the size range from 4 to 20 μm. Absolute collection efficiencies for samplers are impractical to obtain and are not readily available in the open literature. The inefficiency of a particular sampler may be due to losses on the sampler wall, death of the organism on impact, escape from the collecting media, and mechanical limitations of the sampler. Comparisons between different samplers (relative efficiencies) are available in the literature and are listed in Table 7.3. The salient aspects of the various samplers are summarized next.

A sketch of an all-glass impinger is given in Figure 7.1. The air sample is drawn through a critical nozzle into a liquid and then allowed to bubble through. Sonic speed can

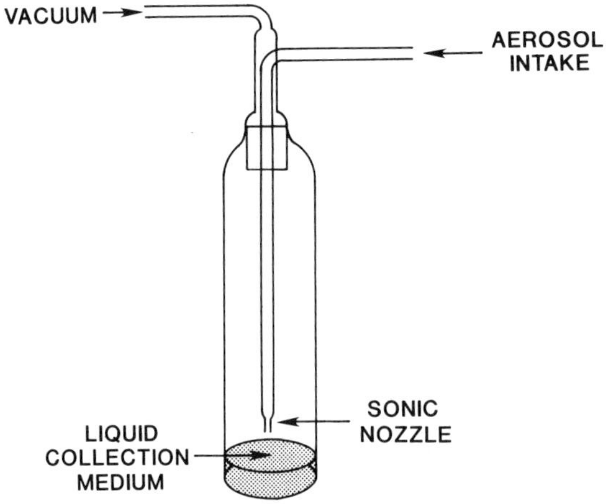

Figure 7.1 *All glass impinger.*

TABLE 7.3 Comparisons of Samplers

Test Sampler	*Reference Sampler*	*Test Organisms %*	*Relative Efficiency*[a] *(%)*	*Comments*
Glass cyclone sampler	AGI-30	*B. subtilis* spores	52–68	Dependent on downwind flow; non discriminant in size without modification
	AGI-30	*S. marcescens*	46	
	AGI-30	f2 Coliphage	76	
	AGI-30	Poliovirus type 1	92	
Porton impinger	h = 30 mm	Bacteria (overall)	99.9[b]	
	Impinger	*S. marcescans*	80.0	Loss of vegetative cells
Midget impinger	AGI	1.3 μm	46 ± 6.2	
	AGI	2.0 μm	37 ± 4.7	
	AGI	Larger aerosols	Same	
Glass midget impinger	AGI	2.0 μm	83 ± 11	—
Spill-proof impinger	AGI	1.3 μm	52 ± 8.1	—
	AGI	2.0 μm	86 ± 12	
Modified personal impinger	AGI	0.8 mm	80 ± 8.2	—
	AGI	1.1 mm	82 ± 9.1	
	AGI	2.0 mm	89 ± 11	
	AI	*B. subtilis*	78	
Multislit impinger	AGI	*S. marcescans*	82	Sterilization is difficult
	AGI	*B. subtilus*	78	
	AGI	Particles 1 μm	50	
Continuous impinger	AGI	*E. coli*	47	Sampling of high volumes for long periods of time
	AGI	*B. subtilis* spores	61	
Shipe impinger	—	Overall (bacteria)	98.2[b]	—
		Coxsackievirus	50[b]	
AGI	—	*B. subtilis* spores (3 μm)	99	Efficiency based on cells collected only and does not include vegetative cell death; good for high-volume collection only
AGI	AI	*B. subtilis*	55	Corrections were made to compensate for difference in sampling methods
		S. marcescens	42	
Personal cascade impactor	AI	*B. subtilis* spores	37	Filters commonly used in collection with the organisms being transferred to media later
	AGI		40	

(*continued*)

TABLE 7.3 (continued)

Test Sampler	*Reference Sampler*	*Test Organisms %*	*Relative Efficiency[a] (%)*	*Comments*
Gelatin Filters	Membrane Filters	Particles 0.5 to 3.0 mm	125	Limited lifetime; difficult to sterilize without damage
PEEP	NA	*B. subtilis* (0.9 μm)	20	Electrostatic precipitators tend to kill vegetative cells; they are built for high volume filtration
REEP	NA	*B. subtilis* (0.8 to 1.0 μm)	50	Evaporation of liquids can cause poor flow of current
SCAP	NA	*B. subtilis* (0.9 μm)	35	—
LVAS	AGI-30	*S. marcescens*	95	Evaporation and problems with quantitation; good for collection with low concentrations of aerosols
	AGI-30	*B. subtilis*	74	

[a]Percent of bacteria collected assuming 100% collection efficiency by reference sampler.

[b]Filtration efficiency refers in this case to the ratio of the number of particles found in the sampling device to the total number of particles entering the sampler, the latter being estimated by the number on an efficient backing filter and the sampler recovery.

AGI, all glass impinger.

AI, Andersen impactor.

PEEP, porous electrode electrostatic precipitator.

REEP, rotating electrode electrostatic precipitator.

SCAP, space charge aerosol precipitator.

LVAS, large volume air sampler.

NA, not available.

Source: Robertson et al., 1990.

be achieved at the nozzle exit with good retention of the microorganism in the liquid. This is done to ensure the microorganism's survival. May and Harper (1957) used two commercially available liquid impingers (an all-glass proton impinger and a Shipe impinger) and four liquid impingers that they designed to collect airborne bacteria. A mixture of *Bacillus subtilis* and *Serratia marcescens* was used as the test aerosol to determine the relative efficiencies of these samplers. The *B. subtilis* was chosen because of its robustness and ability to survive under vigorous agitation, whereas *S. marcescens* is more delicate and sensitive to stress. The authors reported that an impinger operating at sonic speed can damage cells, but at an air velocity of 300 mph or less, no loss of organism viability was observed. Bacteria that were already weakened in their airborne state did not survive even under gentle impingement. The amount of liquid or the liquid depth in the impinger did not affect the collection efficiency as long as the liquid was in circulation. Although the collection efficiency of the proton impinger was nearly 100% for *B. subtilis*, it was only about 76% for *S. marcescens*. A modified impinger, called a subcritical impinger, had a better collection efficiency for both organisms than did the other samplers.

The AGI sampler is inexpensive, is easily sterilized, and with the use of a preimpinger can increase the collection efficiency. However, the collection efficiency of the AGI sampler depends on the size of the aerosol, and it appears to work best in a narrow size range. Generally, the efficiency is a maximum for intermediate sized aerosols, in accordance with the general principles of aerosol mechanics that relate to Brownian and inertial motions. This particular size depends on the impinger design, flow rate, and nozzle to surface separation. Impingers are generally very poor collectors of viruses.

The Andersen impactor is the most commonly used impactor for making bioaerosol studies (Andersen, 1958). It is a six-stage apparatus with approximately 400 holes provided in each stage. Air is drawn directly through the impactor at a flow rate of 1 ft^3/min into a Petri dish that contains an agar. When the velocity imparted to a particle is sufficiently large, its inertia will overcome the aerodynamic drag, and the particle will leave the air stream and deposit on the agar. Otherwise, the particle will continue to the next stage. The quantitative limit on a stage is approximately 2500 particles. The apparatus is designed to minimize the overlapping of colonies. The Andersen impactor provides a differential bacteria count and is competitive in efficiency with the AGI sampler, which gives the total count. The impactor is highly efficient for collecting airborne particulates larger than 0.3 μm in diameter. Collected materials, such as viruses or toxins, can be washed off the agar and used for injection into animals or for chemical analysis. Experimental studies indicate that wall losses are negligible, there is no slippage of bacterial particles, the instrument is extremely sensitive, and the particle size discrimination makes it possible to calculate the particle size distribution of bioaerosols.

Another sampler frequently used for collecting organisms by impaction is the slit impactor. Slit impaction samplers are efficient particle collectors and can provide time-discriminated collections. Rotating-plate slit samplers are especially useful for determining the contribution of specific sources of viable organisms to the total aerosol content. Sieve impactors provide accurate estimates of culturable airborne organisms. A correction for multiple impactions must be considered when estimating the collected number, because multiple impactions affect the collection of fungi more so than bacteria. Although portable-type sieve samplers are useful in inaccessible areas, they are considered to be less efficient than the Andersen impactor.

Conventional filters, including cotton, membrane, and gelatin filters, have been used to collect bioaerosols, but with limited success. Filters are generally very efficient for collecting particles down to their rated pore size. However, the transfer of bioaerosols from the filter to a culture media can be a problem. The microscopic evaluation of the particles is tedious, often resulting in an underestimation of the number of organisms collected. Cell damage and drying of the filter further increase this problem. Washing and culturing of the cells obtained from a filter are possible as long as a proper solution is used. Most vegetative cells are too fragile for this type of collection.

Electrostatic precipitators have also been used to collect bioaerosols. Although these can be used to sample large volumes of air, the dehydration of bacteria is a significant problem.

Harsted (1965) used liquid impingers, filter paper, and fritted bubblers as partial collectors for radioactive submicron T1 bacteriophage. Differences in viable collections between these samplers were attributed to incomplete collection (slippage) and to killing of the phage. High-velocity, all-glass, and capillary impingers gave the highest viable recoveries, but allowed considerable slippage (30% to 48%). The impinger slippage varied inversely with relative humidity. Type 6 filter paper collected nearly 100% of the particles, but killed a significant number of bacteria. Fritted bubblers had a slippage of more than 80%. The collected particle size was slightly larger at 85% relative humidity than at 30% to 55% humidity.

Macher and First (1984) conducted a comparative study of gelatin filter media, impinger samplers, and spiral and cascade impactors with aerosols 2 μm in size. A personal spill-proof impinger was found to collect *E. coli* and the spores of *B. subtilis var. niger* with an efficiency equal to or higher than that of an all-glass impinger. Spiral and cascade impactors were found to be effective for size-discrimination sampling. They were, however, not sufficiently efficient with very small particles to warrant their use in quantitative studies.

Burge and Solomon (1987) reviewed both sampling and analysis techniques using gravity devices, impactors, and suction samplers. They concluded that no single sampler/analysis method is applicable to all microorganisms, but a combination of various sampling techniques can cover a variety of types. Gravity methods, which have very limited use, are neither qualitatively nor quantitatively accurate. Either a liquid impinger or a cascade impactor containing a growth media is suitable for bacteria. For larger fungus spores, a slit sampler and direct visual identification might be useful. Also, a slit sampler in combination with either a culture plate cascade sampler, a filter sampler, or a liquid impinger can cover a broad range of airborne organisms.

Viruses can be collected in the same manner as bacteria. A slit sampler used for a few minutes in combination with a culture plate cascade sampler usually provides a sufficient sample. Generally, the sample should be immediately placed on ice, kept free from contamination, heat, acids, and drying, and taken to the laboratory for identification and quantification. The analysis is carried out by introducing the virus into either the suspected host species (not feasible if it were human) or another susceptible host (for example, chicken embryo), or more generally by tissue cultures of various species. The number of viruses in the infected subjects can be determined using electron microscopy or the opacity of the viral suspension, which is related to the viral number density. Such assays, together with other techniques of modern microbiology, help determine both the type and quantity of the virus collected. The effects of viral infections in humans are often evident, and air sampling is not needed.

The sampling devices described above can also be used for collecting either airborne antigens or the source material (dust, mites, cockroaches) using a specially adapted vacuum cleaner. The samples should be analyzed immediately or freeze-dried for storage and later analysis. The analysis of antigens is generally difficult. The analytical techniques include direct skin tests, bronchial challenge (breathing of aerosolized antigens by the susceptible individual to obtain an immune response), and radioimmune assays. In the assays, radioisotope tracers (for example, Iodine-125) are attached to known purified antigens that must be selected on the basis of which antigen is suspected of being in the sample. These labeled antigens are then placed in wells of an agar gel, along with a known quantity of the sample (the unknown) material and a prescribed monoclonal antibody that is specific to the antigen. Both the labeled antigen and the antigen in the sample will diffuse to the antibody and attach to binding sites. If the labeled antigen, the sample, and the antibody are properly placed, the labeled antigen and the antigen in the sample diffuse toward the antibody at the same rate. The amount of antigen in the sample is related inversely to the radioactivity of the antibody. These tests also require *a priori* knowledge of the suspected antigen.

7.5 SPECIFIC CONTROL STRATEGIES

The reduction of bioaerosols is achieved either by cleaning the air or by disturbing either one or more of the three factors (reservoir, amplification, or dispersal) that lead to their introduction indoors. In army barracks, dispersal was often controlled by using oil sprays on surfaces and floors to reduce resuspension and using glycols in the air to act as a biocide. Neither method was totally effective. Current methods for controlling indoor bioaerosols include (1) source (reservoir) removal, (2) regular maintenance (reservoir, amplification, and dispersal control), (3) humidity control (amplification control), (4) increased ventilation with filtration, and (5) air cleaning. These are discussed next.

Source (Reservoir) Removal

If the source of the airborne microorganism can be identifed, it can be either eliminated or steps can be taken to reduce its strength. For example, if a cat is the source of allergens, the cat can be kept outdoors. Keeping the yard clean may be helpful.

Regular Maintenance

Preventive maintenance is probably the single most important strategy for controlling microorganisms in buildings and residences. Regular maintenance of air-handling systems and fan-coil units, cleaning of drain pans, and the periodic replacement of filters have been found to reduce the microbial contamination of indoor air. Other preventive measures include the cleaning of acoustical and thermal insulation, the use of smooth-surfaced induction units, and the replacement of microbially contaminated porous materials by equivalent nonporous materials.

Humidifiers that employ recirculated water should not be used, since they can rapidly become contaminated. In heating, ventilating, and air conditioning (HVAC) systems, steam should be used instead of a cold vapor for humidification of air. Cold water humidifiers should be inspected regularly. Residential humidifiers should be cleaned daily and, if possible, distilled water should be used along with a biocide.

The preventive measures that can be most effective depend largely on the contents of the building and the source of the microbial contamination. For example, Morey et al. (1986) surveyed five office buildings that were contaminated with microbes. For one building, they recommended thorough cleaning of all building contents, including books, carpets, and the HVAC system ductwork and discarding those items that could not be cleaned. For the second building, where the contamination was due to the flooding of the cafeteria, they suggested changing the plumbing system, discarding all wet carpet and ceiling tiles, and disinfecting the floor with chlorine bleach. Similar preventive measures were employed in the other three buildings.

Disinfectants or biocides are regularly used to control microbial contamination in humidifier water reservoirs, cooling tower water, and any place that contains stagnant water. A biocide can either kill microorganisms directly or it can reduce the microorganisms' ability to reproduce through chemical or physical interactions. An ideal biocide should be simple to use, and it should be highly toxic to organisms, but nontoxic to humans. Numerous biocides and antifoulants are available for the microbial treatment of water. Their types and feed systems must be chosen carefully to maintain efficiency and reliability of cooling systems for long-term operation. Since microorganisms can mutate and change their biochemical and resistance characteristics over time, the use of more than one biocide is recommended. Dosages and frequency of application of biocides are generally determined by the system requirements, but shock dosages provide the best results. The biocide should inhibit the growth of microorganisms and be compatible with the environment. A list of commonly used biocides is provided in Table 7.4, along with their effects on human health that may result from their use.

Cockcroft et al. (1981) evaluated three biocides: a bismethylene chlorophenol com-

TABLE 7.4 Biocides Commonly Used in Cooling Towers

Biocides	*Sporicidal Activity*	*Mechanism*	*Human Health Effects (inhalation or contact)*
Hypochlorites	Yes	Enzyme inactivation	Irritant, corrosive
Hydrogen peroxide solution	?	Hydroxyl-free radicals	None for 3%
Quaternary ammonium compounds	?	Increase cell membrane permeability	Toxic irritants
Alcohols: ethanol, propanol, *iso*-propanol	No	Denatures proteins	None reported
Phenolics	No	Denatures proteins	Odor, toxic irritants, corrosive
Glutaraldehyde	Yes	Protein cross linking	Toxic irritants
Iodine, idophors	Yes	Iodination and oxidation of proteins	Skin, mucous membrane irritant
Formaldehyde	?	Binds DNA cell proteins	Odor, toxic irritants may be carcinogenic

Source: ACGIH, 1989.

pound; a mixture of picloxydine, octyl phenoxy polyethanol, and benzalkonium chloride; and an antiamoebic drug to control humidifier fever antigens. All three biocides were ineffective in killing the bacteria. Ager and Tichner (1983) suggested periodic mechanical cleaning of the cooling tower, along with treatment of the water by chlorine or a hypochlorite solution. However, the authors warned that such a chemical treatment may cause other problems, such as foaming and vaporization of the chemicals into the indoor air.

Various types of biocides have been employed to control *Legionella pneumophilia* in humidifiers and cooling towers. Biocides commonly used to prevent the growth of *L. pneumophilia* include quaternary ammonia compounds, 1-bromo-3-chloro-5,5-dimethyhydantoin, bis(tri-*n*-butyltin) oxide, *n*-alkyl-1, 3-(propanediamine), methylene-bis(thiocyanate), dithiocarbamates, and chlorine (Godish, 1989).

Antifoulants in domestic humidifiers have shown little or no change in the bacterial count. Although cleaning the components of an HVAC system with ammonia and chlorine can reduce mold levels, these compounds do not kill *Legionella* species.

Using laboratory cultures, Pope (1981) tested a total of 11 biocides for their effectiveness in controlling nine strains of *L. pneumophilia* and eight strains of algae. Three concentrations of each biocide were tested for their ability to kill *L. pneumophilia*. Biocides containing bis(tributyltin)oxide (TBTO) were found to be the most effective for controlling all algae and *L. pneumophilia*; the best results were obtained with a mixture of TBTO and quaternary amines. Braun (1982) tested several commercially available biocides (dithiocarbamates, TBTO, *n*-alkyl dimethyl benzol ammonium chlorides) in cooling towers and evaporative condensers, but all were found to be ineffective with *L. pneumophilia*. Pan et al. (1984) evaluated a specific surfactant, cetydimethylbenzyl ammonium chloride (CDBAC), as a disinfectant against a bacterial indicator, *E. coli*, in cooling tower water. Later they formulated several biocides from one or more of the following agents: sodium chromate, sodium dichromate, polyphosphate, sodium nitrate, antifoamants, and CDBAC. The use of CDBAC in the cooling tower water appeared to be the most effective method of controlling *L. pneumophilia*.

Chlorine and chlorine-based compounds have been used to control the growth of *L. pneumophilia* in cooling towers, but the effectiveness of these biocides under field conditions is less than satisfactory. Witherell (1986) sampled the water from 130 cooling towers, 52% of which employed some kind of biocide. He observed no significant difference in the control of *L. pneumophilia* between the treated and untreated cooling tower water.

Humidity Control

A significant reduction of bacteria and viruses can be obtained by controlling the relative humidity of the space. Dunklin and Puck (1948) investigated the effect of relative humidity on the survival of airborne bacteria. They observed that *S. marcescens*, *E. coli*, *S. albus*, *S. aureus*, and *S. salivarius* were more susceptible to death when the relative humidity was around 50%. Microorganisms are well protected by the moisture surrounding them if the humidity is above 50%. However, *in vitro* tests have demonstrated that completely dehydrated bacteria are more resistant to the glycols in the liquid state. A similar conclusion was also drawn by Lester (1948) from his experiments with influenza *A* viruses and pathogenic bacteria. A series of experiments was conducted in which mice were exposed to various concentrations of viruses and relative humidities. Fewer deaths and lesions occurred at a relative humidity of about 50% for all virus concentrations.

Sterling et al. (1985) conducted an extensive literature search to determine the humidity level at which risks to human health will be minimized. Bacterial populations were found to be at a minimum when the humidity was between 30% and 60%. Viruses are more susceptible to death in the humidity range of 50% to 70%, while the maximum growth of fungi occurred above 95% relative humidity. The growth of fungi was almost totally inhibited when the humidity was below 80%. The mite population indoors appeared to increase when the relative humidity was above 60%. Based on findings from the literature, Sterling et al. prepared a nomograph on the optimum humidity ranges for healthy environments. They concluded that a humidity range between 40% and 60% at normal room temperature will minimize the risks to human health. However, Berendt (1980) found that *L. pneumophilia* have a shorter half-life at 30% relative humidity than at 50% and 80%. Jantunen et al. (1987) concluded from their study on fungal growth in apartments that a 40% to 60% humidity range was indeed optimum. In cold climates, relative humidities in that range may be high enough for moisture to condense on surfaces and lead to fungal growth.

Increased Ventilation with Filtration

As mentioned earlier, a number of bacteria and the majority of fungi originate outdoors. Therefore, the proper location of the air intake of a building ventilation system can reduce microbial contamination significantly. Outdoor air intakes should be located upwind and away from cooling towers, evaporative condensers, and vegetation.

Mechanical filters have been used successfully to remove fungi and other microbial agents from outdoor air before it is introduced indoors. Since most microorganism are larger than 1 or 2 μm in diameter, they can be removed effectively by filters. Decker et al. (1951) reported that the combination of an electrostatic precipitator and a spun-glass fiber filter can be used to remove from 99% to 100% of bacteria and viruses from an air stream. Decker et al. (1963) evaluated various filters for the removal of organisms in the size range from 1 to 5 μm. The collection efficiencies of coarse-fiber panel filters, electrostatic precipitators, and HEPA filters were approximately 10%, 60% to 90%, and 90% to 100%, respectively. Silverman and Dennis (1959) tested several glass- and metal-fiber filters that are used in domestic air conditioners to determine their efficiency for removing pollen from air. Removal efficiencies of mineral filters for giant ragweed pollen were in the range of 50% to 92%, depending on the filter thickness. Glass-fiber filters performed as well as mineral fibers. Margard and Logsdon (1965) tested nine types of air filters and reported that electronic air cleaners and dry-type filters were the most efficient for removing airborne bacteria. Filters in which biocides were incorporated during the manufacturing process displayed little bacteriostatic action and did not exhibit an improvement in collection efficiency.

Harstad et al. (1967) reported that HEPA filters were nearly 100% effective at removing viruses and T1 bacteriophage which had mass median diameters of 0.17 μm and greater.

Morey and Jenkins (1987) found from their study of 25 office buildings that the concentration of fungi indoors averaged only 13% to 15% of outdoor levels. They concluded that the lower fungi level indoors was most likely due to effective filtration by the HVAC filtration system and the settling of the larger spores. As noted by Gritschke (1973), HEPA filters are frequently used in hospitals, especially in surgical suites, because of their ability to remove 99% of all particulate matter. Therefore, increased ventilation and new design concepts, along with various filters, can be used effectively to reduce microbial contamination.

Air Cleaning

Various chemical compounds, including sodium hypochlorite, hydroxy acids, and glycols in the vapor phase, have been used to disinfect air. Chemical disinfectants must satisfy a number of criteria, including being lethal in low concentrations to bacteria. These chemicals in the vapor phase should be nontoxic and nonirritating to humans, should not corrode metal or attack fabric, and should be easily vaporized. The mechanism by which aerial disinfectants kill an organism has been the subject of some debate. According to one theory, chemicals that are present in the particulate form in air collide with the bioaerosols and kill them. Experimental evidence confirms that a biocide must be present in the vapor phase for rapid air sterilization to occur (Puck et al., 1943, 1947).

Among the glycols, propylene and triethylene glycols in the vapor phase appear to be the most effective for sterilizing air. They were found to kill up to 75% of all bacteria and viruses in exposed areas. The killing efficiencies for triethylene glycol range from 50% for viruses to 90% for bacteria and fungi. Much of the literature on glycols dates to the middle 1940s, and it has been reviewed by Hines et al. (1991). The use of glycols in the vapor phase is no longer recommended, because of the undesirability of glycols in the indoor air.

Several acids have been studied for their aerial disinfectant properties. These include formic, propionic, lactic, glycolic, and levulinic acids. Ozone and iodine also have been frequently suggested as disinfectants, but they are not currently used because of the side effects.

Desiccant air-conditioning systems can improve indoor air quality by eliminating bioaerosols from an air stream in two ways: (1) by capturing the organism and/or (2) by killing the organism through severe desiccation. A bed of silica gel or molecular sieves can trap organisms and adsorb moisture and nutrients, causing the death of the organism. Although activated carbon is not a desiccant, it has been used effectively in beds to treat contaminated water (Lowry and Brandow, 1985). Unfortunately, bacterial growth occurred on the surface of the activated carbon. Because of moisture accumulation on the activated carbon when treating air, the possible growth of bacteria and fungi on the surface is of concern. Solid desiccants have the capacity to remove chemical pollutants, but their ability to reduce bioaerosols is not as well understood. Several questions remain, including whether the solid desiccant surface will act to amplify the growth of microorganisms once moisture is adsorbed.

Triethylene glycol-water solutions containing from 95% to 98% triethylene glycol and a 40% lithium chloride-water solution are frequently used as liquid desiccants to dehumidify air in liquid desiccant-based air-conditioning systems. These desiccants can be used to improve indoor air quality by reducing the levels of certain bioaerosols. In these systems, air is typically in contact with the liquid for a period of less than one second. Hines et al. (1992) observed that a 95% triethylene glycol solution can kill more than 92% of the airborne microorganisms under these conditions. A similar killing rate (90%) was observed when a 40% lithium chloride solution was employed. The development of an efficient and economically viable system requires further study of parameters such as pH, temperature, concentration of desiccant solutions, contact time, and regeneration of the desiccant.

REFERENCES

ACGIH (American Conference of Governmental and Industrial Hygienists) "Guidelines for the Assessment of Bioaerosols in the Indoor Environment", The Conference, Cincinnati, Ohio, 1989.

AGER, B. P., and TICHNER, J. A., *Ann. Occup. Hyg.*, **27**, 341 (1983).

ANDERSEN, A. A., *J. Bacteriol.*, **76**, 471 (1958).

ANDERSON, D. A., *Introduction to Microbiology*, The C. V. Mosby Company, St. Louis, Mo., 1973.

ATLAS, R. M., *Microbiology*, Macmillan Press, New York, 1984.

BERENDT, R. F., *J. Infect. Dis.* **141**, 689 (1980).

BRAUN, E. B., "Factors Affecting the Distribution, Seasonal Response and Control of Legionella Pneumophila in Aquatic Environments", Ph. D. Dissertation, Rensselaer Polytechnic Institute, Troy, N. Y., 1982.

BURGE, H. A., and SOLOMON, W. R., *Atmos. Environ.*, **21**, 451(1987).

COCKCROFT, A., EDWARDS, A., BEVAN, C., CAMPBELL, I., COLLINS, G., HOUSTON, K., JENKINS, D., LATHAM, S., SAUNDERS, M., and TROTMAN, D., *Br. J. Ind. Med.*, **38**, 144 (1981).

DECKER, H. M., GEILE, F. A., MOORMAN, H. E., GLICK, A., and FREDERICK, M., *Heat Piping Air Cond.*, **23**, 125 (1951).

DECKER, H. M., BUCHANAN, L. M., HALL, L. B., and GODDARD, K. R., *Am. J. Pub. Health,* **53**(12), 1982 (1963).

DUNKLIN, E. W., and PUCK, T. T., *J. of Exp. Med.*, **87**, 101 (1948).

GODISH, T., *Indoor Air Pollution Control*, Lewis Publishers, Chelsea, Mich., 1989.

GRITSCHKE, R. O., *Heat Piping Air Cond.*, **45**, 71 (1973).

HARSTAD, J. B., *Appl. Microbiol.*, **13**, 899 (1965).

HARSTAD, J. B., HERBERT, M. D., LEE, M. B., and FILLER, M. E., *Am. J. Pub. Health*, **57**(12), (1967)

HINES, A. L., GHOSH, T. K., LOYALKA, S. K., and WARDER, R. C., Jr., "Investigation of Co-Sorption of Gases and Vapors as a Means to Enhance Indoor Air Quality", Report No. GRI-90/0194, Gas Research Institute, Chicago, Ill., NTIS Document No. PB91-178806, 1991.

HINES, A. L., GHOSH, T. K., LOYALKA, S. K., and WARDER, R. C., Jr., "Investigation of Co-Sorption of Gases and Vapors as a Means to Enhance Indoor Air Quality-Phase II", Report No. GRI/92-0157, Gas Research Institute, Chicago, Ill., 1992.

HOWORTH, F. H., and HUGH, F., *J. Med. Eng. Technol.*, **11**, 263 (1987).

JANTUNEN, M. J., NEVALAINEN, A., RYTKÖNEN, A-L., PELLIKKA, M., and KALLIOKOSKI, P.,

Proceedings of the 4th International Conference on Indoor Air Quality and Climate, West Berlin, West Germany, **1**, 643, 1987.

LESTER, W., *J. Exp. Med.* **88**, 361 (1948).

LOWRY, J. D., and BRANDOW, J. E., *J. Environ. Eng.*, **111**, 511 (1985).

MACHER, J. M., and FIRST, M. W., *J. Am. Ind. Hyg. Assoc.*, **45**, 76 (1984).

MARGARD, W. L. and LONGSDON, R. F., *ASHRAE J.*, 49, (May 1965).

MAY, K. R. and HARPER, G. J., *Br. J. Ind. Med.* **14**, 287 (1957).

MOREY, P. R., HODGSON, M. J., SORENSON, W. G., KULLMAN, G. J., RHODES, W. W., and VISVESVARA, G. S., *Ann. Am. Conf. Gov. Ind. Hyg.*, **10**, 21 (1984).

MOREY, P. R. and JENKINS, B. A., Proceedings IAQ '87: Practical Control of Indoor Air Problems, American Society of Heating, Refrigeration, Air-Conditioning Engineers, Atlanta, Ga., 67, 1987.

MOREY, P. R., JONES, W. G., CLERE, J. L., and SORENSON, W. G., Proceedings IAQ '86: Managing Indoor Air for Health and Energy Conservation, American Society of Heating, Refrigeration, Air-Conditioning Engineers, Atlanta, Ga., 500, 1986.

MOREY, P. R., FEELY, J. C., and OTTEN, J. A., *Biological Contaminants in Indoor Environments*, STP 1071, American Society of Testing Materials, The Society, Philadelphia, Pa. (1990).

PAN, T. J, WANG, L. K., and WANG, M. H. S., "Prevention of Airborne Legionnaire's Disease by Formulation of a New Cooling Water for Use in Central Air Conditioning Systems', NTIS Document No. PB85-215317 (1984).

POPE, D. H., "Effects of Biocides on Algae and Legionnaire's Disease Bacteria", Ph. D. Dissertation, Rensselaer Polytechnic Institute, Troy, N. Y., (1981).

PUCK, T. T., ROBERTSON, O. H., and LEMON, H. M., *J. Exp. Med.*, **78**, 387 (1943).

PUCK, T. T., *J. Exp. Med.*, **85**, 729 (1947).

ROBERTSON, K. A., GHOSH, T. K., HINES, A. L., LOYALKA, S. K., WARDER, R. C., Jr., and NOVOSEL, D., Proceedings 5th International Conference on Indoor Air Quality and Climate, Toronto, **4**, 565 , 1990.

SILVERMAN, L. and DENNIS, L., *Air Cond. Heat. Vent.*, **56**, 61 (1959).

STERLING, E. M., ARUNDEL, A. and STERLING, T. D., *ASHRAE Trans.* **91**(Pt. 1B), 611 (1985).

WITHERELL, L. E., *J. Environ. Health*, **49**, 134 (1986).

8

Radon

8.1 INTRODUCTION

Radon is a noble gas produced by the alpha decay of radium-226. As radon undergoes further radioactive decay (the half-life of radon is 3.82 days), it produces a series of short-lived radioisotopes, known as radon daughters or progeny. The radon decay chain is shown in Figure 8.1. Radon itself is inert and causes little damage, since most of it is exhaled with the breath. However, the progeny, Po-218 and Po-214, are electrically charged species and can be inhaled either directly or through their attachment to airborne particles. Once inhaled, the radon progeny tend to be retained in the lungs, where they may ultimately cause cancer. The health risks from the exposure to radon and/or radon progeny have been investigated by a number of researchers, and the findings are summarized in a report entitled Biological Effects of Ionizing Radiation (BEIR, 1988).

In the past the contamination of air by radon and the subsequent exposure to radon daughters were believed to be a problem associated only with uranium miners. Workers involved in the mining and processing of other types of ore that have a high radium content, such as phosphate, might also be exposed to radon daughters. Recently it has been recognized that some homes and buildings far away from uranium or phosphate mines or their processing plants can have radon concentrations higher than those found in homes close to the mines or plants. As a consequence, radon and radon progeny have been recognized as indoor pollutants.

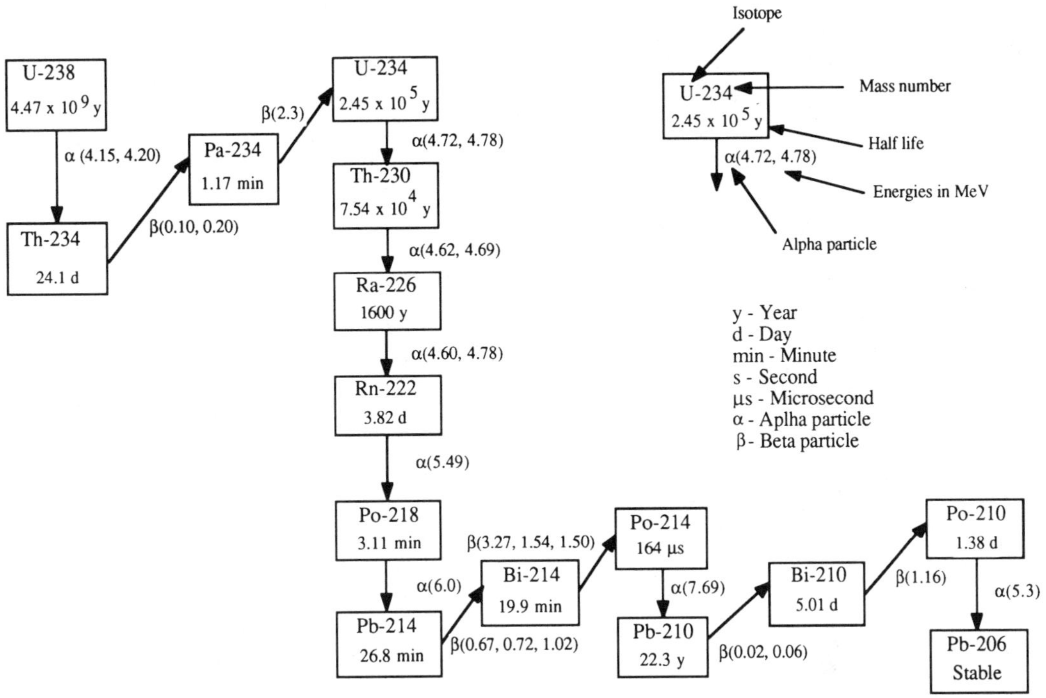

Figure 8.1 *Radon decay chain.*

The National Council on Radiation Protection and Measurement (NCRP) pointed out that indoor radon exposure constitutes the most important public health hazard attributable to radiation. A number of subsequent studies have documented the sources and concentration levels of radon in homes in various regions of the country. A distribution of the concentration of radon in the 50 states of the United States is provided in Table 8.1. A detailed distribution on a county basis is provided in the appendix. In every state, there are some homes with unacceptable levels of radon; and in some regions, a significant fraction of homes have radon concentrations higher than 4 pCi/L, the upper limit recommended by the U. S. Environmental Protection Agency (EPA). Therefore, a better understanding of radon health effects, its standards, sources, and measurement and removal techniques is necessary.

8.2 HEALTH EFFECTS AND STANDARDS

Radon is a colorless and odorless inert gas. Assuming spatial homogeneity, its concentration and those of its daughters in air are determined from the Bateman equation

$$\frac{dC_i(t)}{dt} = \lambda_{i-1}C_{i-1}(t) - \lambda_i C_i(t) + S_i(t) - R_i(t) \tag{8.1}$$

TABLE 8.1 Radon Concentrations in Living Areas and Basements

States	Living Areas (pCi/L)			Basements (pCi/L)		
	No. of Measurements	*Geometric Mean*	*Arithmetic Average*	*No. of Measurements*	*Geometric Mean*	*Arithmetic Average*
Alabama	334	1.23	2.33	181	2.09	3.55
Alaska	59	0.81	1.28	13	1.66	3.48
Arizona	192	1.48	2.17	14	2.54	4.99
Arkansas	168	1.18	2.26	62	2.34	5.41
California	940	0.82	1.19	118	1.29	2.82
Colorado	1657	3.30	6.45	2023	5.49	8.45
Connecticut	1887	1.08	1.75	4750	1.89	3.49
Delaware	75	0.74	0.96	237	1.37	2.20
District of Columbia	483	0.88	1.31	1181	1.48	2.32
Florida	1310	0.95	3.45	59	1.01	2.11
Georgia	623	1.07	1.51	380	2.26	3.26
Idaho	1222	2.48	4.29	249	4.02	7.40
Illinois	2458	1.43	2.19	3585	2.79	4.69
Indiana	636	1.77	2.99	751	3.19	5.71
Iowa	1921	3.25	5.00	7805	5.43	7.97
Kansas	138	2.29	3.44	249	3.64	6.63
Kentucky	355	2.31	6.42	630	4.29	9.31
Louisiana	240	0.62	0.80	7	1.66	3.17
Maine	544	1.28	2.37	752	2.79	5.64
Maryland	7945	1.59	3.16	16,693	2.79	5.66
Massachusetts	2822	1.11	1.97	8130	2.16	3.88
Michigan	1527	1.27	1.99	3308	1.98	3.69
Minnesota	771	1.81	2.61	3507	3.41	4.80
Mississippi	97	0.76	1.17			
Missouri	329	1.56	2.55	465	2.66	5.47
Montana	115	2.20	5.15	114	3.67	6.27
Nebraska	43	2.18	2.77	107	3.87	5.09

(continued)

TABLE 8.1 Continued

	Living Areas (pCi/L)			Basements (pCi/L)		
States	No. of Measurements	Geometric Mean	Arithmetic Average	No. of Measurements	Geometric Mean	Arithmetic Average
Nevada	69	1.15	2.81	27	1.89	4.00
New Hampshire	662	1.37	2.97	1439	2.81	5.35
New Jersey	10,706	1.31	2.57	10,976	2.56	5.77
New Mexico	432	2.64	3.15	88	3.18	5.26
New York	2865	1.09	1.92	4132	1.96	4.06
North Carolina	1221	1.30	2.22	120	3.06	6.42
North Dakota	193	2.30	3.62	456	4.69	7.15
Ohio	1395	2.12	3.68	2416	3.02	5.83
Oklahoma	186	1.00	1.49	54	1.72	3.56
Oregon	149	1.13	1.79	65	1.77	2.84
Pennsylvania	8264	2.30	5.14	16,519	4.17	9.53
Rhode Island	338	1.23	2.09	1304	2.21	5.12
South Carolina	257	0.93	1.48	77	2.43	4.10
South Dakota	75	2.84	4.58	122	5.32	9.00
Tennessee	1605	2.04	3.17	682	3.87	6.72
Texas	896	0.86	1.46	106	1.42	3.60
Utah	41	1.38	2.02	40	1.94	2.80
Vermont	264	0.95	1.95	538	1.80	3.00
Virginia	4810	1.36	2.13	8448	2.40	3.91
Washington	477	1.40	3.10	3.02	3.10	7.92
West Virginia	439	1.75	3.01	527	3.77	8.12
Wisconsin	569	1.65	2.79	3671	3.39	5.43
Wyoming	233	2.54	4.26	285	3.39	5.21

Source: Cohen, B.L., and Shah, R.S., Health Physics, 60(2), 243 (1991) (printed with permission).

where λ is known as the decay constant and has units of time^{-1}, C_i is the concentration per unit volume of the ith isotope at time t, and S_i and R_i are the source (for example, infiltration) and the removal (for example, deposition) rates (volume^{-1} time^{-1}), respectively. The activity, A_i, is defined as the number of disintegrations per unit time per unit volume, and for the ith isotope it is expressed as

$$A_i(t) = \lambda_i C_i(t) \tag{8.2}$$

Thus, the total activity is

$$A(t) = \sum_i A_i(t) \tag{8.3}$$

Since each disintegration emits an energetic particle (ray), the energy activity is defined as

$$Q_i = E_i A_i \tag{8.4}$$

where E_i is the energy of the associated alpha, beta, or gamma particle (ray), and the corresponding total energy activity is

$$Q = \sum_i Q_i \tag{8.5}$$

Various units are used to express the activities. These include the following:

Curie (Ci): 3.7×10^{10} disintegrations/s
Bequerel (Bq): 1 disintegration/s
Picocurie (pCi): 3.7×10^{-2} disintegrations/s = 2.2 disintegrations/min
1 pCi/L = 37 Bq/m^3

The quasi-steady-state concentrations can be expressed in terms of the parent (long-lived) isotope as

$$\lambda_{\text{Po-214}} C_{\text{Po-214}} = \lambda_{\text{Po-218}} C_{\text{Po-218}} = \lambda_{\text{Rn-222}} C_{\text{Rn-222}} \tag{8.6}$$

In the case of radon and its progeny, this state is reached in a matter of few hours. Thus, for all practical purposes, the ambient concentrations are expressed in the above terms. Henceforth, the radon activity level will be expressed with respect to the radon decay only, with the understanding that an additional associated activity due to the progeny is also present. This latter activity for Po-218 and Po-214 is thus twice that of the radon isotope itself (in the quasi-steady state).

Radon progenies can form clusters or attach themselves to other aerosol particles. During the decay of Rn-222, approximately 88% of the daughter Po-218 has a unit positive charge at the end of the recoil path, while the remaining Po-218 are neutral. Some of the charged Po-218 and all of the neutral ones can attach themselves to airborne particles. The attachment of radon and its progeny to particles depends on the specific properties of the particles, such as size, shape, and composition. Attachment can be characterized both theoretically and experimentally, but this is a difficult task and uncertainties exist. Radon and its progeny, both in attached and unattached forms, can find their way to sensitive tissue in the human respiratory system. Depending on the size and the nature of the particles (hygroscopic or nonhygroscopic), they will deposit in various regions of the respiratory tract. Typically, the surface tissue of the bronchi in the lungs is damaged the most by the alpha particles. Approximately 90% of the lung cancers observed in uranium miners arise in the upper bronchial tract, between the second and third generation bifurcation of the respiratory tract (see Chapter 6).

Since radon and its progeny emit energetic radioactive particles that can damage the bronchial cells, radon is recognized as a carcinogen. Trace amounts of it are of significant concern. A concentration of 1 pCi/L (37 Bq/m^3) of radon in air corresponds to 1.7 atoms of radon/cm^3, an amount 19 orders of magnitude less than the approximate 2.5×10^{19} molecules of nitrogen and oxygen that are present in the same volume at standard conditions. Precise quantification of the carcinogenicity, however, is a very difficult problem, which has elicited most of the controversy that is associated with the health effects of low-level radiation. Much of the controversy stems from the epidemiological data, which were obtained on miners who were subjected to radon concentrations that were one to two orders of magnitude higher than the concentrations encountered in homes. Inferences regarding risks to other than miners have been based on extrapolations from the use of physicochemical, mathematical, statistical, and microdosimetric models and animal studies. Since many of the models are rather crude, the estimated risks are continually subject to revision, largely because of improvements in the models and data interpretations. Radon was recognized as an indoor pollutant in the 1970s, yet very little field data exist to verify the dose models.

The interested reader can find extensive discussions of the health effects in the BEIR IV report (BEIR, 1988); however, a short primer on the topic may be helpful here. Radon progeny will deposit along the bronchial surface in a very complicated way during inhalation. Theoretical calculations and experimental data show that deposition in the carina (hot spots) is the highest among all locations at all flow rates. There are a number of other minor hot spots along the daughter tubes of the bifurcation. The alpha emissions from the deposited radioisotopes can traverse to various cells, since they are only a few microns away and well within the range of the particles. Damage to the cells depends on factors such as the source energy and strength, the source-cell geometry, the surrounding tissue, and the nature and mass of the cells and their distribution. The problem is statistical in nature because of the minute amounts of radiation that subject the microscopic cells (micron sized) to damage. As a consequence, if two identical persons are subjected to the same radon exposure for the same amount of time, the radiation damage to their cells may not be identical. Since the basal cells are believed to be the ones most susceptible to damage, the models attempt to estimate the extent of damage for conditions corresponding to a given ambient radon distribution, inhalation rate, and numerous other parameters. The problem is indeed difficult to solve, and the results obtained by different models are neither consistent nor reliable.

Standards for exposure still need to be set. The individual lifetime excess cancer risk for a 70-year lifespan at an exposure level of 0.004 working level month, as interpreted from studies of different organizations, is shown in Figure 2.1. The working level month (WLM) is an arcane unit, common in the mining industry, and it is approximately equivalent to a constant exposure of about 1 pCi/L. These estimates (1.5×10^{-3} to 6.0×10^{-3}) lead

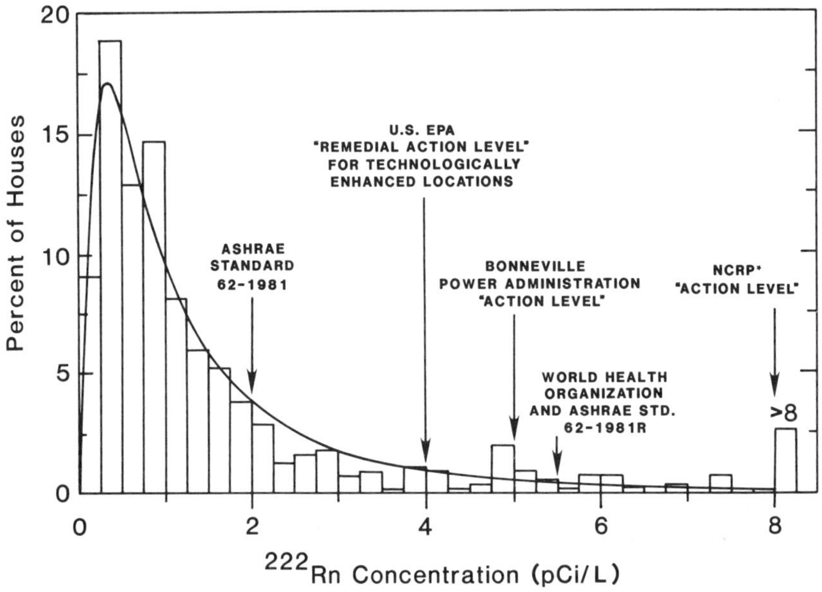

Figure 8.2 *Frequency distribution of Rn-222 concentration in 552 U.S. homes.* (Source: *Modified from Nero, et al., 1986.)*

to an individual annual excess cancer risk of 2.0×10^{-4} to 8.0×10^{-4}, which means that in the United States an estimated 5000 to 20,000 excess cancer deaths will occur annually. The frequency distribution of radon in U. S. houses is shown in Figure 8.2. The EPA standard of 4 pCi/L implies that if the radon concentrations in houses are below this level, radon-related deaths in the United States per year would not exceed about 10,000. Risk assessments from radon exposure, the development of regulatory standards, and the associated policy decisions are complex issues. As can be seen from Figure 2.1, estimates of radon risks vary over a wide range. Even though controversy exists over the standard, the current practice for radon mitigation and control is to maintain the indoor radon concentration below 4 pCi/L. On the basis of current information, the EPA has developed an action guideline for residents when the indoor radon concentration exceeds certain values. These guidelines are provided in Table 8.2.

8.3 SOURCES

Sources of indoor radon include the soil and rock beneath or surrounding a building structure, water supplies, building materials, and the natural gas that is used for cooking and heating purposes. Among these sources, soil is the primary contributor in most locations. In the earth's crust, radon originates from uranium, which is distributed in small quantities within a few meters of the earth's surface. Radon that is formed in rocks and on soil parti-

TABLE 8.2 EPA Action Guidelines for Residents

How quickly should I take action?

In considering whether and how quickly to take action, based on your test results, you may find the following guidelines useful. EPA believes that you should try to permanently reduce your radon levels by as much as possible. Based on currently available information, EPA believes that levels in most homes can be reduced to about 0.02 WL (4 pCi/L).

If your results are about 1.0 WL or higher, or about 200 pCi/L or higher:

Exposure in this range are among the highest observed in homes. Residents should undertake action to reduce levels as far below 1.0 WL (200 pCi/L) as possible. It is recommended that action be taken within several weeks. If this is not possible, you should determine, in consultation with appropriate state or local health or radiation protection officials, if temporary relocation is appropriate until the levels can be reduced.

If your results are about 0.1 to about 1.0 WL, or about 20 to 200 pCi/L:

Exposures in this range are considered greatly above average for residential structures. You should undertake action to reduce levels as far below 0.1 WL (20 pCi/L) as possible. It is recommended that action be taken within several months.

If your results are about 0.02 to about 0.1 WL, or about 4 to 20 pCi/L:

Exposures in this range are considered above average for residential structures. You should undertake action to lower levels to about 0.02 WL (4 pCi/L) or below. It is recommended that action be taken within a few years, and sooner if levels are at the upper end of this range.

If your results are about 0.02 WL or lower, or about 4 pCi/L or lower:

Exposures in this range are considered average or slightly above average for residential structures. Although exposures in this range do present some risk of lung cancer, reductions of levels this low may be difficult, and sometimes impossible, to achieve.

Source: U.S. Environmental Protection Agency, A Citizen's Guide to Radon—What it is and what to do about it, EPA-86-004, August 1986.

cles slowly moves through soil pores and into the atmosphere by diffusion or pressure-induced flow. However, not every radon atom migrates into the soil pores. The ratio of the number of radon atoms that escape from a material to the total number of atoms formed by the decay of radium in that material in unit time is generally referred to as the coefficient of emanation. The ratio is typically 0.05 to 0.3 for most soils, but in some fine-grained soil it may be as high as 0.6. Radon enters buildings through cracks and joints in the foundation, through floor drains and sumps, and through openings around utility lines. A high radon concentration may be found in structures that are built on soil that has a high radium content, or on soil that has a low radium content but is extremely porous. The types of rock present in the soil and its permeability can provide useful information in the selection of construction sites.

Igneous rocks (granites) generally contain two to three times more U-238 and Th-232 than sedimentary rocks such as limestones and sandstones. However, some shales and phosphate rocks may contain even greater amounts of U-238. The NCRP measured the concentration of major radionuclides in selected rock types and soils; these results are summarized in Table 8.3. The typical Ra-226 content of crustal rocks and soil is 40 Bq/kg. All radon atoms do not escape into the soil gas; hence, the effective Ra-226 concentration may vary from 7 to 20 Bq/kg, which would provide an estimated radon concentration in the soil

TABLE 8.3 Radium Content of Various Types of Rocks

Rock Type	*Ra-226 (Bq/kg)*	*Ra-228 (Bq/kg)*
Rhyolite	71 (10–285)[a]	91 (4–470)
Granite	78 (1–372)	111 (0.4–1025)
Andesite	26 (2–64)	27 (2–113)
Dioxite	40 (1–285)	49 (2–429)
Phonolite	368 (24–769)	543 (38–1073)
Syenite	692 (4–8930)	5 (2–3560)
Limestone	25 (0.4–223)	7 (0–45)
Sedimentary rocks (petrital)	60 (1–992)	50 (0.8–1466)
Clay	50 (14–198)	35 (8–223)
Shale	73 (11–992)	66 (21–158)
(Sandstone and conglomerate)	51 (1–770)	39 (3–919)

[a]Concentration range is provided in parenthesis.
Source: Wollenberg, 1984.

gas from 30 to 100 kBq/m^3. The effective radium contents of various soils and the associated radon emanation rates are given in Table 8.4.

As radon diffuses into the atmosphere from the soil, it mixes thoroughly with the prevailing winds. The average concentration of Rn-222 at a distance of one meter above the ground in the United States is in the range of 7 to 15 Bq/m^3, with a mean concentration of about 9 Bq/m^3.

Indoor radon concentrations are primarily dependent on the architecture of the building, the radium content of the underlying soil, and the soil permeability. The National Uranium Resource Evaluation (NURE) program, supported by the Department of Energy, con-

TABLE 8.4 Effective Radon (Rn-222) Content of Soils

Soils	*Range of Emanation Coefficient*	
Crushed rocks	0.005–0.40	
Soil	0.03–0.55	
Soil	0.22–0.32	13% to 20% of dry weight
Sand	0.06–0.18	
Sandy loam	0.10–0.36	
Silty loam	0.18–0.40	
Heavy loam	0.17–0.23	
Clay	0.18–0.40	
Soil	0.09–0.10	Dried at 105°C for 24 h
Uranium ore	0.06–26	Saturated with water
Crushed uranium ore	0.055–0.55	Saturated with water
Tailings from uranium plant	0.067–0.072	Dried at 110°C

Source: Nazaroff et al., 1988.

ducted an aerial survey of the United States to determine the geographic distribution of surface radium sources. The EPA, in collaboration with the United States Geological Survey (USGS), has estimated the indoor radon concentrations at different geographical locations. These projections were based on the results of the NURE survey, regional information on soil composition and permeability from the Department of Agriculture, and architectural information as to whether homes had basements, were slab-on-grade, or were of mixed type construction. Detailed data are available from the EPA on a county basis within any state. The primary results of the EPA study are shown in Figure 8.3. Other factors, such as climate, the lifestyle of building occupants, and their daily activities, can influence the local radon concentration significantly, and the map should be used only as a guideline. Soil testing should precede building construction if there is any question about the presence or concentration level of radon.

The transport of radon from soil gas to air depends on several factors, including soil permeability, porosity, water content, temperature, and the pressure differential between the soil and the building structure. Radon migration through soil occurs primarily by molecular diffusion and convective transport, which can be described mathematically by Fick's and Darcy's equations, respectively. The radon concentration in the soil gas (30 to 100 kBq/m^3) is several thousand times greater than that in the atmosphere (7 to 15 Bq/m^3). Therefore, a large concentration difference or driving force exists between the two. The air permeability of the soil is on the order of 10^{-8} to 10^{-16} m^2/h, depending on the soil type, and must be considered in radon migration calculations. Air permeability data for various types of soil are provided in Table 8.5.

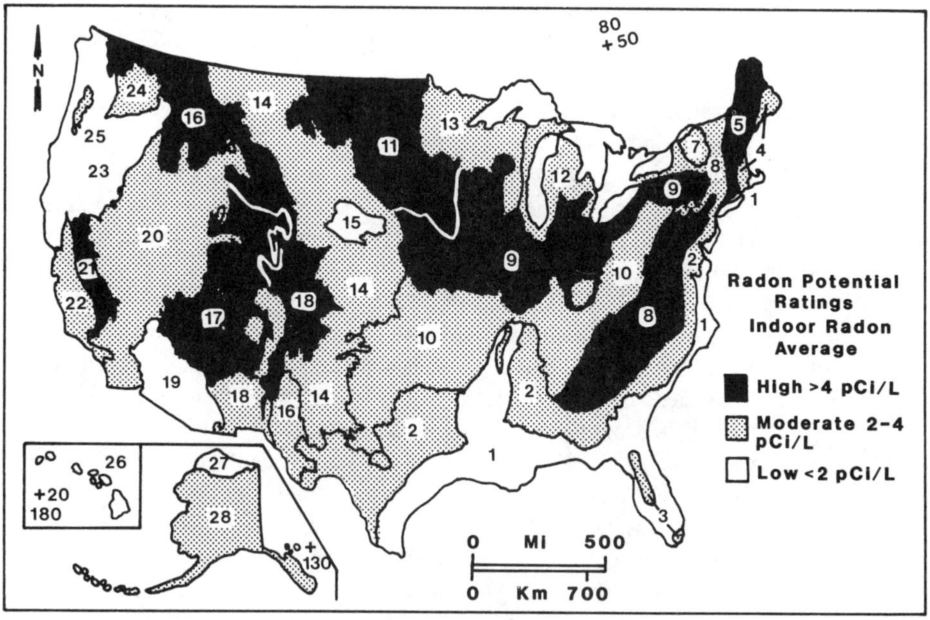

Figure 8.3 *Geologic radon potential of the United States.* (Source: *IAR, 1992.*)

TABLE 8.5 Permeability of Soils

Soil Type	*Permeability (m^2/h)*
Clay	1×10^{-16}
Sandy clay	5×10^{-15}
Silt	5×10^{-14}
Sandy silt and gravel	5×10^{-13}
Fine sand	5×10^{-12}
Medium sand	1×10^{-10}
Coarse sand	5×10^{-10}
Gravel	1×10^{-8}

Source: Terzaghi, 1967; Tuma, 1973.

The presence of water in the soil pore space tends to significantly reduce radon diffusion through soils. The distance traveled by a recoiling radon atom is approximately 60 μm in air; however, it is only 70 nm in water. Thus, the diffusion coefficient for radon in water is about 1000 times smaller than it is in air.

There are daily and seasonal variations in the radon concentrations within buildings and homes. These are due to several factors, including the daytime heating of the ground's surface by the sun and human activities. Figures 8.4 and 8.5 depict these concentration variations.

Generally, the pressure inside of a building is subatmospheric, particularly in the summer, with the pressure differential being typically in the range of 5 to 10 Pa. Clements and Wilkening (1974) studied the effect of barometric pressure change on the emanation of radon from uncovered soil in New Mexico. A barometric pressure change of 1000 to 2000 Pa over a period of one to two days can produce convective velocities on the order of 10^{-6} m/s at the surface of a soil that has a permeability of 10^{-12} m^2/h. This is a 20% to 60% increase in the radon flux over molecular diffusion alone. A number of other factors influence radon migration from soil into the building. This is shown schematically in Figure 8.6.

Domestic water supplies are a potential source of radon. The solubility of radon in water is rather high, approximately 0.24 L of radon/kg of water under standard conditions. Water supplies can be divided into three categories: surface water, which serves approximately 50% of the U. S. population; public groundwater supplies, which serve approximately 32% of the population; and private groundwater, which serves the remaining 18% of the population. Surface water generally has a smaller radon concentration than the underground water supplies, particularly those that come from deep wells. Underground water is usually surrounded by rock, and radon produced from radium in the rock dissolves in the water, increasing its radon content. The radon content of surface water is approximately 2000 Bq/m^3, whereas that from underground aquifers may vary between 20 and 50,000 Bq/m^3. Private groundwater supplies, which are provided primarily by deep wells in rural areas, have the greatest radon content, reaching a level of about 360,000 Bq/m^3 based on a sampling of 44 sources. The EPA conducted extensive tests to determine the radon concentrations in public ground water supplies, and a summary of these results is shown in Figure 8.7.

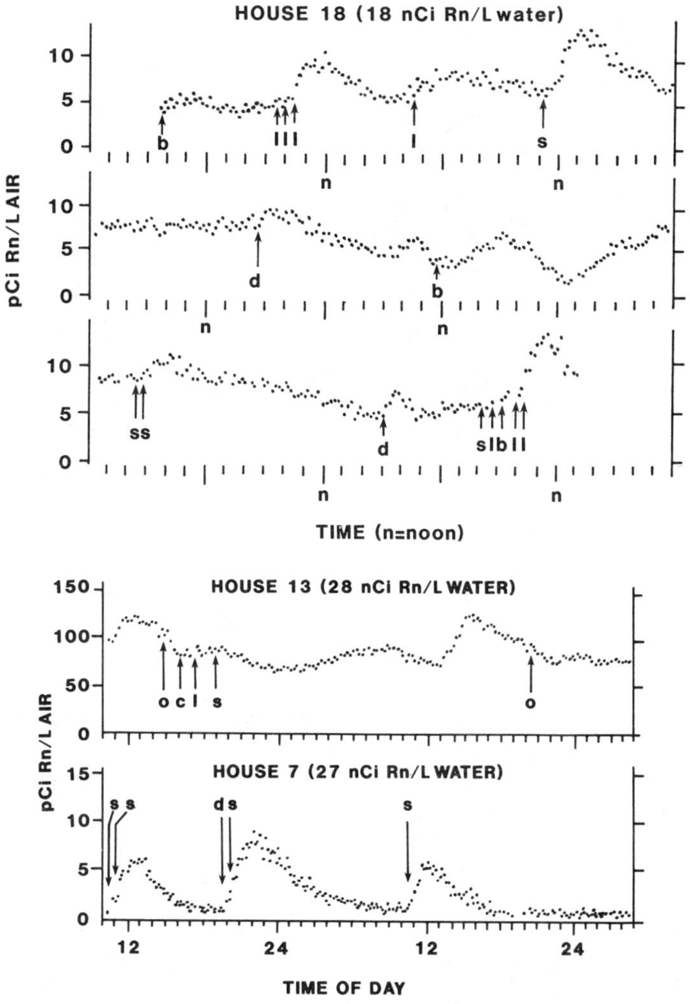

Figure 8.4 *Radon concentration in selected houses for 7 days. (*Source: *Hess et al., 1982; Activities are marked as b: bath, c: window closing, d: dishwasher, l: laundry, o: window opening, s: showers.)*

Although water may contain a high concentration of radon, the fraction released to indoor air depends on the daily water use rate and the transfer of radon from water to air. The transfer rate of radon also depends on the volume of the building and the air exchange rate. Generally, a large concentration difference exists between the radon in air and water, resulting in a long time period to reach equilibrium between these two phases. The quantity of radon released to air is dependent on the activities taking place within the home. Nazaroff et al. (1988) compiled data on the transfer coefficient of radon during various

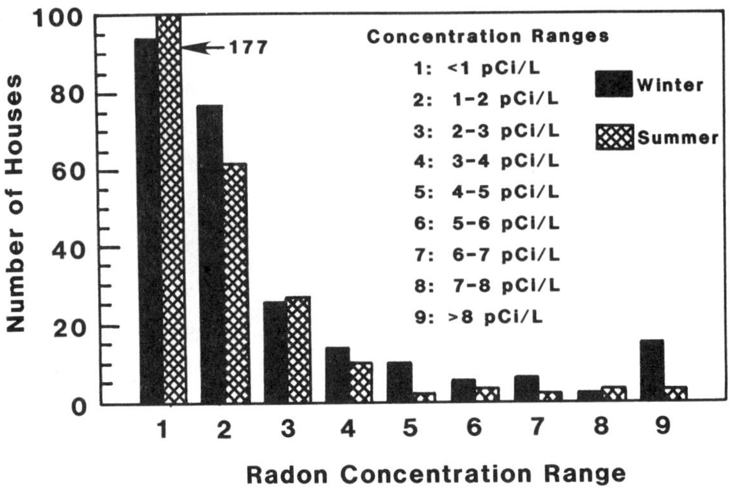

Figure 8.5 *Distribution of radon levels in homes.* (Source: *Hawthorne et al., 1984.)*

activities, such as dishwashing, laundry, and bathing. The coefficient can range from 0.3 when cleaning the home or drinking water to 0.98 when washing dishes. Transfer coefficients for these and other activities are given in Table 8.6.

Building materials also contribute to indoor radon levels. Natural materials, such as granite, clay bricks, marble, and sandstone, have a wide range of radium contents and may

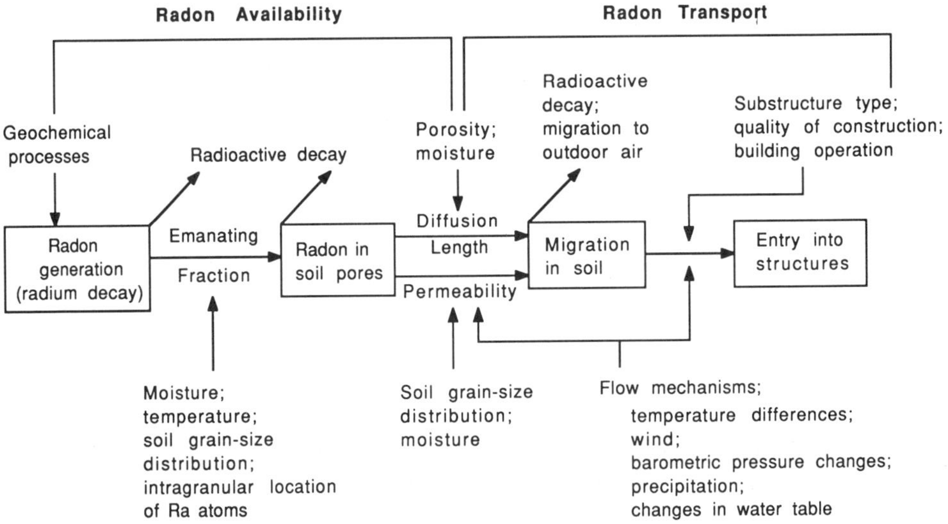

Figure 8.6 *Schematic representation of radon production and migration in soil and its entry into buildings.*

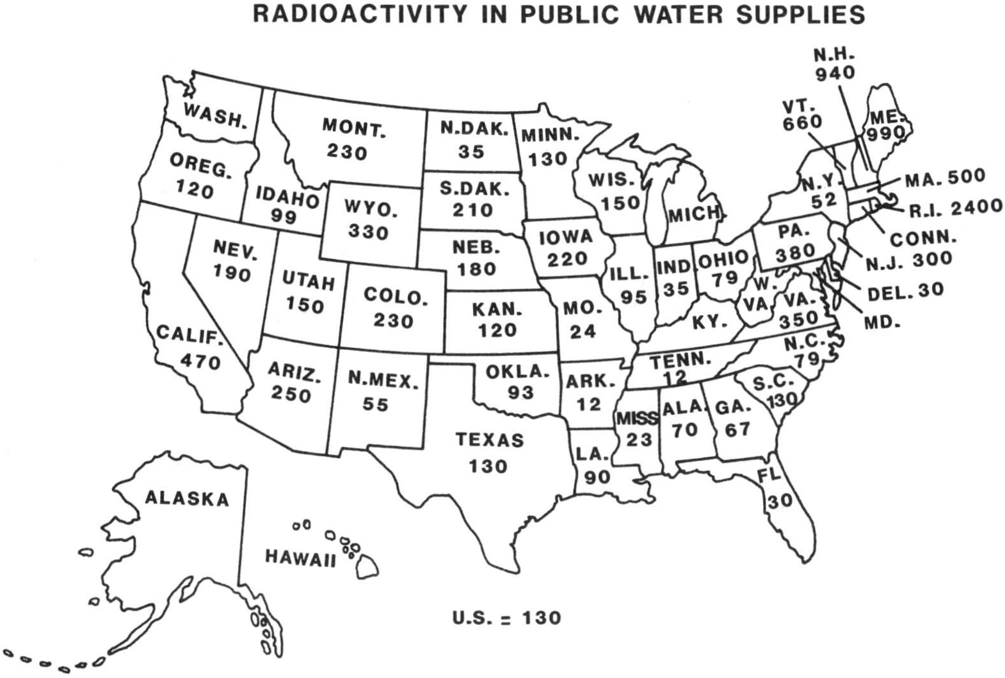

Figure 8.7 *Geometric average Rn concentration in pCi/L for public ground-water supplies in the United States.* (Source: *Hess et al., 1985.*)

have a high radon emanation rate. In recent years, wastes from a number of industries have been used widely as backfill around building foundations, and many of these buildings have now been identified as having elevated radon levels. Fly ash from coal-fired power plants, which can emanate up to several hundred picocuries per liter of radon, has been used in the manufacture of concrete and cement. By-product gypsum from the phosphate indus-

TABLE 8.6 Transfer Coefficient of Radon from Water to Air During Household Activities

Activity	*Transfer Coefficient (% released of total radon content)*
Dishwashing	0.98–0.90
Laundry	0.95–0.90
Shower	0.71–0.63
Bath	0.50–0.47
Toilet	0.29–0.30
Cleaning	0.28–0.45

Source: Partridge, et al., 1979; Gesell and Prichard, 1980.

try, which also contains radium, is utilized in the manufacture of plasterboard and concrete. The contribution of radon from these building materials depends on the total volume of the material used in the structure and on their emanation rate, which in turn depends on environmental factors such as temperature, pressure, and humidity. There is a strong correlation between the pressure difference between the interior and exterior of buildings and the emanation rate (Jonassen and McLaughlin, 1980). This rate for a concrete wall can increase by a factor of five if the pressure is decreased by 25 mmHg. A general rise and fall in the radon level with a respective fall and rise in humidity and temperature have also been observed. Figure 8.8 shows the effect of humidity variations. Air movement through and around building materials can increase indoor radon levels by a factor of two to three.

The use of new energy conservation methods, such as thermal energy storage in rocks or concrete in solar-heated houses, can also introduce radon into houses and elevate radon levels, especially in combination with low air exchange rates. The radon emanation characteristics for selected building materials are given in Table 8.7.

Because it originates in the earth's crust, natural gas contains a significant amount of radon that can be released indoors during unvented combustion. The contributions of the various sources described above to indoor radon concentrations are provided in Table 8.8.

8.4 SAMPLING AND MEASUREMENT

The techniques used to measure indoor radon concentrations are based on the detection of alpha, beta, or gamma emissions during the decay of radon-222. The sampling methods that have been developed using this principle can be classified into three categories: grab sampling, continuous sampling, and integrated sampling. Depending on the method used for measuring the radon concentration, the techniques may involve direct measurement of the activity of Rn-222 only, of Rn-222 and its short-lived daughters, or of daughter products only. After the radon sample is collected, a certain period of time must be allowed for

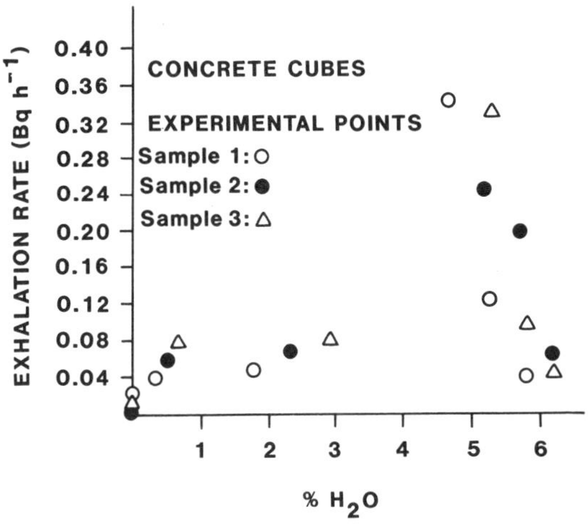

Figure 8.8 *Effect of the moisture content of concrete on radon exhalation.* (Source: *Stranden et al., 1984.)*

TABLE 8.7 Radon Emission Rates from Building Materials

Material	*Radon Flux ($pCi/m^2/h$)*
Concrete (ordinary)	270.3
	18.0 (10 cm thick)
Lightweight concrete	54.1
	72.0 (20 cm thick)
Alum shale concrete	5405.4
Clay brick	108.1
By-product gypsum	540.5
	36.0 (wall 7.6 cm thick)
	3.6 (ceiling 1.3 cm thick)
Heavy concrete	36.0 (8 cm thick)
Soil	1800.0
Brick	
Solid type	0.46
Cavity type	0.19
Chipboard	0.068
Gypsum board	0.122

Source: Jonassen and McLaughlin, 1982; UN, 1982.

equilibrium to be achieved among the radon and its daughters prior to making an activity measurement (counting).

Grab Sampling

The grab sampling method consists of measuring the radon concentration in air over a short period of time. Samples are collected instantaneously or over a time period in an airtight container coated on the interior with a silver-activated zinc sulfide phosphor. One surface of the container is flat, transparent, and usually made of an optical glass. When an alpha particle that is produced during the decay of radon strikes the phosphor, visible radiation is

TABLE 8.8 Approximate Contributions of Various Sources to Indoor Radon Concentrations

Source	*Single-family houses*[a,b] *(pCi/L)*	*High-rise apartments*[a,b] *(pCi/L)*
Soil potential (based on flux measurements)	1.5	>0
Water (public supplies)	0.01	0.01
Building materials	0.05	0.1
Outdoor air	0.25	0.25

[a]In each case, the arithmetic mean is shown.

[b]Based on an air exchange rate of $0.5\ h^{-1}$.

Source: Nero, 1988.

emitted and then transmitted through the glass window to where it is detected with a photomultiplier tube and associated electronics. This device, called a scintillation flask, is referred to as a Lucas cell in recognition of its inventor, H. F. Lucas. The container has one or two sampling ports fitted with a valve for evacuation or filling of the chamber. Usually three or four hours are allowed for growth of the radon daughters Po-218 and Po-214 before the counting of the alpha particles is started. The entire measurement is completed within 10 minutes thereafter. The advantages of the grab sampling method include both its sensitivity and the rapidity with which the results can be obtained. However, its accuracy is limited by the detection efficiency of the scintillation cell, which usually is in the range of 75% to 80%. The error in the measurement can be more than 30% if the radon concentration is below 10 Bq/m^3.

Continuous Sampling

Continuous sampling devices provide real-time measurements at short-time intervals over a long period. The time intervals required for a measurement can vary between 10 minutes and 4 hours, depending on the measurement techniques used. Two types of devices are available for continuous monitoring of indoor radon concentrations: flow-through scintillation chambers (a two-port Lucas cell) and solid-state detectors (Wrenn chambers). The Wrenn chamber is more widely used for continuous radon measurements. The working principle of the detector is based on the electrostatic collection of Po-218. Radon diffuses passively into a hemispherical chamber, whose walls consist of foam rubber supported on a wire screen. An optical fiber coated with zinc sulfide and covered with aluminized Mylar is located in the center of the chamber near the flat-bottomed surface. As radon diffuses into the chamber, positively charged Po-218 and Po-214 are drawn to the Mylar by an electric field, formed by applying a high voltage between the Mylar (–) and the support screen (+). The alpha particles produced from the decay of Po-214 and Po-218 strike the phosphor, producing a light pulse that is detected by the photomultiplier tube assembly. The Wrenn chamber is capable of measuring radon concentrations down to 10 Bq/m^3. However, the efficiency decreases significantly in the presence of moisture because of the neutralization of Po-218. The effect of moisture can be minimized by placing a desiccant material inside the chamber.

The Lucas cell can also be used in a continuous flow mode by using one of the two cell ports as the inlet and the other as the outlet. Time integration is a problem with the scintillation cells. The Po-214 and Po-218 deposited on the wall of the cell emit alpha particles upon decay and contribute to the observed count.

Integrated Sampling

Integrated sampling methods provide the average concentration of indoor radon over time periods ranging from several days to one year. The detectors that are used to measure these average radon concentrations include alpha track detectors, electret ion chambers, and charcoal canisters.

An alpha track detector consists of a material that becomes damaged or etched when exposed to alpha particles. Generally, the detector uses CR-39, a polymer of oxydi-2, 1-

ethanediyl dia-2-propenyl diester of carbonic acid, mounted inside a plastic cup. The plastic cup is covered with a filter or a membrane to prevent radon daughters from entering. As a result, only radon enters the cup by diffusion and reaches equilibrium with the surrounding air. The detector is placed in the radon atmosphere for the time period of interest, which may be from several weeks to a year. The penetrating alpha particles actually attack the chemical bond of the polymer and leave a permanent footprint on the material. Following the exposure period, the damage tracks are chemically etched by a basic solution (usually KOH or NaOH). The etching extends to the region of damage around each track, and the separate tracks become visible as small holes on the surface of the material. These tracks are then counted visually, using either a microscope or with the aid of a computer. Although the alpha track sampling method provides a true integrated measurement under the average exposure conditions, several studies have shown that a high degree of uncertainty is associated with the measurement. Some concerns have been expressed that a static charge on the plastic cup that contains the detector may affect the response of the device by altering the deposition pattern of alpha particles. This may account for up to 20% of the error associated with the measurements made by this method.

The electret ion chamber consists of a dielectric material that is given an electric charge and sealed from the air in a chamber which is coated with a material of low atomic number. By measuring the difference in the charge (voltages) before and after deployment in a radon-containing atmosphere, and with the aid of a suitable calibration sample, the radon concentration can be estimated.

Another method for measuring indoor radon concentrations is a charcoal canister. This method is widely used and is recommended by the EPA, as well as by numerous other public agencies. Charcoal canisters can be distributed to the user and collected by mail for later radon analysis, and they can be used by persons who do not have technical training. Activated charcoal has long been known to have the ability to adsorb radon (Rutherford, 1906). The adsorbed radon is retained by the charcoal until it is desorbed by heating. Typically, from 25 to 100 g of charcoal are packed in a canister. The canister, which is impervious to radon diffusion, is opened to the indoor air for a specified period of time. During this time period, radon is adsorbed onto the surface of the charcoal, and the canister is then sealed for future radon analysis. The gamma activities of both Pb-214 and Bi-214 are measured by a gamma ray spectrometer, using a scintillation detector. The corresponding radon concentration in the indoor air is then calculated from the measured activity, after correcting for the efficiency of the radon detector, the water gain by the charcoal in the canister, and the decay of radon with time. An assumption that is frequently made with this method is that charcoal never reaches equilibrium with atmospheric radon. Therefore, it serves as a sink for radon and provides an integrated measurement. As will be explained later, this assumption is incorrect. The indoor radon concentration is calculated using the equation

$$Rn = \frac{\text{net } CPM}{(T_S)\,(E)\,(CF)\,(DF)} \tag{8.7}$$

where Rn is the radon concentration in the indoor air, net CPM is the net counts per minute from the exposed charcoal canister, T_s is the exposure time, E is the efficiency of the detector, CF is a calibration factor for the water gain of the charcoal, and DF is the decay factor for radon. Although *CF* is called a calibration factor, it has units of liters per minute and is physically interpreted as an effective sampling rate. The purpose of *CF* is to correct for the relative humidity of the air and the exposure time.

The calibration factor is determined by exposing the canister to known concentrations of radon for a predetermined period of time in a test chamber maintained at various humidity levels. Although calibration factors are determined with respect to water gain, the EPA recommends that each batch of charcoal be calibrated before it is packed in a canister. The reasons for this are multifold. Radon adsorption capacity is a function of various environmental factors, including temperature, humidity, and barometric pressure, and variations in these factors can cause significant changes in the accuracy of the measurements. For example, the water vapor adsorption capacity of charcoal depends on the physical characteristics of the charcoal, including the surface area, total pore volume, and pore size distribution. The water gain by an activated charcoal measured by different investigators is shown in Figure 8.9. This figure demonstrates the importance of calibrating each batch of charcoal prior to use.

Charcoal canisters are calibrated by several methods. One method is to use a filter paper to collect the radon daughters while charcoal adsorbs radon from the air. The concentration is calculated from the activity of the radon daughters collected on the filter paper. Shleien (1963) compared the radon measuring efficiency of Millipore type AA filters with the charcoal adsorption method. The filter paper consistently provided lower radon concentration values than did the charcoal adsorption method. The radon concentration in the charcoal was determined by measuring the activity of Bi-214. Equilibrium was established

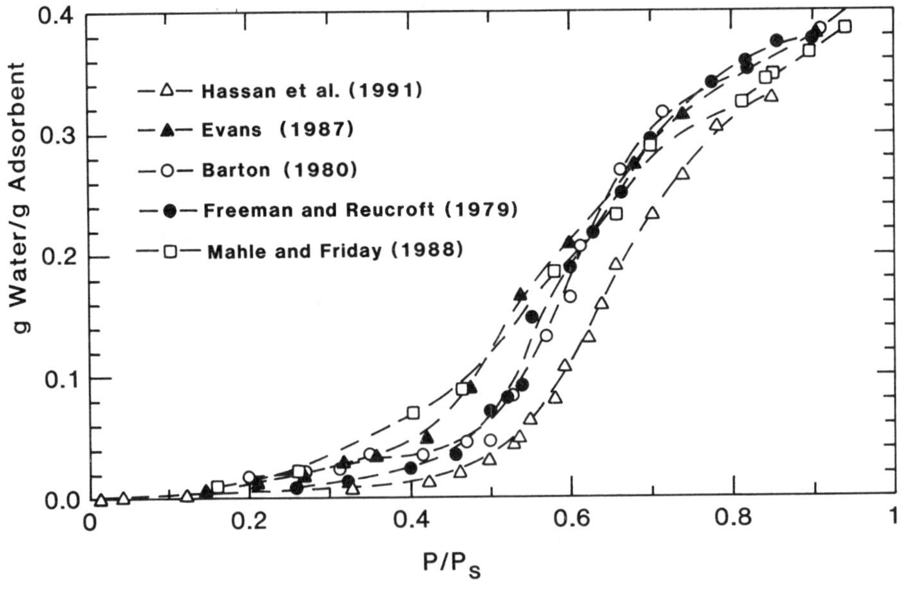

Figure 8.9 *Comparison of water vapor isotherm data on BPL activated carbon with the literature at 298 K. (*Source: *Hassan et al., 1991a.)*

between radon and its daughters in the charcoal within three hours, and the radon concentration did not change for the next five hours.

Countess (1976) used standard M11 Army charcoal canisters to measure the indoor radon concentration. Air was drawn through the canister instead of letting radon diffuse passively into the canister from the air. To prevent radon breakthrough from the charcoal bed, the total volume of air passing through the charcoal was less than 100 L. The radon concentration in the charcoal was measured by counting the gamma activity of the radon daughters. The counting efficiency was found to depend significantly on the counting time; and, the error was reduced from ± 20% to ± 7% when the count time was increased from 10 to 100 minutes.

The effect of a varying radon concentration on the accuracy of a measurement can be determined with a control study in a test chamber. A charcoal canister is exposed in a chamber to varying radon levels at different relative humidities and temperatures for different periods of time. The concentration of radon in the chamber during the test is varied deliberately, either from low values to high values or vice versa. The average radon concentration measured with the canister is compared with the time-weighted average concentration measured with a continuous radon monitor. George (1984) varied the concentration of radon in a chamber from 5 to 55 pCi/L, the relative humidity from 15% to 100%, and temperature from 291 to 300 K. The mean radon concentration calculated from the canister was 0.6 times that of the known time-weighted mean value when the canister was exposed first to a high radon concentration followed by a lower one. In the reverse situation, the concentration was 1.25 times higher. Radon that is adsorbed on the charcoal can desorb from the same charcoal if it is exposed to an atmosphere that contains radon at a lower concentration than the one in which the adsorption study was carried out. This may be one of the reasons for the inaccurate results. Therefore, the exposed canister must be made airtight at the end of the exposure period to obtain the most accurate estimate of the radon concentration. From these standard tests, calibration factors are determined as a function of radon concentration and are plotted versus exposure time for different relative humidities.

A number of researchers have noted similar results when the radon concentration varied during the sampling period. Michaels et al. (1987) reported that canisters exposed to constant radon concentrations performed much better; the measured concentrations were within ± 25% when a 96-hour sampling period was employed. Their findings were confirmed by Ronca-Battista and Gray (1987), who exposed charcoal canisters to radon-laden air at varying radon concentrations for one to six days at different relative humidities and temperatures. The average concentration was underestimated by 75% when the initial radon concentration was decreased continuously until the final concentration was only one-tenth of the original value. When the testing conditions were reversed, the average concentration was overestimated by 64%. The sampling period was 96 hours. The radon concentration was underestimated by 54% when these canisters were exposed for 48 hours to a decreasing radon atmosphere. Ronca-Battista and Gray also concluded that radon can desorb from charcoal if the radon concentration of the surrounding air is less than that in the charcoal. Based on the above findings, the exposure time recommended by the EPA was changed from 96 to 48 hours. However, Pearson (1989) noted that canisters gave a 40% overestimation of radon concentrations for decreasing radon levels and a 30% underestimation when the conditions were reversed. He observed that charcoal canisters provide results that were lower than the actual radon concentrations at any radon level, when exposed to higher temperatures, regardless of the relative humidity. Also, separate lots of activated charcoal differed in performance.

While the charcoal canister method is sufficiently sensitive for measuring concentrations at or above the EPA guideline level of 4 pCi/L, large errors occur for concentrations below 1 pCi/L. Since charcoal canisters are affected by varying indoor radon concentration levels, they do not provide a true time-integrated measurement. Most researchers have found that the adsorption efficiency of charcoal canisters decreased with an increase in temperature and humidity, although there is a great deal of controversy over the extent to which these factors affect the measurements. Cohen (1986) stated that the correction due to the change in relative humidity was not more than 6%. However, Ronca-Battista and Gray (1987) reported that radon adsorption changed up to 50% when the humidity changed from 20% to 100%. George (1984) and Ren and Lin (1987) observed that the amount of radon adsorbed by canisters did not change significantly when temperature was changed from 17° to 27°C. Similar results were also reported by Megumi and Mamuro (1972), who noted that the efficiency of radon adsorption on charcoal was a function of temperature, but it was independent of the relative humidity in the range of 40% to 90%. On the other hand, Cohen (1986) observed a change of 1.5% in adsorption capacity for each 1°F change in temperature. Moisture in the air appears to exert the greatest influence on the measurement efficiency.

The manner in which a canister is exposed to radon and the canister configuration are believed to affect the measurement. With proper design of the exposure path, the measurements could be made less sensitive to humidity and varying radon concentrations. However, canisters still should be calibrated by exposing them to a range of radon concentrations. When the above procedure was followed in a test chamber for a three-day integrating time measurement, the error ranged from 2% to 17%. The measurements should be repeated over a long time period to obtain a better estimate of concentrations.

To improve the reliability of radon measurements and to make them less sensitive to moisture and varying radon concentrations, a new type of charcoal canister, the diffusion barrier charcoal canister, was developed. The canister is packed with charcoal in the usual manner, but is completely sealed. An air diffusion path is provided by attaching a tube to a hole cut in the lid. It is expected that the performance of these canisters will be less dependent on the physical properties of the charcoal. In principle, a shorter integration time can be used. Time periods varying from one to seven days have been used in field tests. Using a shorter integration time (eight hours), the uncertainties associated with back diffusion and radon decay may be minimized.

Prichard and Marien (1985) used an integration time of up to 24 hours. To ensure complete adsorption of radon by charcoal, only a small volume of air was drawn into the sampler. Cohen (1987) optimized the dimensions of the diffusion barrier and the integrating time by using a one-dimensional diffusion equation. The model yielded an effective length for the diffusion path of 2.28 cm for a fixed barrier area and an integration time constant of about three days. The humidity problem was avoided by placing a bag filled with a desiccant material inside the diffusion barrier. The EPA has also developed a diffusion barrier charcoal canister.

Although diffusion barrier canisters are expected to provide better measurement accuracy than open-face canisters, various independent studies have shown that the error can still be significant. The errors in the measurement associated with the canisters were determined by exposing them to radon at 50 pCi/L for 24 hours and then keeping them in an atmosphere containing less than 0.5 pCi/L of radon for one week. A 1-week measurement of radon concentration by this method gave the average monthly concentration within a standard deviation of about 15%. The effect of humidity was taken into account by using a correction factor. The largest source of error associated with this detector was attributed to the user. Once a measurement was complete, these canisters were regenerated by applying heat for repeated use. The adsorption capacity for radon decreased by 5.5% with an increased number of adsorption-desorption cycles.

Lee and Sextro (1987) tested the model of Cohen and Nason (1986a, 1986b) with experimental data and found that diffusion-limited charcoal canisters can have errors of more than 20% and should not be relied on if accurate measurements are needed. Lee and Sextro also evaluated the performance of modified diffusion-barrier EPA canisters. The charcoal in the canisters was preadsorbed with radon that was then allowed to desorb from the canisters over a period of a few days into an atmosphere with low radon concentration. The experimental data were fitted to a two-dimensional diffusion model, which gave close agreement between the predicted and the actual experimental values. Later, Sextro and Lee (1989) studied both open-faced and diffusion-barrier charcoal canisters at various radon concentrations. The errors ranged from 10% to 105% for the open-faced canisters, while they were in the range of 77% to 115% for the diffusion-limited canisters. The data were taken by using two to six canisters for each set of conditions. Pojer et al. (1990) assessed the effects of temperature and humidity, both theoretically and experimentally, on the performance of a diffusion barrier charcoal canister. The adsorption capacity decreased by 30% when the temperature was increased from 13° to 35°C; the amount of radon adsorbed decreased by a factor of three when the relative humidity was increased from 15% to 90% at 35°C.

The physical characteristics of charcoal are often ignored in determining the efficiency of charcoal canisters. Even the amount of charcoal used in the canister and how it is spread in the canister seem to influence the outcome of the measurement. Li and Hopke (1991) calculated the distribution coefficients for radon on various types of charcoal. The coefficients expressed as cm^3 STP air/g of carbon ranged from 2000 to 9650 (see Table 8.9). Scarpitta and Harley (1990) investigated the adsorptive and desorptive characteristics of canisters that contained a petroleum-based charcoal, Witco 6-12 mesh, under controlled conditions of temperature, relative humidity, and radon concentration. Two sets of experiments were conducted. In the first, the charcoal was spread in one layer. In the second, a packed bed of charcoal was used. The exposure times in both cases were varied from one to seven days. The results demonstrated that radon adsorption and desorption depend on the bed depth and on the amount of water adsorbed. The adsorption of radon decreased by an order of magnitude when water started condensing in the pores of the charcoal. Conven-

TABLE 8.9 Distribution Coefficients of Radon Between Air and Activated Carbons

Carbon Type	*Distribution Coefficients (cm^3 STP in air/g of carbon)*	*Temperature (°C)*
SKT-2M	9650	18
No. 2	6300	18
SKT-1	9200	18
No. 4	7600	18
MSKT (5)	7400	18
SKT (82)	6000	18
No. 7	4500	18
SKT (84)	5700	18
Ag-2	4250	18
No. 12	2000	18
Witco AC-337	5660	20
Peat	2250	20
Sutcliffe-207	3530	25
Norit RFL 3	4610	25
Norit RFL 111	4660	25
Ultrasorb	5000	25
Pittsburgh PCB	5690	25
Lime wood	7000	20

Source: Bocanegra and Hopke, 1987.

tional charcoal canisters can become saturated in less than four days at 70% relative humidity if exposed in the fully opened configuration. The diffusion characteristics of radon in a half-filled and a completely filled charcoal bed were studied by Wilson (1989). The diffusion coefficient was found to be independent of exposure time in the half-filled bed, but decreased with time in the filled canister. The humidity did not affect the radon concentration profile in either of the beds.

Charcoal canisters are not effective for measuring radon concentrations below 10 Bq/m^3, and large errors may be expected due to the low efficiency of the gamma detector at these low concentrations. Researchers have tried to develop other methodologies to measure radon concentrations after it is adsorbed on charcoals. Shimo et al. (1983) transferred the adsorbed radon into an ionization chamber by heating the charcoal; the activity of radon was determined from the current density necessary to ionize radon in the chamber. The collection efficiency of this system was found to be 88% ± 0.5% for radon concentrations in the range of 0.05 to 1.2 pCi/L, although a high temperature was required to desorb radon from the charcoal.

Prichard and Marien (1983) extracted radon from activated charcoal by a chemical solvent, followed by the measurement of alpha and beta activities with a liquid scintillation counter. Low background counts, greater counting efficiency, and lower limits of detection permitted a better estimate of the radon concentration. Organic solvents such as carbon disulfide, *n*-hexane, ethyl ether, acetone, ethanol, and a mixture of *n*-hexane containing 10% toluene were used to extract radon from charcoal.

All three detector types (alpha track detectors, electret ion chambers, and charcoal canisters) are frequently used indoors. The General Accounting Office conducted a study to evaluate the performance of various radon detectors (GAO, 1989). A summary of their study is provided in Table 8.10. The average errors in the measurements were 19% for charcoal canisters, 25% for alpha track detectors, and 31% for electret ion chambers.

The working principle of charcoal canisters is based on the adsorption process. The amount of radon that can be adsorbed by a certain quantity of charcoal depends on the radon's gas phase concentration. If charcoals are exposed to radon for a sufficient period of time, equilibrium will be established between radon in the two phases as long as the gas phase concentration remains constant and other conditions, such as temperature, pressure, and humidity, do not change. This equilibrium is different from the radioactive equilibrium, which is established between radon and its daughter products. Once the adsorption equilibrium is reached, charcoal will not adsorb additional radon. But if the radon concentration changes in the atmosphere, charcoal will reach a new equilibrium based on the new concentration. Therefore, it is not surprising that investigators have observed an underestimation of radon concentration by charcoal canisters that were exposed to a decreasing concentration. The opposite phenomenon could be expected in an increasing radon atmosphere. This type of behavior can be predicted from adsorption isotherm data. Knowledge of the time required to reach adsorption equilibrium is extremely important, since the exposure time will depend on this information.

Hassan et al. (1991b) designed an experimental apparatus to measure the radon concentration simultaneously in the gas and solid phases at any given time. They observed that it took less than 60 minutes to reach adsorption equilibrium when a 5-g sample of activated carbon was used and was spread in two to three layers in the adsorption apparatus. The amounts of radon adsorbed by charcoal at low concentrations were rather small, but the charcoal's capacity rose sharply as the gas phase concentration increased. This type of behavior is common when adsorbent-adsorbate interactions are relatively weak. Once an

TABLE 8.10 Efficiency of Radon Detectors Recommended by the EPA

Method	*Number of Tests*	*Average Error (percent)*[a]	*Range of Company Error (percent)*[b]
Alpha track detector	10	25	11–55
Activated-charcoal adsorption detector	256	19	1–133
Continuous radon monitor	99	25	0–658
Continuous working level monitor	75	40[c]	0–1353[c]
Electret ion chamber	127	31	5–486
Grab sampling radon	66	18	3–75
Grab sampling working level	58	29	3–328
Radon progeny integrated sampling unit	4	27	1–80

[a]The average error was computed by taking the mean of all errors of the individual devices of a given type.

[b]The average error for devices of a given type was computed for each company for the test and also for the retest. The range represents the low and the high average company error.

[c]If the company whose average error was 1353 is excluded, the average error becomes 22% and the range of company error becomes 0 to 518%.

Source: GAO, 1989.

atom or a molecule is adsorbed, adsorbate-adsorbate (here radon to radon) interactions promote the adsorption of further atoms or molecules so that the isotherm becomes convex to the pressure axis (see Chapter 10). A similar behavior was observed by researchers almost 30 years earlier. Therefore, the calibration factors should be determined on the basis of the radon concentration in the gas phase, rather than on the water gain of the charcoal. The adsorption capacity of charcoal or any solid adsorbent is a strong function of temperature as demonstrated in Figure 8.10 for radon. The presence of moisture changes the capacity for radon, as shown in Figure 8.11.

The relationship between the gas phase and the solid phase radon concentrations when no moisture is present can be described by a modified Freundlich equation written as

$$q = k'\left(\frac{P}{P_0}\right)^n \tag{8.8}$$

or

$$\ln q = \ln k' + n \ln\left(\frac{P}{P_0}\right) \tag{8.9}$$

where q is the amount of radon adsorbed (g/cm^3 of the adsorbent), P is the partial pressure (mmHg) of radon in the air, P_0 is a reference pressure, set at 10^{-14} mmHg for convenience, k' is a measure of the adsorbent capacity, and n is the intensity of adsorption. From experimental data taken at three temperatures, a value of n equal to 1.75 was obtained. The temperature dependence of the parameter k' was found to be linear and is represented by

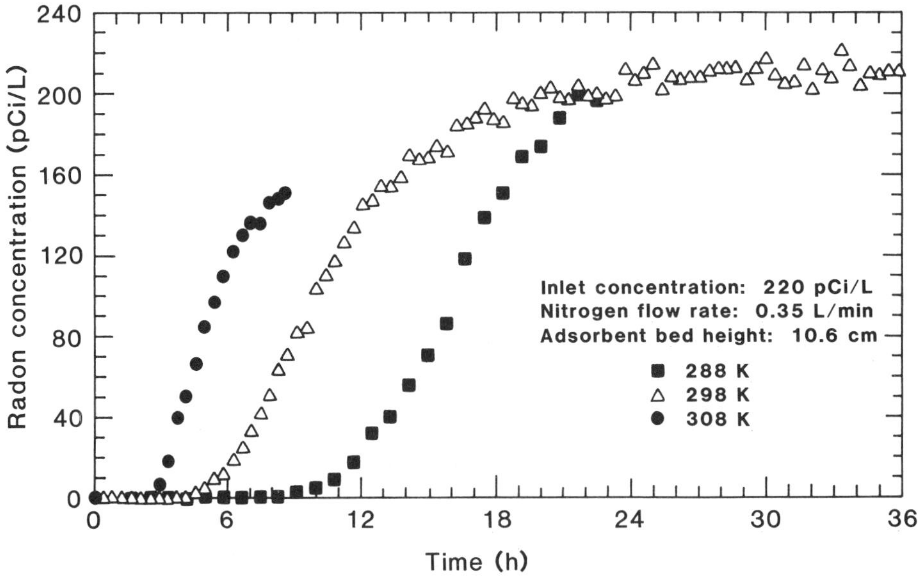

Figure 8.10 *Experimental breakthrough curves for radon adsorption on BPL activated carbon. (Source: Hassan et al., 1991c.)*

$$k' = 2.10 \times 10^{-11} - 6.58 \times 10^{-14}\, T \tag{8.10}$$

Thus, the adsorption of radon on charcoal can be expressed as

$$q = (2.10 \times 10^{-11} - 6.58 \times 10^{-14}\, T)\left(\frac{P}{1.0 \times 10^{-14}}\right)^{1.75} \tag{8.11}$$

where P is in millimeters of mercury and T is in Kelvins. The radon concentration in air, C, can be obtained in picocuries per liter from the expression

$$C = \frac{5.48 \times 10^{3}\, q^{0.571}\, T^{-1}}{(2.10 \times 10^{-11} - 6.58 \times 10^{-14}\, T)^{0.571}} \tag{8.12}$$

The parameter n is a weak function of the physical properties of the charcoal. The parameter k' can be obtained for each batch of charcoal by conducting experiments at three temperatures.

An interesting phenomenon was observed during radon adsorption in the presence of moisture. Although it is expected that the capacity will decrease due to the presence of moisture, a large change in the capacity may occur depending on how the charcoal is exposed to the radon-moisture mixture. Hassan et al. (1992b) conducted two sets of experiments; in the first case, charcoal was preequilibrated with moisture at various relative humidities (40%, 60% and 80%). The charcoal was then exposed to radon-laden air whose relative humidity was also the same as that at which the charcoal was preequilibrated. In the second case, dry charcoal was exposed to mixtures of water vapor and radon. Significant differences in the capacities were found (see Figure 8.11). The temperature of the charcoal increased slightly due to the exothermic heat of adsorption, but this was not large enough to

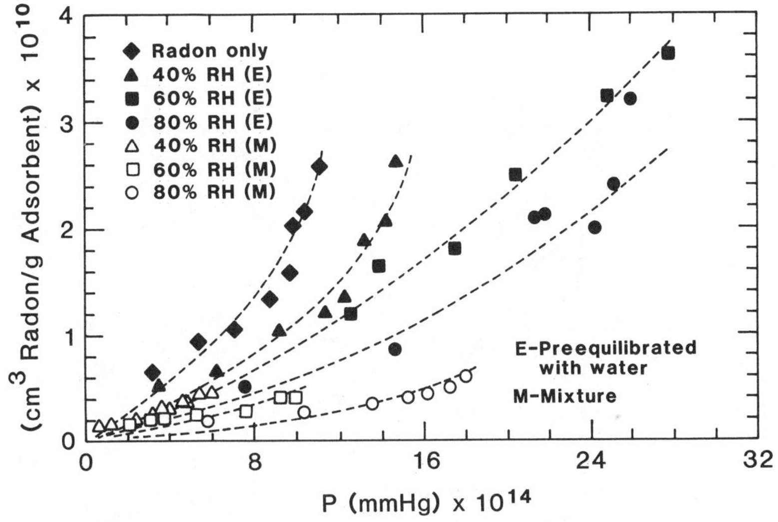

Figure 8.11 *Comparison of radon uptake from water vapor mixture with that on pre-equilibrated BPL activated carbon at 298 K.* (Source: *Hassan et al., 1992b.*)

account for the change in adsorption uptake. A similar observation was reported by Thomas (1974), who investigated the adsorption of radon on activated charcoal in a dynamic system. A rise in temperature of about 7°C, which was attributed to the heat of adsorption of water, was assumed to be responsible for the decrease in uptake capacity of radon by the charcoal. However, the greater uptake capacity of radon by the charcoal preequilibrated with water vapor may involve radon absorption in the water that is condensed in the micropores of the charcoal. The solubility of radon in water is relatively high, approximately 24 cm^3/100 cm^3 of water at 298 K. The water contained in the charcoal's pores can hold up to 44 times more radon than would have been adsorbed on the charcoal surface. Since both radon adsorption and absorption are occurring when the charcoal is preequilibrated with water vapor, the uptake capacity of radon on activated charcoal should be higher when the charcoal was first preequilibrated with water than when radon was adsorbed from the mixture. If a longer period of time is allowed, the amount adsorbed from the mixture should eventually equal the preequilibrated value.

Water usually exhibits a large hysteresis upon desorption from charcoal. As a result, when water is adsorbed, it will not desorb from the pores under normal conditions. Therefore, care must be taken when packing the charcoal canisters. Manufacturers generally ship the material in a dried condition in an airtight container. Once the container is opened, charcoal will slowly adsorb moisture from the air; but the water gain will differ, depending on how quickly the charcoal is packed in the canister. This may ultimately lead to inaccurate radon measurements.

8.5 SPECIFIC CONTROL STRATEGIES

Strategies to control radon in residences can be broadly classified into three categories and several subcategories as follows:

1. Source removal (New construction considerations)
2. Source control
   a. Sealing of radon entry paths
   b. Subslab ventilation
   c. Basement pressurization
3. Air cleaning
   a. Filtration and/or electronic air cleaner
   b. Increased ventilation
   c. Adsorption

These strategies can be employed during the retrofitting of old structures or incorporated during the construction of new ones. The choice of a particular method and its effectiveness, particularly during retrofitting, depends primarily on the radon source, its entry path, and the structural condition of the building. Some of these methods, such as the removal of high-radium-content soil from the construction site, may not be convenient and economical for older structures, but they may be more cost efficient during new construction than any of the post construction remedial techniques.

Source Removal

Since soil is the main source of indoor radon in most cases, selecting a construction site where the radium content of the soil is low will result in a building with a low radon concentration. Knowledge of the local soil characteristics, such as radium content, permeability, and moisture content, is necessary when making such a selection. This information is usually available from the Department of Agriculture Soil Conservation Service. Site modification prior to construction is another approach. The removal of high-radium-content soil from a perimeter of 3 m from the building foundation and replacing it with a low-radium-content soil can substantially reduce radon concentrations in new structures. One study showed that the radon emanation rate decreased by almost 80% when a phosphate fill was replaced with a low-radium-content soil to a depth of 10 ft (Culot et al., 1973).

The removal and replacement of soil from beneath an old structure and the subsequent backfill around the foundation and the concrete floor slab are frequently too costly, since much of the excavation must be done by hand. Although the costs are site specific, they can range from $5000 to $20,000.

New Construction Considerations

A substantial reduction in radon concentration can be achieved by adopting alternative construction techniques. The objective is to reduce the direct contact between the soil and the building foundation by placing a physical barrier between them. Generally, provisions are made for soil gas to permeate into the subslab and crawl spaces so that a ventilation system can be effective. Often a passive stack vent is installed to develop differential pressures between the basement and the subslab fill material. The EPA has developed guidelines to assist home builders in constructing radon-resistant residential homes and to facilitate radon removal after construction (EPA, 1988).

The permeability beneath the subslab can be increased by placing a minimum of four inches of aggregate (or stone chips) under the slab. The aggregate is covered by a plastic sheet that prevents radon migration into the basement area. The highly permeable aggregate serves two purposes. A post construction active or passive subslab ventilation system can be installed. Second, radon accumulated in the aggregate can flow out through any control openings in the barrier to the outside air. The efficiency of the method depends on the effectiveness of the barrier in containing the radon. Fifteen homes were built in New York state employing the above technique, but required active subslab ventilation to reduce indoor radon concentrations below 4 pCi/L. Kunz (1991) suggested that a better result might be obtained if the barrier is placed between the soil and the aggregate. In some instances, the use of two barriers may be needed: one between the soil and the aggregate and another between the aggregate and the slab. The barrier at the soil-aggregate interface should be resistant to both diffusion and convective flow of radon. Radon gas accumulated in the aggregate can be vented passively to the outdoor air through a hollow-block wall. This will also allow outdoor air to flow freely into the aggregate. However, to have more control of the airflow in and out of the aggregate, a small pipe can be installed in the subslab. Using outdoor air, the space can be pressurized or depressurized alternatively through the pipe to remove radon from the aggregate. A basement with a sump can also be used effectively to accumulate soil gas, which can then be vented outside. This is illustrated in Figure 8.12. The sump ensures the effective drainage of water from the permeable space between the barriers. A piping network much like a drainage system can also be installed in the bed of crushed aggregate beneath the slab to draw radon from all sides of the house.

The double-barrier approach can be used for slab-on-grade and crawl space construction. As shown in Figures 8.13 and 8.14; provision can be made to depressurize or pressurize the permeable space and to drain water. Several field studies have shown that the radon concentration indoors can be effectively reduced below 4 pCi/L by adopting the new design approaches.

A housing project with 828 residential units in 251 buildings was built by incorporating the newer design concepts (Harris et al., 1991). Since the soil of the construction site was known to have a high radium content, an active soil depressurization system was built with assistance from the Radon Mitigation Branch of EPA's Air and Energy Research Laboratory to reduce the indoor radon levels to below 4 pCi/L. The construction of the housing

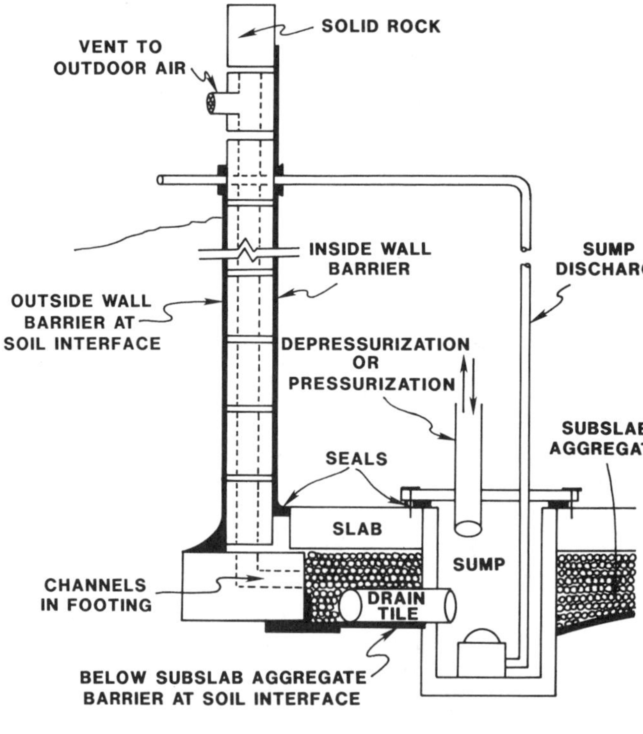

Figure 8.12 *Double-barrier construction for a basement with sump.* (Source: *Kunz, 1991.)*

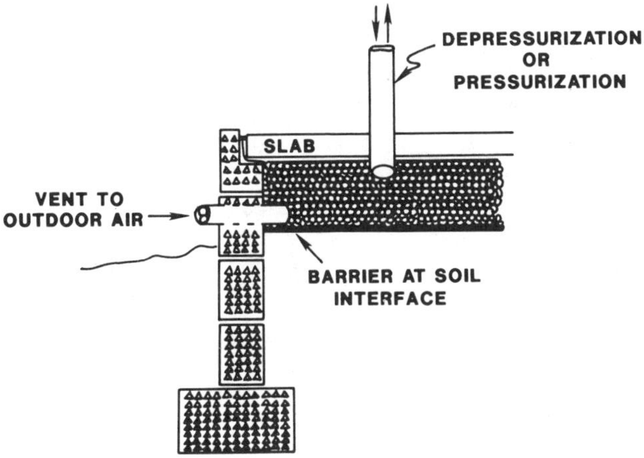

Figure 8.13 *Double-barrier system for slab-on-grade ventilation.* (Source: *Kunz, 1991.*)

units on expansive soils required that the lower floors be constructed of treated plywood over joists with a crawl space below. The mitigation technique involved direct suction of the airtight crawl space. The radon concentration in the crawl space increased to 100 pCi/L when the mitigation fan was off for several days, but decreased to virtually zero when it was on. Grisham (1991) evaluated two houses in which new radon-resistant residential construction methods were adopted. The new construction strategies included installing a piping network in the permeable subslab during construction, placing a radon barrier on the subslab aggregate, and installing a passive stack vent to induce differential pressures between the basement and the subslab fill material. The total costs for these measures were \$1615 and \$1992 for the two houses. The difference in the costs was due primarily to the use of different subslab fill materials. Radon concentrations in the two houses were below 0.7 pCi/L, while those in surrounding structures exceeded 20 pCi/L.

Indoor radon concentration levels can be controlled by modifying the HVAC system to pressurize the basement. When using combustion appliances, the movement of hot gases

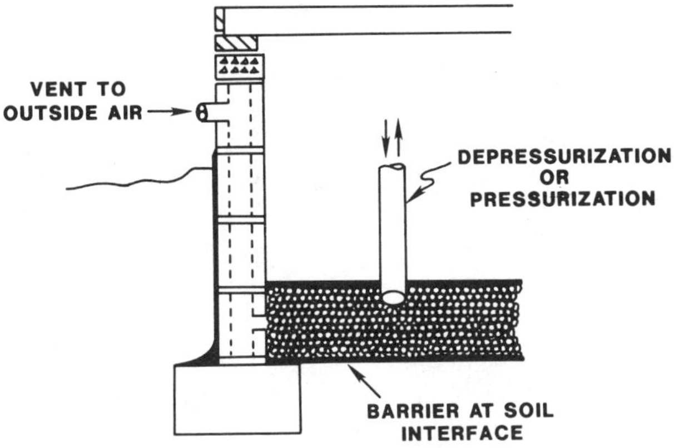

Figure 8.14 *Double-barrier system for crawl space ventilation.* (Source: *Kunz, 1991.*)

through vents and chimneys creates a draft or a negative pressure inside the house, which leads to radon-laden soil gas entering the structure. Brennan et al. (1991) installed an HVAC system in a newly constructed house that could slightly pressurize the basement while depressurizing the upper floors. When the air handler was on, there was a positive air pressure difference between the basement and the outdoor air of about 4 Pa. This method was found to be as effective as an active soil depressurization method in controlling indoor radon levels. During the cooling season, the average indoor radon concentration was 12 pCi/L in the basement and kitchen when the HVAC system was off, but the average radon concentration dropped to 1.4 pCi/L and 2 pCi/L in the basement and kitchen, respectively, when the HVAC system was on.

Source Control

Sealing of radon entry paths.

One of the most common radon control strategies, particularly in old structures, is to seal openings that permit radon entry. As noted earlier, radon can enter a building by convection due to the negative pressure differential between the structure and the outdoors. Therefore, it is important to identify openings or other radon entry pathways in the structure. The areas that should be checked for such openings include the following:

1. floor drains and sumps that are connected to the drainage system
2. openings around utility lines (gas and water service lines)
3. hollow concrete block walls
4. junctions between walls and floors and junctions between floor slab sections
5. cracks in building materials due to thermal expansion or shrinkage of the material themselves and in basement floors
6. exposed soil and rock surrounding the structure
7. unpaved crawl space

Most of these openings occur as a result of conventional construction practices. During construction of concrete basements, the concrete floor is generally poured after the surrounding walls have been erected. Although the concrete floor contacts the wall when it is first poured, as the concrete sets and dries, the floor shrinks and separates from the wall, leaving small gaps for radon entry. If the building support columns are made from hollow concrete blocks or hollow steel, they can provide major soil gas entry routes. Radon from soil gas slowly diffuses from the block into the interior or through the open block cavities at the top of the wall. A number of openings are also created in the structure during the installation of plumbing fixtures and service conduits. Once these openings are identified, a variety of sealing agents can be employed to close these entry routes.

Sealing agents.

Materials that are used to control the influx of radon gas through cracks and joints can be classified as caulking agents, paints, membranes, and cement-type materials. These sealants should be flexible and able to expand or shrink in response to climate changes to accommodate changes in the building materials. Although the foundation of a building is placed at a specific depth to minimize movement, a significant structural movement can still take place, depending on the soil characteristics and environmental factors. In the northern part of the United States, the annual summer-winter cycle changes the volume of the upper soil layer. Similarly, in regions where there is a great deal of clay soil, the volume change occurs because of rainfall variations. Inside the house the service conduits are subjected to thermal expansion and contraction cycles relative to the concrete wall.

Sealants should be resistant to moisture, since concrete slabs are in contact with soil and are usually damp. They are typically tested in the laboratory because of the difficulties associated with field tests. Bedrosian et al. (1974) tested four types of gel for use as a sealing material. The gels remained in the fluid state at about 50°C, but solidified at lower temperatures. They retained their flexibility even at 5°C. On the basis of laboratory tests, a petroleum gel was judged to be the best. A 5-cm thick gel prevented approximately 99% of the radon from diffusing from a test chamber during a 30-day test. A glass container with two leaky radium sources sealed by a 10-cm thick plain gel did not release radon to the environment after three years of use.

Eight potential caulking materials were laboratory tested by Fleischer (1992) for their ability to obstruct radon flow. These included two types of GE silicone Silpruf weatherproofing sealants, three varieties of GE white acrylic latex silicone caulk, a Bondex (polyurethane) radon gas barrier, and a Geocel radon shield. A 0.2-cm thick GE acrylic latex silicone caulk reduced the radon flow by 98% to 99.9%, compared to 75% to 90% for two commercial radon barriers.

Radon emanation from concrete blocks can be reduced by painting the blocks. The radon emanation rate from a stuccoed block coated heavily with an epoxy paint decreased by a factor of four (Auxier et al., 1974). In their study, the concentration of radon in the surrounding air was reduced from 113 to 28.6 pCi/L after painting. The effectiveness of a surface coating in reducing the radon emanation rate from a phosphate slag concrete block was investigated by Eichholz et al. (1980). A section of a wall was covered by various radon barriers, including a block filler, a polyethylene sheet, an epoxy paint, an epoxy paint on the filler, a latex paint, a latex paint on the filler, and wallpaper. The polyethylene sheet and the epoxy paint on a filler appeared to be the most effective barriers, giving 97% and 87% reductions, respectively. The epoxy materials are good sealants, but some are rather rigid and cannot withstand thermal fluctuations. Cracks have been observed to develop in epoxy sealants after two to three years of use.

Barriers used at the soil-concrete interface are frequently made from plastic sheets. A 10-day test at temperatures of 0.5°, 16°, and 40°C and at relative humidities between 30% and 50% showed no detectable loss of radon from a container made of polyamide foil.

If radon-sealant materials are to be effective, they must adhere strongly to the concrete block or other surfaces. Often concrete blocks are covered with dust or a thin layer of cement and sometimes with a paint. A damp surface can also prevent good bonding between the concrete and the sealant. Therefore, care must be taken in preparing the surface before applying a sealant. To ensure proper bonding, the surface must be thoroughly cleaned and dried and treated with a primer. The primer penetrates the concrete pores and forms a thin layer on the surface. When the sealant is applied over the primer, it adheres to the primer through a strong chemical bond. Similar precautions should be taken when applying sealants around openings such as drainpipes, service conduits, and wall-floor joints.

Subslab ventilation. The design of a subslab ventilation system is house specific and depends on the nature of the foundation. If the structure has a subfloor filled with aggregate, an exhaust pipe is inserted through the basement floor into the aggregate. A fan capable of creating a suction of 50- to 100-Pa is installed on the end of the pipe outside the house. This method is most effective if the basement walls are poured concrete rather than concrete blocks. The suction fan draws the soil gas and a small portion of house air through cracks in the basement floor and through openings at the wall-floor joints. If the suction is not adequate to prevent soil gas flow through the block walls, several suction ports are installed directly inside the wall. Multiple collection ports and an exhaust fan for each wall may be required.

The weeping tile system or underground drainpipes can be used effectively to vent soil gas from beneath the structure by connecting an exhaust fan to the piping network. Outdoor air is drawn through the soil adjacent to the walls, which reduces the radon concentration. This also creates a negative pressure in the basement by drawing air from the house through cracks in the floor and preventing radon entry. This method is extremely effective if drain tiles surround the entire house. However, an open water discharge pipe can reduce its effectiveness. If the weeping tiles are drained into a sump inside a building, the sump can be vented outside by placing an airtight cover over it and then exhausting the contents to the exterior of the house by a fan.

Subslab ventilation is an effective radon removal method for older structures, but the effectiveness of subslab ventilation is diminished by the existence of too many cracks in the basement wall, the presence of other gaps, and openings in the slab. Major openings in the basement and block walls must be closed, while small openings can be sealed by using one of the sealants described earlier. Larger openings can be closed by using polyurethane foam. A critical evaluation of different soil gas ventilation systems has been carried out by Henschel and Scott (1986) using the performance data on 18 concrete block basement homes. In most of these, radon levels were reduced by 84% to 99%. For houses that showed no significant decrease, it was found that some major openings were not properly sealed, or improper mitigation techniques were installed.

A significant percentage of the 40 houses with basements in Pennsylvania that installed the EPA-sponsored radon-mitigation system had radon concentrations higher than 4 pCi/L, even after four years (Henschel and Scott, 1991). The elevated levels were attributed to failure of the suction system, reentry of radon from the exhaust of the mitigation system into the house, and the release of radon from building materials and the water supply that were not considered in the design. Therefore, as with any other system, the periodic replacement of system components and regular maintenance must be an integral part of the total mitigation strategy. Despite these problems, subslab ventilation is still one of the most effective mitigation techniques, and is used successfully in other countries, particularly in Canada and Sweden.

Basement pressurization. Basement pressurization can significantly reduce the entry of radon into a building, but the method is effective only if the basement is relatively airtight. The reversal of the indoor-outdoor pressure differential prevents the flow of soil gas into the house. Turk et al. (1986) studied the effect of basement pressurization in the Spokane River Valley region. The overpressurization of the basement by 3 to 4 Pa appeared to be sufficient to reduce the radon concentration below 4 pCi/L. The results were encouraging, but problems such as first-floor depressurization, a subsequent increase in energy expense, and a change in human comfort level, due to an increase in the ventilation rate, were encountered. The normal opening and closing of doors and windows can easily disrupt the mitigation system.

Air Cleaning

Air cleaning as a means of reducing indoor concentrations of radon and radon progeny has received considerable attention in the last 10 years. These methods include increasing the natural ventilation or air exchange rates through the HVAC system, using mixing fans, electrostatic precipitators and negative ion generators, and using filtration or an activated carbon adsorber. Most of these methods remove both airborne particulates and attached radon progeny from indoor air, leaving a greater fraction of unattached radon in the air. Only increased ventilation and activated carbon beds can remove both radon gas and its daughter products. Unattached radon progenies are believed to deliver a greater alpha dose than attached progeny to lung tissue.

Electronic air cleaners. Electrostatic precipitators are capable of reducing indoor particle concentrations and the potential alpha energy concentration (PAEC) by a factor ranging from 2 to 20. Maher et al. (1987) measured the steady-state concentrations of Po-218, Pb-214, and Bi-214 in a 78-m^3 room before and after air treatment. Air was cleaned by using an electrostatic precipitator, a ceiling fan, a high-efficiency filter, a negative ion generator, a positive ion generator, and a combination of a negative ion generator and a positive ion generator with a ceiling fan. The combination of a positive ion generator and a ceiling fan provided the best result, with an estimated 87% reduction in the dose of PAEC to the bronchial tissue. Two electrostatic precipitators tested by Offermann et al. (1984) in an airtight chamber reduced the PAEC by a factor of 13 and the particle concentration from 40,000 particle/cm^3 to a range of 200 to 500 particles/cm^3. However, the field evaluation of one electronic air cleaner by Li and Hopke (1991) did not show a significant reduction in the radon level inside a single-family home. Although the air cleaner reduced the dose from radon progeny, the reduction was not enough to lower the concentration below the risk level of 4 pCi/L. A median dose reduction of only 17.4% was reported.

Another mechanism that has been explored to remove radon from indoor air is forced plate-out. The unattached charged progenies are forced toward walls or floors by fans to facilitate the plate-out procedure. However, the plate-out activity depends on the indoor concentrations of particles. If the particle concentration is high, radon progenies will tend to attach to airborne particulates first. Other factors that may influence deposition on the walls include the flow pattern near the walls (laminar or turbulent), wall roughness, and environmental parameters such as the temperature and relative humidity of the air. The plate-out activity was found to be greater on walls than on floors. The convective currents near the wall may have induced turbulence, resulting in a higher deposition rate.

Studies in environmental chambers showed a reduction of radon progeny in the range of 40% to 70%. Kinsara (1991) reported that a mixing fan can reduce the working level of radon progenies in a chamber by a factor of two to three. However, the effectiveness of using a mixing fan in homes or buildings has not been fully demonstrated. Several factors should be carefully considered: the speed of the fan, its operation during winter months when people tend to switch it off, the location of the fan in the ceiling or in a free standing position in a corner of the room, and the eventual particulate/radon resuspension from wall surfaces or floors.

Increased ventilation. A very simple, but effective, method for removing radon from the interior of a building is to increase the ventilation rate. Quantitative information on the effectiveness of building ventilation is limited, but the available data suggest that increasing the ventilation in a building can reduce the indoor radon concentration by a factor of three. The ventilation of crawl spaces requires no knowledge of the foundation structure, soil conditions, or identification of entry routes. Structures with crawl spaces often have high radon concentrations due to a combination of a large area of exposed soil and low ventilation rates. Direct introduction of fresh outside air dilutes the concentration of radon in the crawl space and slightly pressurizes the space, which reduces the emanation rate of radon from the soil. Mechanical ventilation of crawl spaces is highly effective in reducing radon entry into a building. The ventilation of the crawl space can be increased by installing a small fan that delivers unconditioned outside air into the building. The amount of air that can be handled is, however, limited by the capacity of the heating and cooling system. If only a small reduction in radon concentration is required, an air-to-air heat exchanger can be used without altering the pressure in the building. The air circulation strategy is effective only when the airborne particle concentration is low enough so that a significant fraction of the radon progeny will not attach to the particles. Reductions in radon concentrations ranging from a factor of 2 to 20 and a fourfold decrease in the concentration of radon progeny have been reported. When crawl space ventilation areas were increased (Scott and Findlay, 1983), the indoor radon concentration dropped to almost one-half of its original concentration. Ericson et al. (1984) observed that homes built on uranium-rich soil required no additional measures to maintain the indoor radon concentration at acceptable levels when the crawl-space was equipped with forced ventilation.

Adsorption. A review of the literature on radon adsorption is provided by Hines et al. (1991), and a discussion of adsorption processes and various design criteria are discussed in Chapter 10. However, the nature of the available data that are useful in the design of removal equipment are described briefly below.

During the 1960s the removal of radon by a carbon bed was considered to be a viable method for special applications, such as the removal of radon from the lunar sample laboratory and from uranium and lead mines. Lucas (1964) investigated the removal of radon from an air stream by adsorbing it on activated charcoal at temperatures ranging from 193 to 233 K, using an underground concrete chamber. The efficiency of the charcoal bed for radon removal was determined at a flow rate of 9 ft^3/min for a fixed radon concentration. To improve the efficiency of the charcoal bed, moisture and CO_2 were removed from the air by passing it through a molecular sieve bed prior to its introduction into the charcoal bed. The temperature of the bed was set at –70°C, but it varied with the flow rate of air. Lucas suggested that for a room with a volume of 7200 ft^3, a charcoal bed of 150 lb could be used for several months to maintain the radon concentration in the room below 10^{-2} pCi/L, provided that the radon concentration of the intake air was between 0.03 and 5 pCi/L. Later, Lucas (1967) designed a radon removal system for the NASA lunar sample laboratory using an activated charcoal bed that would be operated at –60°C. The system could process 20 ft^3/min of air and provide radon-free air (radon concentration of less than 10^{-4} pCi/L) for more than 24 hours before needing to be regenerated. The method was never tested in houses or other buildings. One obvious reason is that the low temperature of the adsorption bed cannot be maintained economically in residences. Anderson and Palmer (1975) designed a radon retention system that might be employed for residential applications. The system was capable of processing 0.1 m^3/min of air containing 3×10^9 pCi/m^3 of radon. The retention system contained two activated-charcoal beds, each consisting of 100 kg of charcoal packed in a column 31.5 cm in diameter and 145 cm long. While one bed was exposed to radon, the other bed was regenerated by allowing the radon to decay. Since the decay products were allowed to stay on the charcoal, a lead shield was used to reduce the gamma radiation to the atmosphere.

The adsorption of radon on activated charcoal depends on radon concentrations, air flow rates, and relative humidities. The influence of these parameters must be investigated thoroughly before designing a system. Boncanegra and Hopke (1988, 1989) conducted a series of experiments to evaluate the technical feasibility of carbon-based radon removal systems. The radon was adsorbed from air-radon mixtures on six types of activated charcoal. Radon rapidly desorbed from the charcoals at temperatures close to 90°C. A decrease of approximately 25% in the adsorption capacity was observed after two adsorption-desorption cycles. However, the cumulative effect on the degradation of the charcoals was not clear even after five cycles. The presence of isooctane, ethylene chloride, and formaldehyde (which are very common indoor air pollutants) in the radon gas mixtures had a detrimental effect on the adsorption capacity. By contrast, the presence of toluene increased the radon adsorption capacity. The number of theoretical stages required to remove radon from air was calculated using typical values for the dynamic distribution coefficient (5 m^3/kg). The time required to reach steady-state conditions and to reduce the radon concentration below the EPA recommended level was predicted by a two-box building model in which the test chamber was divided into several small zones, and each zone was divided into two sections. It was found that when the air exchange rate was low and the ratio of the radon concentration in the basement to that on the ground level of the house was high, a charcoal bed could not be used to decrease the radon concentration in the house.

Hassan et al. (1992a) also conducted extensive adsorption experiments using activated carbon, silica gel, and molecular sieve 13X packed in a glass column. Using 50 g of activated carbon in the column and an air flow rate of 2 L/min with a radon concentration

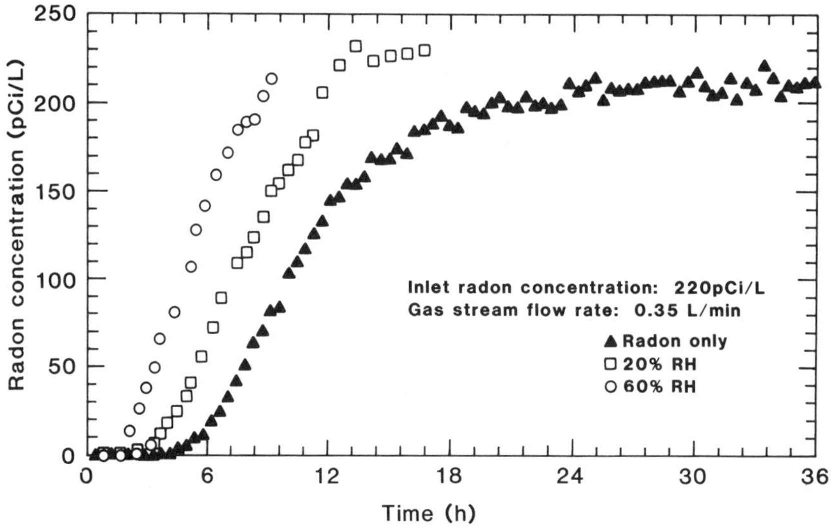

Figure 8.15 *Breakthrough curves for radon from a moist nitrogen stream on BPL activated carbon at 298 K.* (Source: *Hassan et al., 1992a.*)

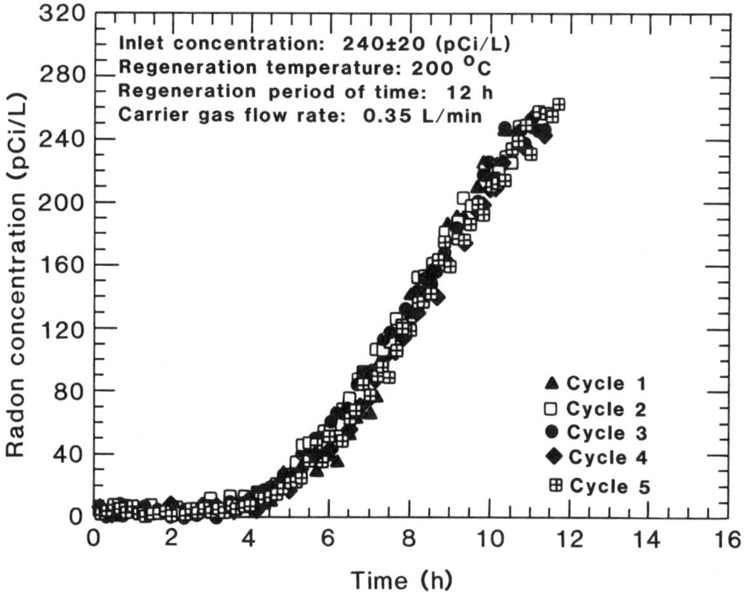

Figure 8.16 *Adsorption/regeneration cycles of radon on BPL activated carbon at 298 K.* (Source: *Hassan et al., 1992a.*)

of 220 ± 20 pCi/L, the bed adsorbed more than 99.9% of the radon for the first 4.5 hours at 25°C. The capacity decreased dramatically when moisture was present in the stream and radon broke through the column within 30 minutes (see Figure 8.15). It can be seen from Figure 8.16 that the adsorption capacity of carbon for radon did not change significantly after five regeneration cycles. It appears that the design of such a unit may range from a single packed column containing an activated carbon section downstream from a desiccant bed, to a complex system of multiple columns. Considering that desiccant humidity control devices are becoming more competitive with other humidity control systems, the integration of a carbon-based removal system may be economical.

For a carbon-based radon removal system to be economically feasible, it should have a high capacity for radon and minimum interference from moisture and other volatile organic pollutants. Since the bed must be regenerated for repeated use, the two-bed concept discussed by Anderson and Palmer (1975) could be quite attractive.

REFERENCES

ANDERSON, B. V., and PALMER, R. F., "Cleanup of Radon Inhalation Study Effluents," BNWL-SA-5541, Battelle Pacific Northwest Laboratory, Richland, Wash., 1975.

AUXIER J. A., SHINPAUGH, W. H., KERR, G. D., and CHRISTIAN, D. J., *Health Phys.*, **27**, 390 (1974).

BEDROSIAN P. H., SENSINTAFFAR, E. L., GELS, G. L., NORWOOD, D. L., and CULVER, A. A., *Health Phys.*, **27**, 387 (1974).

BEIR IV: *Health Effects of Radon and Other Internally Deposited Alpha-Emitters,* National Academy Press, Washington, D. C., 1988.

BONCANEGRA, R., and HOPKE, P. K., "The Feasibility of Using Activated Charcoal to Control Indoor Radon," in *Radon and Its Decay Products*, (ed. P. K. Hopke), *ACS Symp. Ser.*, No. 331, 560 (1987).

BONCANEGRA, R., and HOPKE, P. K., *The Science of the Total Environment*, **76**, 193 (1988).

BONCANEGRA, R., and HOPKE, P. K., *J. Air Pollut. Control Assoc.*, **39**, 305 (1989).

BRENNAN, T., CLARKIN, M. DYESS, T. M., and BRODHEAD, W., "Preliminary Results of HVAC System Modification to Control Indoor Radon Concentrations," in The 1991 International Symposium on Radon and Radon Reduction Technology, 2, VIIIP-2, Philadelphia, Pa., 1991.

CLEMENTS, W. E., and WILKENING, M. H., *J. Geophys. Res.*, **79**, 5025 (1974).

COHEN, B. L., *Health Phys.*, **50**, 828 (1986).

COHEN, B. L., and NASON, R., *Health Phys.*, **50**, 457 (1986a).

COHEN, B. L., and NASON, R., *Health Phys.*, **51**, 135 (1986b).

COHEN, B. L., "Studies of Our Diffusion Barrier Charcoal Adsorption Detectors," paper presented at the Technical Exchange Meeting on Passive Radon Monitoring, Department of Energy CONF-8709187, Technical Measurements Center, Grand Junction, Colo., 1987.

COHEN, B. L., and SHAH, R. S., *Health Phys.*, **60**, 243 (1991).

COUNTESS, R. J., *Health Phys.*, **31**, 455 (1976).

CULOT, M. V. J., OLSON, H. G., and SCHIEGER, K. J., *Health Phys.*, **34**, 498 (1973).

EICHHOLZ, G. G., MATHENY, M. D., and KAHN, B., *Health Phys.*, **39**, 301 (1980).

EPA (Environmental Protection Agency), "Radon Reduction Techniques for Detached Houses," Technical Guidance, 2nd ed., Report No. EPA 625/5-87/019, NTIS Document No. PB88-184908, 1988.

ERICSON, S. O., SCHMIED, H., and CLAVENSJO, B., *Rad. Prot. Dosim.*, **7**, 233 (1984).

FLEISCHER, R. L., *Health Phys.*, **62**, 91 (1992).

GAO (U. S. General Accounting Office), Report GAO/RCED-90-25 Air Pollution, Washington, D. C., 1989.

GEORGE, A. C., *Health Phys.*, **46**, 867 (1984).

GEORGE, A. C., "Recent Studies with Activated Carbon for Measuring Radon," paper presented at the Technical Exchange Meeting on Passive Radon Monitoring, U. S. Department of Energy CONF-8709187, Technical Measurements Center, Grand Junction, Colo., 1987.

GESELL, T. F., and PRICHARD, H. M., "The Contribution of Radon in Tap Water to Indoor Radon Concentrations," in *Natural Radiation Environment III*, (eds. T. F. Gesell and W. M. Lowder), U. S. Department of Energy CONF-780422, 1347, 1980.

GRISHAM, C. M., "Radon Prevention in Residential New Construction: Passive Designs That Work," in The 1991 International Symposium on Radon and Radon Reduction Technology, **2**, VIIIP-1, Philadelphia, Pa., 1991.

HARRIS, D., CRAIG, A. B., and HAYNES, J., "Building Radon Mitigation into Inaccessible Crawlspace New Residential Construction," in The 1991 International Symposium on Radon and Radon Reduction Technology, **2**, VIII-6, Philadelphia, Pa., 1991.

HASSAN, N. M., GHOSH, T. K., HINES, A. L., and LOYALKA, S. K., *Carbon,* **29**(4), 681(1991a).

HASSAN, N. M., GHOSH, T. K., HINES, A. L., LOYALKA, S. K., and KETRING, A., *Ind. Eng. Chem. Res.*, **30**(9), 2205 (1991b).

HASSAN, N. M., GHOSH, T. K., HINES, A. L., and LOYALKA, S. K., "Radon and Water Vapor Coadsorption on Solid Adsorbents," in The 1991 International Symposium on Radon and Radon Reduction Technology, Philadelphia, Pa., **3,** IIIP-4, 1991c.

HASSAN, N. M., GHOSH, T. K., HINES, A. L., LOYALKA, S. K., and NOVOSEL, D., *ASHRAE Trans.*, **98**(Part 1), 699 (1992a).

HASSAN, N. M., GHOSH, T. K., HINES, A. L., and LOYALKA, S. K., *Sep. Sci. Technol.,* **27**(14), 1955 (1992b).

HAWTHORNE, A. R., GAMMAGE, R. B., DUDNEY, C. S., HINGERTY, B. E., SCHURESKO, D. D., PARZYCK, D. C., WOMACK, D. R., MORRIS, S. A., WESTLEY, R. R., WHITE, D. B., and SCHRIMSHER, J. M., "An Indoor Air Quality Study of Forty East Tennessee Homes," Report No. ORNL-5965, Oak Ridge National Laboratory, Oak Ridge, Tenn., 1984.

HENSCHEL, D. B., and SCOTT, A. G., "The EPA Program to Demonstrate Mitigation Measures for Indoor Radon: Initial Results," in *Indoor Radon*, Air Pollution Control Association International Specialty Conference Proceedings, Pittsburg, Pa., 1986.

HENSCHEL, D. B., and SCOTT, A G., "Causes of Elevated Post-Mitigation Radon Concentrations in Basement Houses Having Extremely High Pre-Mitigation Levels," in The 1991 International Symposium on Radon and Radon Reduction Technology, **2**, IV-1, Philadelphia, Pa., 1991.

HESS, C. T., MICHEL, J., HORTON, T. R., PRICHARD, H. M., and CONIGLIO, W. A., *Health Phys.,* **48**(5), 553 (1985).

HESS, C. T., WEIFFENBACH, C. V., and NORTON, S. A., *Environ. Int.*, **8**, 59 (1982).

HINES, A. L., GHOSH, T. K., LOYALKA, S. K., and WARDER, R. C., Jr., "Investigation of Co-

Sorption of Gases and Vapors as a Means to Enhance Indoor Air Quality," Report No. GRI-90/01934, Gas Research Institute, Ill., NTIS Document No. PB91-178806/GAR, 1991.

IAR (Indoor Air Review), "EPA Develops Preliminary Radon Potential Map of the U. S.," 27 (February, 1992).

JONASSEN, N., and MCLAUGHLIN, J. P., "The Effect of Pressure Drops on Radon Exhalation from Walls," in *Natural Radiation Environment III*, (eds. T. F. Gesell and W. M. Lowder), U. S. Department of Energy CONF-780422, 1225, 1980.

JONASSEN, N., and MCLAUGHLIN, J. R., *Environ. Int.*, **8**, 71 (1982).

KINSARA, A. A., "Deposition of the Radon Daughter Po-218 in a Lung Bifurcation," Ph. D. Dissertation, Nuclear Engineering, University of Missouri-Columbia (1991).

KUNZ, C., "Radon Reduction in New Construction: Double-Barrier Approach," in The 1991 International Symposium on Radon and Radon Reduction Technology, **2**, VIII-3, Philadelphia, Pa., 1991.

LEE, D. D., and SEXTRO, R. G., "The Response of Charcoal Canister Detectors to Time-Variant ^{222}Rn Concentrations," paper presented at the Technical Exchange Meeting on Passive Radon Monitoring, Department of Energy CONF-8709187, Technical Measurements Center, Grand Junction, Colo., 1987.

LI, C. S., and HOPKE, P. K., *Health Phys.*, **61**, 785 (1991).

LUCAS, H. F., Jr., "The Low-Level Gamma Counting Room: Radon Removal and Control," Report No. ANL-6938, Argonne National Laboratory, Argonne, Ill., 1964.

LUCAS, H. F., Jr., "A Radon Removal System for the NASA Lunar Sample Laboratory: Design and Discussion," Report No. ANL-7360, Argonne National Laboratory, Argonne, Ill., 1967.

MAHER, E. F., RUDNICK, S. N., and MOELLER, D. W., *Health Phys.*, **53**, 351(1987).

MEGUMI, K., and MAMURO, T., *J. Geophysical Res.*, **77**, 3052 (1972).

MICHAELS, L. D., VINER, A. S., and BRENNAN, T., "A Comparison of Laboratory and Field Measurements of Radon," paper presented at the Technical Exchange Meeting on Passive Radon Monitoring, U. S. Department of Energy CONF-8709187, Technical Measurements Center, Grand Junction, Colo., 1987.

NAZAROFF, W. W., and DOYLE, S. M., *Health Phys.*, **48**, 265 (1985).

NAZAROFF, W. W., MOED, A. B., and SEXTRO, R. G., "Soil as a Source of Indoor Radon: Generation, Migration, and Entry," *Radon and its Decay Products in Indoor Air*, (eds. W. W. Nazaroff and A. V. Nero, Jr.), Wiley, New York, 1988.

NERO, A. V., SCHWEHR, M. B., NAZAROFF, W. W., and REVZAN, K. L., Science, 234, 992 (1986).

NERO, A. V., JR, "Radon and Its Decay Products in Indoor Air: An Overview," *Radon and its Decay Products in Indoor Air*, (eds. W. W. Nazaroff and A. V. Nero, Jr.), Wiley, New York, 1988.

OFFERMANN, F. J., SEXTRO, R. G., FISK, W. J., NAZAROFF, W. W., NERO, A. V., REVZAN, K. L., and YATER, J., "Control of Particles and Radon Progeny with Portable Radon Cleaners," LBL-16659, Lawrence Berkeley Laboratory, Berkeley, Calif., 1984.

PARTRIDGE, J. E., HORTON, T. R., and SENSINTAFFAR, E. L., "A Study of Radon-222 Released from Water During Typical Household Activities," Eastern Environmental Radiation Facility, Montgomery, Ala., NTIS Document No. PB-295881, 1979.

PEARSON, M. D., "Evaluation of the Performance Characteristics of Radon and Radon-Daughter Concentration Measurement Devices Under Controlled Environmental Conditions," Report UNC/GJ–44(TMC), UNC Geotech, Grand Junction, Colo., 1989.

POJER, P. M., PEGGIE, J. R., O'BRIEN, R. S., SOLOMON, S. B., and WISE, K. N., *Health Phys.*, **58**(1), 13 (1990).

PRICHARD, H. M., and MARÏEN, K., *Analyt. Chemistry*, **55**, 157 (1983).

PRICHARD, H. M., and MARÏEN, K., *Health Phys.*, **48**(6), 797 (1985).

REN, T., and LIN, L., *Radiat. Prod. Dosim.*, **19**, 121 (1987).

RONCA-BATTISTA, M., and GRAY, D., "The Influence of Changing Exposure Conditions on Measurements of Radon Concentrations with the Charcoal Adsorption Technique," paper presented at the Technical Exchange Meeting on Passive Radon Monitoring, U. S. Department of Energy CONF-8709187, Technical Measurements Center, Grand Junction, Colo., 1987.

RUTHERFORD, E., *Nature*, **74**, 634 (1906).

SCARPITTA, S. C., and HARLEY, N. H., *Health Phys.*, **59**, 393 (1990).

SCOTT, A. G., and FINDLAY, W. O., "Demonstration of Remedial Techniques Against Radon in Houses on Florida Phosphate Lands," EPA 520/5-83-009, Environmental Protection Agency, Eastern Environmental Radiation Facility, Montgomery, Ala., 1983.

SEXTRO, R. G., and LEE, D. D., "The Performance of Charcoal Based Radon Under Time-Varying Radon Conditions: Experimental and Theoretical Results," Proceedings of the 1988 Symposium on Radon and Radon Reduction Technology, **2**, 81, 1989.

SHIMO, M., IKEBE, Y., MAEDA, J., KAMIMURA, K., HAYASHI, K., and ISHIGURO, A., *J. Atom. Energy. Soc. Jpn.*, **25**(7), 562 (1983).

SHLEIEN, B., *Am. Ind. Hyg. Assoc. J.*, **24**, 180 (1963).

STRANDEN, E., KOLSTAD, A. K., and LIND, B., *Health Phys.*, **47,** 480 (1984).

TERZAGHI, K., *Soil Mechanics in Engineering Practice*, 2nd ed., Wiley, New York, 1967.

TUMA, J. J., *Engineering Soil Mechanics in Engineering Practice*, Prentice-Hall, Englewood Cliffs, N. J., 1973.

THOMAS, J. W., "Evaluation of Activated Carbon Canisters for Radon Protection in Uranium Mines, Report No. HASL-280, Health and Safety Laboratory, New York, 1974.

TURK, B. H., PRILL, R. J., FISK, W. J., GRIMSRUD, D. T., MOED, B. A., and SEXTRO, R. G., "Radon and Remedial Action in Spokane River Valley Residences," LBL-21399, Lawerence Berkeley Laboratory, Berkeley, Calif., 1986.

UN (United Nations), "Sources and Effects of Ionizing Radiation," A report to the General Assembly, Scientific Committee on the Effects of Atomic Radiation, 1982.

WILSON, O. J., *Nucl. Instr. Meth. Phys. Res.*, **A275**, 163 (1989).

WOLLENBERG, H. A., "Naturally Occurring Radioelements and Terrestrial Gamma-Ray Exposure Rates: An Assessment Based on Recent Geochemical Data," Report LBL-18714, Lawerence Berkeley Laboratory, Berkeley, Calif., 1984.

9

Absorption Applications

9.1 INTRODUCTION

Numerous investigations have shown that indoor air contains a variety of different types of pollutants, including inorganic gases, organic vapors, aerosols, and particulates. Because of the wide range of types, no single mitigation process could be expected to be effective in reducing or eliminating all pollutants from indoor air. One of the more important mitigation methods is absorption, which is used to remove moisture and certain accompanying pollutants from indoor air, including volatile organic carbons, particulates, bacteria, and fungi. Liquid solutions that have the property of absorbing moisture from air, are referred to as liquid desiccants and include aqueous solutions of lithium chloride [40% by weight (w/w)], calcium chloride (25% w/w), lithium bromide (40% w/w), and triethylene glycol (98.5% w/w). Among these solutions, lithium chloride and triethylene glycol are most frequently employed in desiccant systems. More recently, however, mixtures of lithium chloride and lithium bromide have been shown to perform better than either lithium chloride or lithium bromide alone (Grover et al., 1989).

Absorption involves the transfer of one or more materials from the gas phase to a liquid solvent. The material that is transferred from the gas to the liquid phase is referred to as the solute. Although absorption typically is a physical phenomenon and often involves no change in the chemical species present in a system, absorption processes have been developed in which chemical reactions do occur. Absorption may involve the use of a given portion of solvent more than one time. More frequently, however, the condensible vapor (solute) is separated from the solvent in a stripping process, and the solvent is recirculated to the absorber for further use.

Absorption (also referred to as *scrubbing*) is used throughout industry in processes in which various gaseous contaminants are removed from effluent streams by absorbing them in a liquid solution. Several devices have been developed to provide good contact between the gas stream and the absorbing liquid, and many of these have been employed successfully in a number of industrial applications. These devices include venturi, cyclone, packed bed, and sieve plate scrubbers, but applications of these systems in the treatment of air in indoor environments (except in submarines) have not been reported in the open literature. In this chapter we will discuss the application and design of absorption systems because of their enormous potential.

9.2 FUNDAMENTALS

Three approaches have been employed to develop the equations used to predict the performance of absorbers and absorption equipment:

1. The first approach makes use of mass transfer coefficients and depends on the molecular and eddy diffusivities of the solute for the equipment in which the operation is being carried out.
2. The second technique is a graphical solution method generally attributed to Lewis (1927).
3. The third method is described as the absorption factor or overall approach and is generally attributed to Kremser (1930).

The graphical solution is simple, direct, and easy to use for one or two components. It has the advantage of giving the thinking user an explicit graphical presentation of the interrelationships of the variables and parameters in an absorption process. It has the disadvantage, however, that it becomes very tedious when several solutes are present that must be considered. The absorption factor approach can be utilized for either hand or computer calculations. The principles involved are the same, but the equations and solution techniques differ. Because of the nature of the systems of interest in this work, only the approach utilizing mass transfer coefficients will be considered here. Furthermore, this discussion will be limited to the transfer of a single component. The reader desiring more detailed design information is referred to the work of Holland (1963) and Hines and Maddox (1985).

Material Balances for Countercurrent Operations

Most absorption operations are carried out in countercurrent flow processes, in which the gas phase is introduced in the bottom of the absorber and the liquid solvent is introduced in the top. The contact tower may be equipped with either trays or filled with an inert packing. From the standpoint of the mathematical analysis, the two are equivalent.

Figure 9.1 shows schematically a countercurrent absorption process. The overall material balance for this is

$$L_b + V_a = L_a + V_b \tag{9.1}$$

where
V = moles of gas phase per unit time
L = moles of liquid phase per unit time
a,b = top and bottom of the contactor, respectively

If a single solute, A, is being transferred from the gas phase to the liquid phase, the component material balance is

$$L_b\, x_{A,b} + V_a\, y_{A,a} = L_a\, x_{A,a} + V_b\, y_{A,b} \tag{9.2}$$

where
y_A = mole fraction of A in the gas phase
x_A = mole fraction of A in the liquid phase

In many instances, more convenient expressions can be derived for evaluating the absorption process if a solute–free basis is used for compositions rather than mole fractions. The solute–free concentration in the liquid phase is

$$\overline{X}_A = \frac{x_A}{1-x_A} = \frac{\text{mole fraction of } A \text{ in the liquid}}{\text{mole fraction of non–}A \text{ components in the liquid}} \tag{9.3}$$

If the carrier gas is considered to be completely insoluble in the solvent and the solvent is considered to be completely nonvolatile, the carrier gas and solvent rates remain constant throughout the absorber. Using $\overline{L}$ to describe the flow rate of the nonvolatile solvent and $\overline{V}$ to describe the carrier gas flow rate, the material balance for solute A becomes

$$\overline{L}\,\overline{X}_{A,b} + \overline{V}\,\overline{Y}_{A,a} = \overline{L}\,\overline{X}_{A,a} + \overline{V}\,\overline{Y}_{A,b} \tag{9.4}$$

or

$$\overline{Y}_{A,a} = \frac{\overline{L}}{\overline{V}}\overline{X}_{A,a} + \left(\overline{Y}_{A,b} - \frac{\overline{L}}{\overline{V}}\overline{X}_{A,b}\right) \tag{9.5}$$

Obviously, Eqs. (9.4) and (9.5) are not restricted in application to the two column terminals but can be used to relate compositions and flow rates between any two points in the countercurrent absorber. When plotted on $\overline{X}_A - \overline{Y}_A$ coordinates, Eq. (9.5) gives a straight line with a slope of $\overline{L}/\overline{V}$ and an intercept of $(\overline{Y}_{A,b} - \overline{L}\,\overline{X}_{A,b}/\overline{V})$.

Gas solubilities in liquids are frequently given in terms of the Henry's law constants.

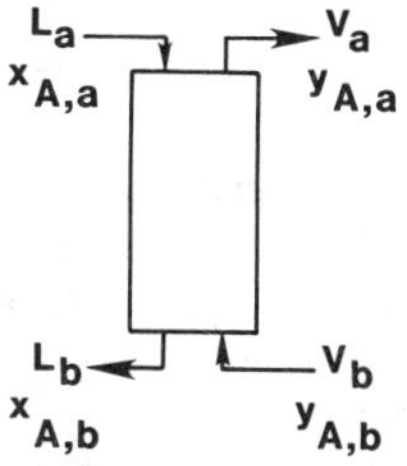

Figure 9.1 *Countercurrent absorption process.*

Henry's law states that the quantity of gas that dissolves in a given quantity of solvent is directly proportional to its partial pressure over the solution. This is expressed by

$$\overline{P}_A = mx_A \tag{9.6}$$

where $\overline{P}_A$ is the partial pressure of A and x_A is its mole fraction. The Henry's law constant, m, has units of pressure per mole fraction. When the solubility is expressed in volumes of gas per volume of solution, the units will be pressure units per [(gas volume)/(liquid volume)]. Values of m for a number of gases in water are given in Table 9.1. In many cases, the equilibrium between a gas and liquid will be expressed in terms of an equilibrium constant or the vapor–liquid equilibrium ratio (K_A) as

$$y_A = K_A x_A \tag{9.7}$$

If m in Eq. (9.6) or K_A in Eq. (9.7) may be assumed constant throughout the absorber, they can be plotted as straight lines on $x_A - y_A$ coordinates. Example 9.1 illustrates the use of the techniques that are necessary for making the design calculations described in subsequent examples. In the following example, water is used as the absorbing liquid. For the simultaneous removal of moisture (dehumidification process) and indoor pollutants, liquid desiccants would be used and are discussed later in this chapter.

TABLE 9.1 Henry's Law Constants for Several Gases in Water as a Function of Temperature (mmHg/mol fraction)

T °C	Air $m \times 10^{-7}$	SO_2 $m \times 10^{-4}$	H_2S $m \times 10^{-4}$	NO $m \times 10^{-6}$	CO $m \times 10^{-7}$	CO_2 $m \times 10^{-6}$	CH_4 $m \times 10^{-6}$	C_2H_4 $m \times 10^{-6}$
20	5.044	—	36.66	20.055	4.070	1.079	28.531	7.74
25	5.468	—	41.38	21.806	4.401	1.243	31.36	8.67
30	5.858	3.640	46.29	23.511	4.711	1.414	34.08	9.62
35	6.249	4.259	51.40	25.170	5.007	1.588	36.95	—
40	6.611	4.954	56.59	26.750	5.285	1.771	39.46	—
45	6.916	5.716	61.83	28.277	5.540	1.954	41.83	—
50	7.188	6.53	67.19	29.638	5.784	2.153	43.85	—

Source: Perry, Chilton, and Kirkpatrick, 1963.

Example 9.1

Solute A is to be removed from an air stream by absorbing it in a water–liquid desiccant solution. The air mixture entering the absorber flows at a rate 100 kgmol/h with the concentration $y_A = 0.03$. The mole fraction of A in the air stream leaving the absorber is $y_A = 0.001$. The solvent enters the absorber at a rate of 300 kgmol/h with a concentration of $x_A = 0.0001$. The equilibrium relationship between the concentration of A in the air mixture and water solution is $y_A = 2.5\,x_A$. The carrier stream, air, may be assumed to be insoluble in the water solvent, and the solvent is nonvolatile. Construct the $x - y$ plots for the equilibrium curve and operating line using both mole fraction and solute–free coordinates.

Solution: We first calculate the flow rates of the solvent and air streams without the presence of the solute.

For the solvent

$$\bar{L} = L(1 - x_A) = 300\,(1 - 0.0001)$$
$$= 299.97 \text{ kgmol/h}$$

For air

$$\bar{V} = V(1 - y_A) = 100\,(1 - 0.03)$$
$$= 97.0 \text{ kgmol/h}$$

We must now determine the concentration of A in the solvent stream leaving the absorber. The quantity of A leaving with the air is

$$y_{A,a} = 0.001 = \frac{\text{moles } A \text{ in } V_a}{\text{moles } A \text{ in } V_a + 97}$$

$$\text{moles } A \text{ in } V_a = 0.097097$$

$$\text{moles } A \text{ in } L_b = (100 \times 0.03) - 0.097097 + 300\,(0.0001)$$
$$= 2.9329$$

$$x_{A,b} = \frac{2.9329}{299.97 + 2.9329} = 0.00968$$

and

$$\bar{X}_{A,a} = \frac{0.0001}{1 - 0.0001} = 0.00010$$

$$\bar{X}_{A,b} = \frac{0.00968}{1 - 0.00968} = 0.00978$$

$$\bar{Y}_{A,a} = \frac{0.001}{1 - 0.001} = 0.001001$$

$$\bar{Y}_{A,b} = \frac{0.03}{1 - 0.03} = 0.0309$$

To calculate points on the equilibrium curve, we assume arbitrary values for x_A and calculate equilibrium compositions for y_A using the equilibrium relationship.

x_A	y_A	$\bar{X}_A = \dfrac{x_A}{1 - x_A}$	$\bar{Y}_A = \dfrac{y_A}{1 - y_A}$
0.0001	0.00025	0.0001	0.0002501
0.0004	0.001	0.0004	0.001001
0.001	0.0025	0.0010	0.002506
0.003	0.0075	0.0030	0.007557
0.005	0.0125	0.0050	0.012658
0.006	0.015	0.0060	0.015228
0.008	0.020	0.0081	0.020401

Points along the operating line are calculated by using Eq. (9.5) and assuming arbitrary values for $x_{A,a}$. Thus

$$\overline{Y}_{A,a} = \frac{299.97}{97.0} \frac{x_{A,a}}{1 - x_{A,a}} + \left(0.0309 - \frac{299.97 \times 0.00978}{97.0}\right)$$

$$= 3.092 \frac{x_{A,a}}{1 - x_{A,a}} + 0.000664$$

Values for the operating line are given below.

x_A	y_A	$\overline{X}_A = \frac{x_A}{1 - x_A}$	$\overline{Y}_A = \frac{y_A}{1 - y_A}$
0.0001	0.0010	0.0001	0.001001
0.0004	0.0019	0.0004	0.001924
0.0010	0.0038	0.0010	0.00378
0.0030	0.0099	0.0030	0.00996
0.0050	0.0159	0.0050	0.01615
0.0060	0.0189	0.0060	0.01924
0.00968	0.0300	0.00978	0.03090

Figure 9.2 shows the equilibrium and operating lines plotted on mole fraction coordinates. The operating line is curved very slightly, and the equilibrium line is straight. Normally, the compositions and/or temperatures will change from point to point in the absorber, which will bring about corresponding changes in the equilibrium constant (Henry's law constant) and result in a curved equilibrium line. Figure 9.3 shows the equilibrium and operating lines using solute–free coordinates. In this case, the equilibrium line is also curved. Since the equilibrium line is going to be curved anyway, there is an advantage in using solute–free coordinates because the operating line will always be straight. It is important to note, however, that for dilute solutions and low solute concentrations, the equilibrium and operating curves are nearly straight on a mole fraction basis and can be used when making calculations without introducing serious error. This is demonstrated by the previous example and is the case for most indoor air pollutants.

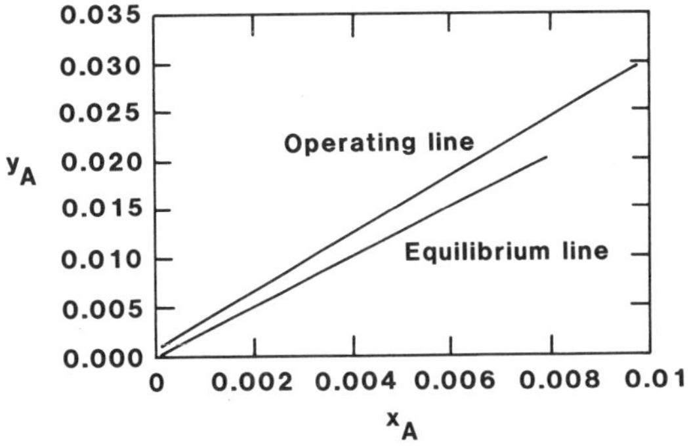

Figure 9.2 *Equilibrium and operating lines on mole fraction coordinates.*

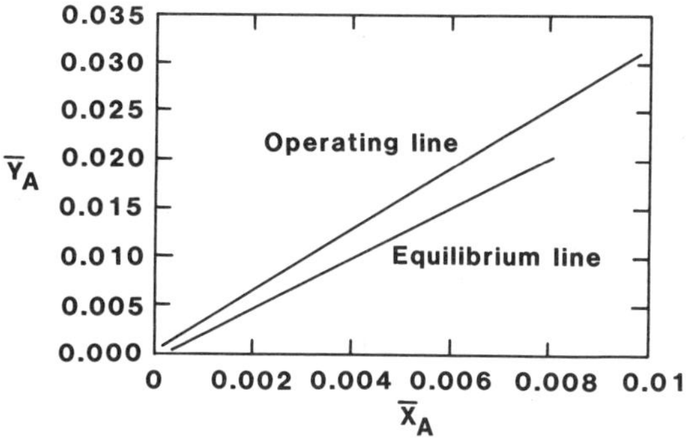

Figure 9.3 *Equilibrium and operating lines on solute free coordinates.*

Minimum Solvent Rates

A minimum solvent rate exists when the driving force for mass transfer from the vapor to the liquid phase becomes zero. This is indicated on the graph by the intersection of the operating and equilibrium lines. The various minimum rates are illustrated in Figures 9.4 through 9.6. Figures 9.4 and 9.5 have straight equilibrium and operating lines for simplicity. In Figure 9.4, the intersection of the equilibrium and operating lines occurs at the bottom of the absorber. This condition defines the minimum solvent rate needed to recover a specified quantity of solute.

In Figure 9.5, the intersection occurs at the top of the absorber. This condition represents the solvent rate required to remove the maximum possible amount of solute. For the case illustrated in Figure 9.5, in which a solvent denuded of solute enters the absorber, this solvent rate would be sufficient to remove all the solute from the carrier gas. If the equilibrium line is curved, the same two minima in solvent rates exist. The minimum solvent-to-vapor ratio for the case shown in Figure 9.4 can be calculated from the following expression:

$$\left(\frac{\overline{L}}{\overline{V}}\right)_{\min} = \frac{\overline{Y}_b - \overline{Y}_a}{\overline{X}_b - \overline{X}_a} \tag{9.8}$$

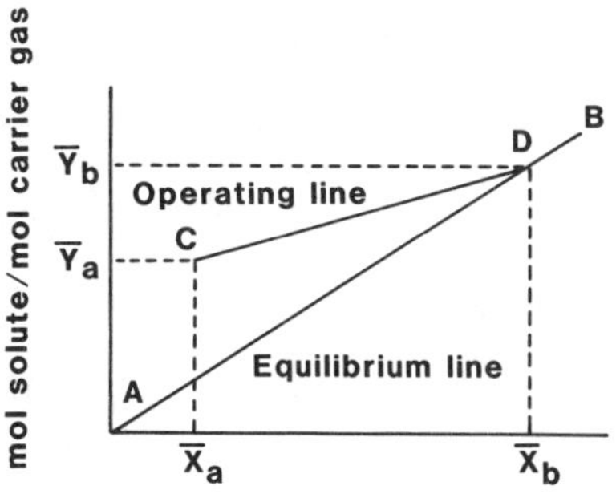

Figure 9.4 *Graphical construction for determination of minimum solvent rate to recover specified amount of solute.*

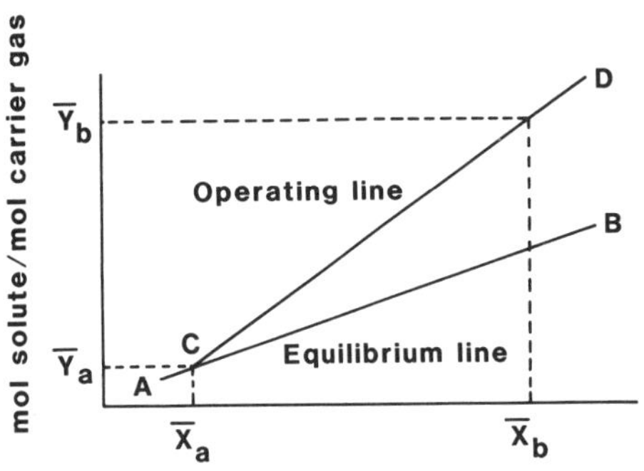

Figure 9.5 *Graphical determination of minimum solvent rate to recover all the solute.*

Figure 9.6 shows a curved equilibrium line that could become tangent to the operating line at a sufficiently low solvent rate. Since this is a condition that cannot be represented by straight lines, the slope of the tangent must be determined from the graph. The minimum liquid–to–vapor ratio for this case can be determined from Eq. (9.8) by replacing $\overline{Y}_b$ and $\overline{X}_b$ with $\overline{Y}_c$ and $\overline{X}_c$, respectively.

9.3 COLUMN DESIGN

A typical absorption tower is shown in Figure 9.7. The top of the tower contains a liquid distributor to distribute the liquid evenly over the packing. A demister, which may consist of extra packing or several inches of wire mesh above the liquid inlet, is installed to remove entrained liquid from the exiting gas stream. Quite often additional liquid distributors are installed in a column to prevent channeling with the accompanying decrease in packing effectiveness. Redistributors are also used at points in the column where either additional feed streams are introduced or side streams are withdrawn.

The design of a packed absorption tower requires a determination of the tower diameter and the height of packing necessary to make a desired separation. The tower diameter can be determined by knowing the gas and liquid flow rates along with the physical properties of the solute and the solvent, and information about the type of packing to be used in the column. Determining the height of packing in the column requires knowing the mini-

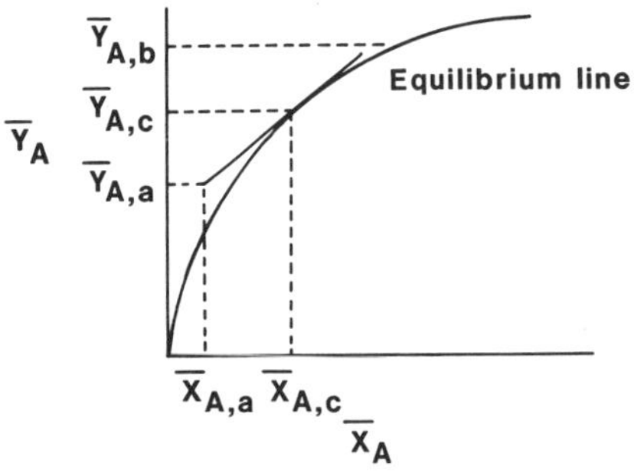

Figure 9.6 *Minimum L/V for equilibrium curve concaved away from operating line.*

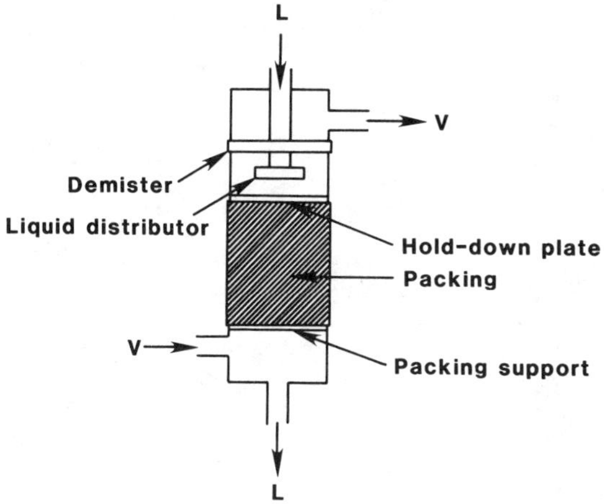

Figure 9.7 *Packed tower.*

mum solvent flow rate, which depends on the solubility of the solute in the solvent and the rate of mass transfer, which requires information about the mass transfer coefficient. In this section, we will first calculate the column diameter and complete the design by calculating the column height.

Tower Packing

A wide variety of packing types are presently in use; several of these are shown in Figure 9.8. Although Raschig rings and Berl saddles were the most popular packings for many years, these have been often replaced by higher-capacity and more efficient packings, such as Pall rings, Intalox and Super Intalox saddles, Flexipak, and more recently, structured packing.

The type of packing selected for a process depends on several factors. Desirable properties of the packing are as follows:

1. Large void volume to decrease pressure drop
2. Chemically inert to the fluids being processed
3. Large surface area per unit volume of packing
4. Lightweight but strong
5. Good distribution of fluids
6. Good wettability

Tower packings are usually available in a variety of materials, including ceramic, metal, plastic, and carbon. In addition to the desirable properties of a packing, another limitation is that its size must be no greater than one-eighth of the tower diameter. If the size of the packing for a particular tower is too large, a decrease in operating performance will result because of channeling along the column wall and through the packing.

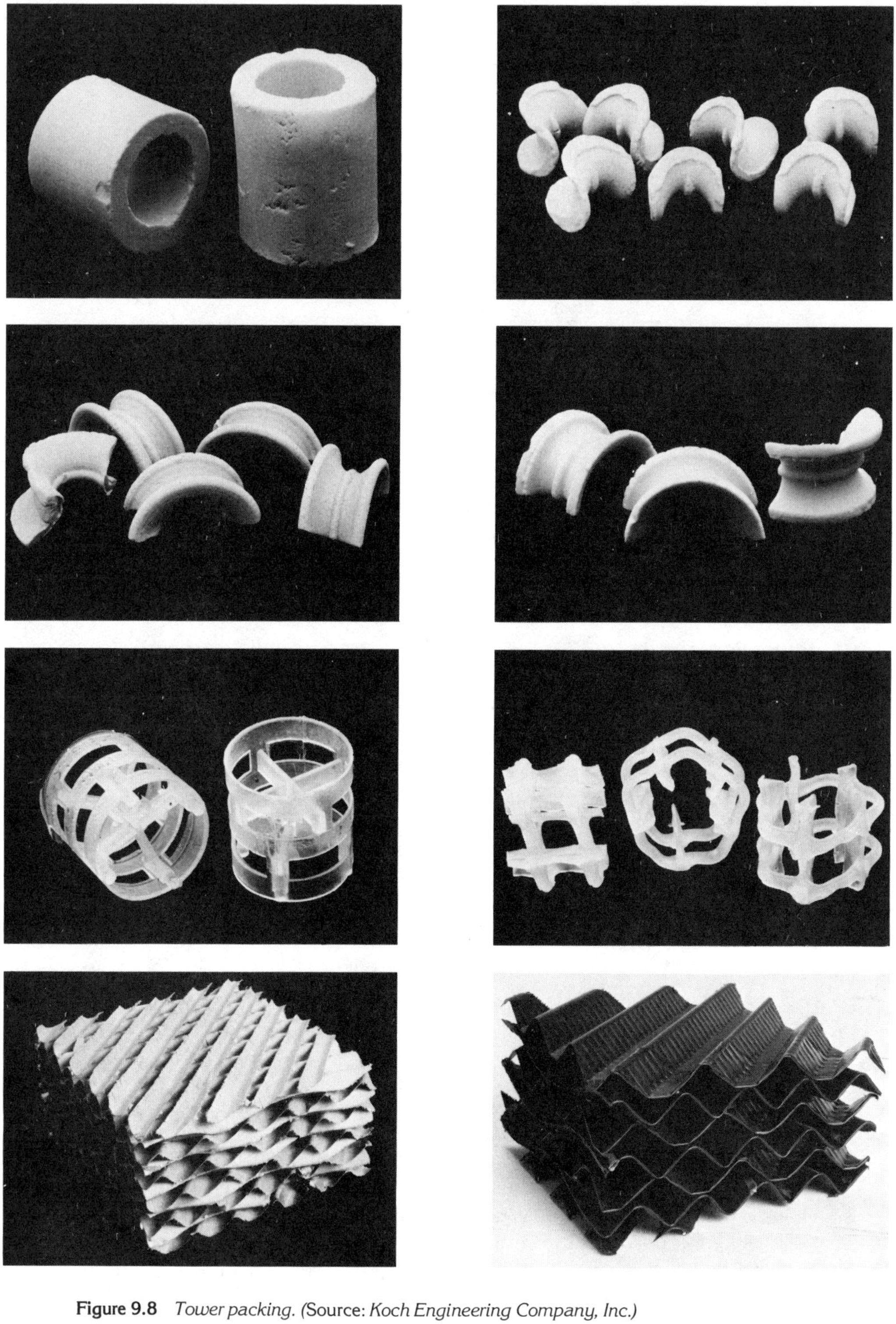

Figure 9.8 *Tower packing.* (Source: *Koch Engineering Company, Inc.*)

Flow Rate and Pressure Drop in Packed Columns

A packing can be loaded into a tower either by stacking it in layers or by filling the tower with water and dumping the packing into the water. The pressure drop through the random packing is usually several times greater than through stacked packing. As indicated by Leva (1953), the primary factors that influence the pressure drop are the percentage of void space in the packed tower, packing size and shape, and the densities and mass velocities of the gas and liquid streams. The primary effect of the liquid flow rate is to fill the void space and thus decrease the space available for gas flow. The liquid that is retained in the voids of the packing is called the dynamic holdup. For a constant liquid rate, the pressure drop increases with increasing gas flow rate. If, however, the gas rate becomes sufficiently large, a condition is reached in which a significant holdup of liquid occurs in the packing. This is defined as the region of loading. From a log–log plot of pressure drop versus gas mass velocity, as shown in Figure 9.9, the loading point can be defined as a point on the curve where the slope is greater than 2.0. After reaching the loading point in the column, the pressure drop increases more rapidly with increasing gas flow until the flooding point is reached. Flooding is defined as the condition at which all the void space in the packing fills with liquid, and the liquid will not flow through the column. This gives a pressure drop of 2 to 3 in of water per foot (approximately 1.6 to 2.5 kPa/m) of packing depth. A column cannot be operated under these conditions. Packed towers are typically operated at a gas velocity that corresponds to about 50% to 80% of flooding. This condition is usually near the loading point and will result in a pressure drop that ranges from 0.5 to 1.0 in of water per foot of packing (approximately 0.4 to 0.8 kPa/m). A large decrease in flow rate, which is typically referred to as the *turn–down ratio*, will cause channeling through the packed bed and will frequently result in poor column performance. This, of course, depends on the operating range.

Two different approaches are currently being used to determine the tower diameter and flow rates through an absorption tower. One approach is to select an allowable pressure drop in the bed, and the other is to select some fraction of the flooding capacity. The preferred method, however, is to design the column to operate at a specific flooding capacity. This is due in part to the uncertainty in the generalized pressure drop correlations for the tower packing. The pressure drop and gas flow rate can be calculated by applying the correlations shown in Figure 9.10. The calculation of pressure drop in a packed tower and the determination of tower diameter are demonstrated in Example 9.2, because of their importance in a column design.

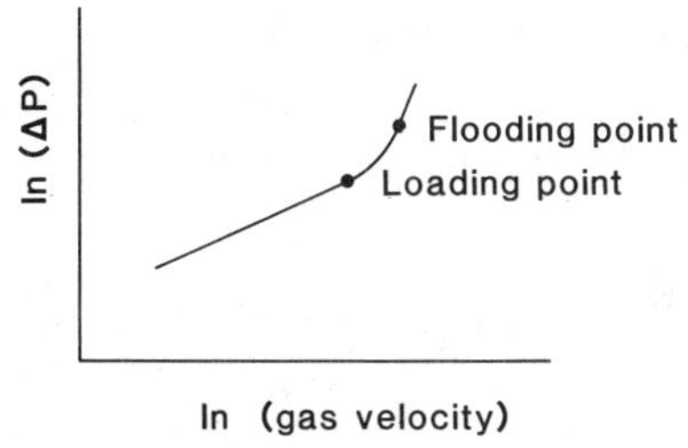

Figure 9.9 *Pressure drop in a packed bed.*

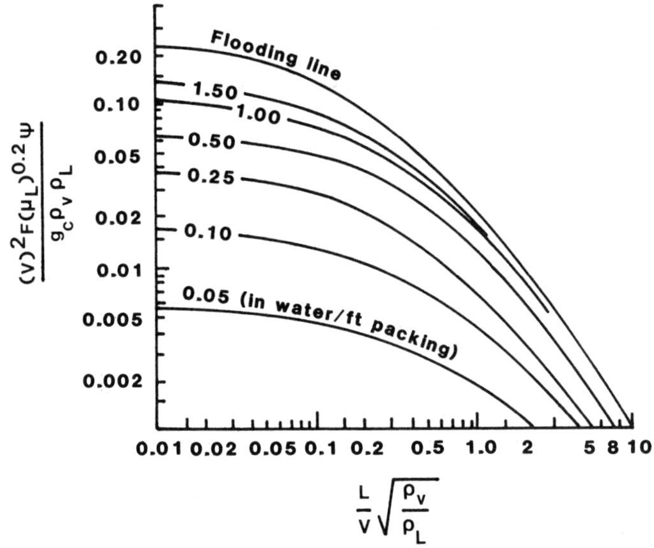

Figure 9.10 *Flooding and pressure drop correlation for random packings. (Source: Eckert, 1970). L = liquid rate (lb/s·ft²); V = gas rate (lb/s·ft²); ρ_L = liquid density (lb/ft³); ρ_V = gas density (lb/ft³); F = packing factor; μ = viscosity of liquid, cP; Ψ = density of water/density of liquid; g_c = gravitational constant, 32.2.*

Example 9.2

A tower packed with 1–in Raschig rings is to be used to absorb NH_3 from air by contacting the air stream with water. The air flow rate entering the bottom of the column is 40 lbmol/h and contains 5.0% NH_3. Ninety percent of the NH_3 is to be removed from the air. The entering water flow rate is 3200 lb/h. Absorption is to be carried out at a pressure of 1 atm absolute and 20°C. For these conditions, (a) calculate the tower diameter for 70% of flooding, and (b) determine the pressure drop per foot of packing. Assume the tower will be operated in countercurrent flow.

Solution: (a) When calculating the tower diameter, conditions at the bottom of the tower should be used since this corresponds to the maximum flow rates. At the bottom of the tower, the molecular weight of the air-ammonia mixture is

$$
\begin{aligned}
MW_{\text{mixture}} &= 0.05(17) + 0.95(29) = 28.4 \text{ lb/lbmol} \\
NH_3 \text{ removed from the air} &= 0.9(0.05)(40) = 1.8 \text{ lbmol/h} \\
&= 30.6 \text{ lb/h} \\
\text{gas flow rate at the bottom} &= (40 \text{ lbmol/h})(28.4 \text{ lb/lbmol}) \\
&= 1136 \text{ lb/h} = 0.316 \text{ lb/s} \\
\text{liquid flow rate at the bottom} &= 3200 + 30.6 = 3230.6 \text{ lb/h}
\end{aligned}
$$

Since there is 359 lbmol/ft³ of gas at 0°C (273 K) and the molecular weight of the gas mixture is 28.4 lb/lbmol, the density of the air stream at the bottom of the tower is calculated from the expression

$$\rho_V = \frac{1}{359 \text{ lbmol/ft}^3} \frac{273 \text{ K}}{293 \text{ K}} (28.4 \text{ lb/lbmol}) = 0.0737 \text{ lb/ft}^3$$

The density and viscosity of the solvent can be assumed to be that of pure water since the concentration of the solute (ammonia) in the water is very small.

$$\rho_L = 62.3 \text{ lb/ft}^3$$
$$\mu_L = 1.0 \text{ cP}$$

The abscissa for Figure 9.10 is calculated as

$$\frac{L}{V}\left(\frac{\rho_V}{\rho_L}\right)^{0.5} = \frac{3230.6}{1136}\left(\frac{0.0737}{62.3}\right)^{0.5} = 0.0978$$

Using the above value for the abscissa on Figure 9.10 and extending a line to where it intersects the flooding line, a value of 0.14 is obtained for the ordinate. Thus,

$$\frac{V^2 F \mu_L^{0.2} \psi}{g_c \rho_V \rho_L} = 0.14$$

From Table 9.2, $F = 155$. Since the absorbing liquid is water, $\psi = 1.0$.

TABLE 9.2 Packing Factors, F (ft^2/ft^3)

Type of Packing	Material	Nominal Packing Size (in) 1/4	3/8	1/2	5/8	3/4	1	1.25	1.5	2	3	3.5
Super Intalox	Ceramic						60			30		
Super Intalox	Plastic						33			21	16	
Intalox saddles	Ceramic	725	330	200		145	98		52	40	22	
Hy-Pak rings	Metal						42			18	15	
Pall rings	Plastic				97		52		40	25		16
Pall rings	Metal				70		48		28	20		16
Berl saddles	Ceramic	900*		240*		170^{+}	110^{+}		65^{+}	45*		
Raschig rings	Ceramic	$1600^{a,*}$	$1000^{b,*}$	580^{c}	380^{c}	255^{c}	155^{d}	$125^{e,*}$	95^{e}	65^{f}	$37^{g,*}$	
Raschig rings 1⁄32-in wall	Metal	700*	390*	300*	170	155	115*					
Raschig rings 1⁄16-in wall	Metal			410	290	220	137	110*	83	57	32*	
Tellerettes	Plastic						40			20		
Maspak	Plastic									32	20	
Lessing exp.	Metal								30			
Cross-partition	Ceramic										70	

[a] $\frac{1}{32}$ wall; [b] $\frac{1}{16}$ wall; [c] $\frac{3}{32}$ wall; [d] $\frac{1}{8}$ wall; [e] $\frac{3}{16}$ wall; [f] $\frac{1}{4}$ wall; [g] $\frac{3}{8}$ wall.

* Extrapolated.

+ Packing factor obtained in 16- and 30-in ID towers.

Source: Eckert, 1975.

$$V = \left[\frac{0.14 g_c \rho_V \rho_L}{F \mu_L^{0.2}}\right]^{0.5}$$

$$= \left[\frac{(0.14)(32.2)(0.0737)(62.3)}{155\ (1.0)^{0.2}}\right]^{0.5}$$

$$= 0.365\ \text{lb/ft}^2 \cdot \text{s}$$

For a design based on 70% of flooding, we obtain

$$V = 0.7(0.365) = 0.256\ \text{lb/ft}^2 \cdot \text{s}$$

The tower cross–sectional area is found by dividing the total gas flow rate at the bottom of the tower by the flow rate per unit area at the desired flooding condition of 70%. Thus,

$$S = \frac{0.316}{0.256} = 1.23\ \text{ft}^2$$

Therefore, the tower diameter is

$$d = \left(\frac{4S}{\pi}\right)^{0.5} = \left[\frac{4(1.23)}{\pi}\right]^{0.5} = 1.25\ \text{ft} = 15.0\ \text{in}$$

(b) To determine the pressure drop, we must first calculate a value for the ordinate using the actual gas flow rate and the cross–sectional area of the tower. Thus,

$$V = \frac{0.316}{1.23} = 0.257\ \text{lb/ft}^2 \cdot \text{s}$$

and

$$\frac{V^2 F \mu_L^{0.2} \psi}{g_c \rho_V \rho_L} = \frac{(0.257)^2(155)(1.0)^{0.2}(1)}{(32.2)(0.0737)(62.3)} = 0.069$$

Using an abscissa of 0.0978, which was used earlier, and a value of 0.069 for the ordinate in Figure 9.10, the pressure drop is 0.95 in of water per foot of packing.

Determination of Packing Height

The rate of mass transfer and subsequently the height of packing in a separation process depend on the area of contact between the phases. Intimate contact of passing streams is provided in a packed tower as a result of the distribution of the phases over the packing surface. While a variety of packings is available for accomplishing this goal, differences in their characteristics result in different mass transfer rates. The distribution of the phases over the surface and the wettability of the packing are a function of the type of material from which the packing is constructed and the type of fluid being processed. Thus, for different packing materials, the difference in wettability causes a change in the mass transfer coefficient and, subsequently, the degree of fluid separation.

The height of a packed tower depends on the properties and flow rates of the contacting streams as well as on the type of packing being used. Therefore, equations that utilize mass transfer coefficients for determining contactor heights are required. The equation used to determine the height of packing in an absorber relates the molar flux to a mass transfer coefficient and to the concentration difference in the column. Because of the limited availability of mass transfer data for indoor air pollutants, in this presentation the molar flux will be related to the overall mass transfer coefficient. This is done by expressing the molar flux as the product of an overall mass transfer coefficient and the difference between the bulk and equilibrium concentrations. For a gas phase mass transfer coefficient and a gas phase concentration difference, we write the molar flux as

$$N_A = \frac{\text{kgmol}}{\text{m}^2\cdot\text{h}} = \frac{d(Vy_A)}{a\,dZ} = \frac{K_y}{\beta_V}(y_A - y_A^*) \tag{9.9}$$

where β_V = bulk flow correction factor related to the composition of the bulk phase and the equilibrium composition

y_A = mole fraction of A in the bulk gas phase

y_A^* = mole fraction of A in the gas phase in equilibrium with the concentration of A in the bulk liquid phase

K_y = overall mass transfer coefficient for the entire diffusional resistance in both phases but related to the gas phase driving force

The factor a is introduced to represent the mass transfer area per unit volume of packing and has typical units of m^2/m^3. This is done because the surface area in a packed column is difficult to obtain, and it changes with packing size. The form taken by β_V depends on the nature of the transfer and can be simplified for the limiting cases of transfer through a stagnant film and equimolar counter transfer.

The relationship between y_A and y_A^* is shown in Figure 9.11. For any gas phase concentration, y_A, a line is extended horizontally from the ordinate to intersect the operating line. A vertical line is then extended from the operating line to intersect the equilibrium

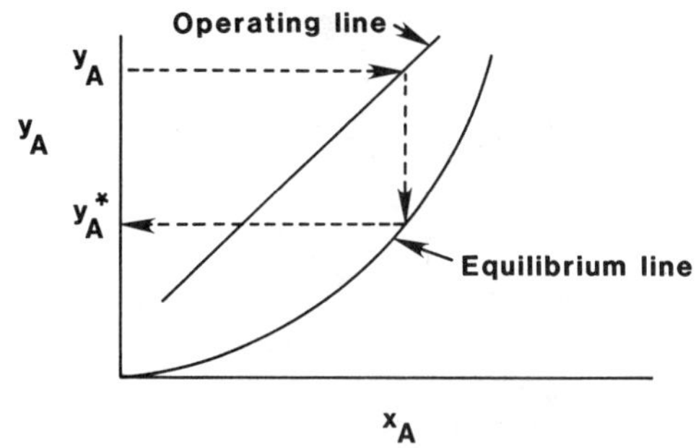

Figure 9.11 *Relationship between bulk and equilibrium concentrations.*

curve. A line drawn from this intersection back to the ordinate gives the equilibrium concentration, y_A^*.

The height of the tower packing is obtained by separating the variables in the above equation and integrating over the concentration range, which corresponds to concentrations at the top and bottom of the tower, as shown.

$$Z = \int_{y_{A,a}}^{y_{A,b}} \frac{\beta_V}{K_y a} \frac{d(Vy_A)}{y_A - y_A^*} \tag{9.10}$$

In the case of transfer through a stagnant film, Eq. (9.10) is written as

$$Z = \int_{y_{A,a}}^{y_{A,b}} \frac{V}{K_y a} \frac{(1 - y_A)_{*M}}{(1 - y_A)} \frac{dy_A}{(y_A - y_A^*)} \tag{9.11}$$

where $(1 - y_A)_{*M}$ is defined as the log mean concentration difference and is given as

$$(1 - y_A)_{*M} = \frac{(1 - y_A^*) - (1 - y_A)}{\ln\left(\dfrac{1 - y_A^*}{1 - y_A}\right)} \tag{9.12}$$

For conditions under which the overall mass transfer coefficient can be used, the term $V/K_y a$ is nearly constant over the height of the packing. Therefore using an average value, we get

$$Z = \left(\frac{V}{K_y a}\right)_{avg} \int_{y_{A,a}}^{y_{A,b}} \frac{(1 - y_A)_{*M}}{(1 - y_A)} \frac{dy_A}{(y_A - y_A^*)} \tag{9.13}$$

The above equation is frequently expressed as

$$Z = H_V N_V \tag{9.14}$$

where H_V is the height of the transfer unit and is given as $H_V = (V/K_y a)_{avg}$. The number of transfer units is

$$N_V = \int_{y_{A,a}}^{y_{A,b}} \frac{(1 - y_A)_{*M}}{(1 - y_A)} \frac{dy_A}{(y_A - y_A^*)} \tag{9.15}$$

The height of a transfer unit has been shown to be less dependent on flow rate than the mass transfer coefficient.

The bulk flow correction factor for equimolar counter transfer is one. Thus, the height of packing is given by

$$Z = \left(\frac{V}{K_y a}\right)_{avg} \int_{y_{A,a}}^{y_{A,b}} \frac{dy_A}{(y_A - y_A^*)} \tag{9.16}$$

It should be noted that if the log mean concentration difference is replaced by the arithmetic mean over the column length, Eq. (9.11) can be written as

$$Z = \left(\frac{V}{K_y\, a}\right)_{\text{avg}} \int_{y_{A,a}}^{y_{A,b}} \frac{dy_A}{(y_A - y_A^*)} + \frac{1}{2} \ln\left(\frac{1 - y_{A,a}}{1 - y_{A,b}}\right) \tag{9.17}$$

For dilute solutions (mole fraction less than about 5%), Eq. (9.17) can be reduced to Eq. (9.16).

Analogous equations for the liquid phase and for local mass transfer coefficients are given by Hines and Maddox (1985), along with a detailed derivation of the above equations.

9.4 MASS TRANSFER CORRELATIONS

Transfer Unit Heights

Mass transfer data are frequently correlated in terms of the height of a transfer unit, since it changes less over the length of a column than does the mass transfer coefficient. The first extensive investigation of transfer unit heights was conducted by Sherwood and Holloway (1940) for the desorption of hydrogen, oxygen, and carbon dioxide from water. They found that the height of the liquid phase transfer unit, H_L, was independent of the gas flow rate. Sherwood and Holloway proposed an empirical relationship to correlate the height of a transfer unit for the liquid phase as a function of liquid flow rate and the Schmidt number. The Sherwood and Holloway correlation is

$$H_L = \beta\left(\frac{L}{\mu_L}\right)^n (\text{Sc})^{0.5} \tag{9.18}$$

where
H_L = height of transfer unit, ft
L = liquid flow rate, $\text{lb/ft}^2 \cdot \text{h}$
μ_L = viscosity of the liquid, $\text{lb/ft} \cdot \text{h}$
Sc = Schmidt number for the liquid phase
β, n = constants for different packings (see Table 9.3)

The liquid phase correlation applies for liquid flow rates that range from 400 to 15,000 $\text{lb/ft}^2 \cdot \text{h}$. A correlation was proposed by van Krevelen and Hoftijzer (1948) in which H_L was related to the Schmidt number raised to the 2/3 power and the Reynolds number raised to the 1/3 power. The Reynolds number was based on the effective interfacial area. Onda et al. (1959) used a similar relationship to show that the Schmidt number varied with the 1/2 power and the Reynolds number varied with the 0.51 power.

A large body of literature data for absorption, stripping, and distillation was correlated by Cornell et al. (1960 a, b) and extended by Bolles and Fair (1979). They proposed an empirical relationship for the height of a transfer unit for rings, saddles, and spiral tile.

A number of relationships have been proposed for correlating the number of gas phase transfer units in terms of the gas and liquid flow rates. Some of these, however, do not take into account the liquid holdup in the voids of the packing and, subsequently, are of

TABLE 9.3 Constants for determining H_L

Packing	β	*n*
Raschig rings		
⅜ in	0.00182	0.46
½ in	0.00357	0.35
1 in	0.0100	0.22
1.5 in	0.0111	0.22
2 in	0.0125	0.22
Berl saddles		
½ in	0.00666	0.28
1 in	0.00588	0.28
1.5 in	0.00625	0.28

Source: Sherwood and Holloway, 1940.

limited use. Another difficulty in obtaining accurate gas phase correlations results from mass transfer data being taken on gas–liquid systems in which the resistance to mass transfer is attributed to both phases. The absorption of NH_3 in water provides a system in which nearly all the resistance to mass transfer lies in the gas phase. Fellinger's (1941) absorption study of NH_3 in water provides the primary source of data for gas phase correlations.

Correlation of Mass Transfer Coefficients

For a system in which the solute is very soluble in the liquid, the overall gas-phase mass transfer coefficient can be approximated with the local gas-phase coefficient. A correlation of the gas phase mass transfer coefficient was developed by Onda et al. (1968). The correlation is expressed as

$$k\left(\frac{RT}{aD_V}\right) = 5.23\left(\frac{G_V}{a\mu_V}\right)^{0.7} (\mathrm{Sc}_V)^{1/3}\,(ad_p)^{-2.0} \tag{9.19}$$

where

k = mass transfer coefficient in the gas phase, lbmol/ft^2 · s · atm
a = surface area to volume ratio of the packing material, ft^2/ft^3
μ_V = viscosity of the gas, lb/ft · s
d_P = diameter (or nominal size) of packing, ft
G_V = mass velocity of gas based on column cross section, lb/ft^2 · s
Sc_V = Schmidt number, $\mu_V/\rho_V D_V$
D_V = diffusion coefficient for the key component, ft^2/s
T = gas temperature, °R
R = universal gas constant, 0.7302 atm · ft^3/lbmol · °R

Correlations for the gas phase mass transfer coefficient for an aqueous calcium chloride solution–air contact system were developed by Gandhidasan et al. (1986) for dehumidification of air. Their correlations are expressed as

1–in ceramic Raschig rings:

$$k_G\, a = 0.147(L')^{0.2}(G')^{0.7} \exp(0.00045\ T) \tag{9.20}$$

2–in ceramic Raschig rings:

$$k_G\, a = 0.081(L')^{0.23}(G')^{0.64} \exp(0.00034\ T) \tag{9.21}$$

1–in Berl saddles:

$$k_G\, a = 0.133(L')^{0.3}(G')^{0.69} \exp(0.00041\ T) \tag{9.22}$$

1.5–in Berl saddles:

$$k_G\, a = 0.072(L')^{0.35}(G')^{0.64} \exp(0.0003\ T) \tag{9.23}$$

where
k_G = mass transfer coefficient for the gas phase, kmol/$m^2 \cdot$ s
a = specific interfacial surface for contact of air with liquid, m^2/m^3
L' = superficial liquid mass velocity, kg/$m^2 \cdot$ s
G' = superficial air mass velocity, kg/$m^2 \cdot$ s
T = gas temperature, °C

According to Gandhidasan et al., these correlations are valid for air flow rates from 0.3 to 1.1 kg/$m^2 \cdot$ s, liquid flow rates from 0.7 to 6.7 kg/$m^2 \cdot$ s, air temperatures from 30° to 70°C, and liquid temperatures from 30° to 60°C.

A more recent correlation was developed by Chung (1991) for the dehumidification of air by aqueous lithium chloride solutions. That equation is given as

$$K_G a\left(\frac{Md_p^2}{D_V\rho_V}\right) = 1.326(10^{-4})(1 - x)^{-0.94}\left(\frac{L_m}{V_m}\right)^{0.27}(\mathrm{Sc}_V)^{1/3}(\mathrm{Re}_V)^{1.16} \tag{9.24}$$

where
K_G = overall mass transfer coefficient for the gas phase, kmol/$m^2 \cdot$ s
a = surface area to volume ratio of the packing material, m^2/m^3
d_p = diameter (or nominal size of packing), m
V = gas flow rate, m/s
L_m = liquid mass flow rate, kg/$m^2 \cdot$ s
V_m = gas mass flow rate, kg/$m^2 \cdot$ s
Sc_V = Schmidt number, $\mu_V/D_V\rho_V$
Re_V = Reynolds number, $\rho_V V d_c/\mu_V$
D_V = diffusion coefficient for the key component, m^2/s
μ_V = viscosity of gas, kg/m $\cdot$ s
ρ_V = density of gas, kg/m^3
d_c = column diameter, m
M = molecular weight of the transferred component, kg/kgmol
x = mass fraction of the solute in the liquid solution

This equation has been tested successfully for air flow rates from 0.9 to 1.9 kg/$m^2 \cdot$ s, liquid flow rates from 8.6 to 15.2 kg/$m^2 \cdot$ s, air temperatures from 73° to 78°F (room temperature), and liquid temperatures from 55° to 73°F. The data were obtained with the column operating between 50% to 80% of flooding. A comparison of the lithium chloride experimental data of Chung et al. (1991) and the values calculated by the above correlations is shown in Figure 9.12.

Equation (9.24) fitted the experimental data within a 3% error. The predictions made with the correlations of Gandhidasan et al., Onda et al., and Bolles and Fair were two to six times greater than the experimental values for this system. The correlations of Onda et al. and Gandhidasan et al. did not take into account the liquid solution properties used in calculating the overall gas phase mass transfer coefficient. Therefore, the estimated values for the operating condition in the 30% and 40% LiCl solutions are approximately the same. The Bolles and Fair correlation was derived by using distillation, absorption, and stripping data. Although one correlation could be used to predict the mass transfer coefficients for all three different processes, the deviation between the experimental data and the calculated values could be very large. In absorption, the mass transfer coefficient is typically related to transfer of a single component; whereas, in distillation it is more closely related to equimolar counter transfer.

It should be noted that Eq. (9.24) is a function of the properties of the liquid absorbent and the packing material. Therefore, the correlation can be extended to other packed column operations if the parameters are available. The values calculated by this equation are the same order of magnitude as the experimental data reported by Bichowsky and Kelley

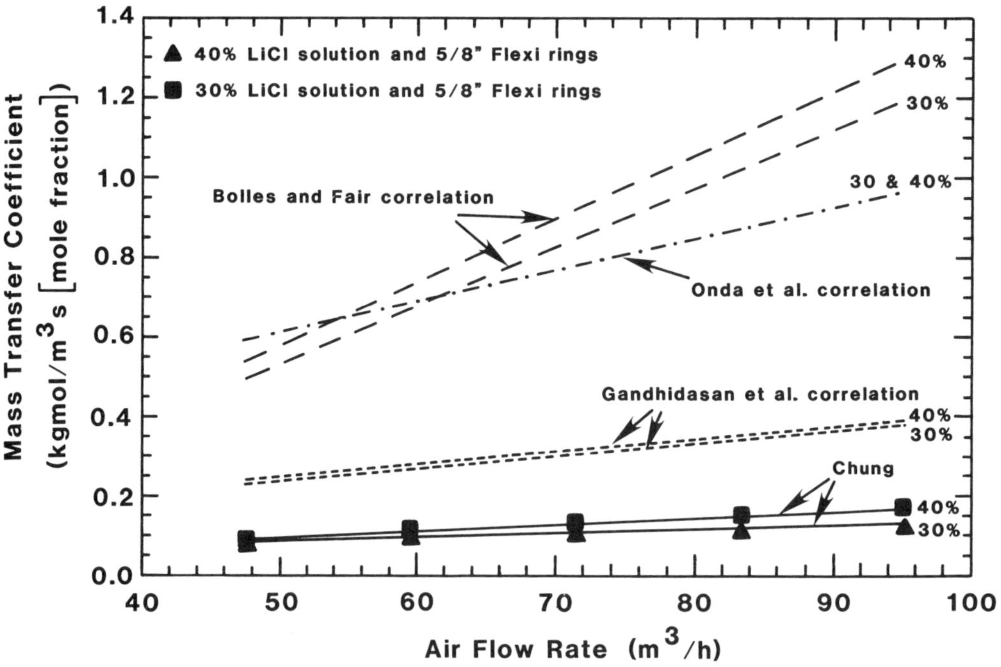

Figure 9.12 *Comparison of the experimental data of Chung (1991) with predictions by different correlations. (*Source: *Chung, 1991.)*

(1935). The values calculated by the correlations of Bolles and Fair, Onda et al., and Gandhidasan et al. did not compare well with the experimental data. The comparisons are shown in Figure 9.13.

9.5 APPLICATION TO DEHUMIDIFICATION AND POLLUTANT CONTROL

Two key factors used in determining the comfort of an environment are the temperature and humidity of the space. Temperatures in the range of 68° to 80°F and relative humidities in the 20% to 80% range, depending on climatic conditions, are considered to be an indicator of a comfortable environment. The increase in the cost of energy over the last two decades has led to the design of energy–efficient homes and buildings. The structures are made airtight, with 80% to 90% of the air being recirculated in the buildings, causing the pollutants that are generated indoors to remain there. As a result, the control of only temperature and humidity is not sufficient to maintain a healthy indoor environment. Fortunately, some types of heating, ventilating, and air conditioning (HVAC) systems have the potential to clean the air as well as modify temperature.

Depending on the application, various types of air–conditioning systems are available, including liquid desiccant–based systems. In liquid desiccant–based systems, air is generally introduced at the bottom of a column, and a liquid desiccant, such as an aqueous solution of 40% lithium chloride or 95% triethylene glycol (TEG), is fed from the top. To enhance the mass transfer rate between the air and the liquid, a number of methods can be

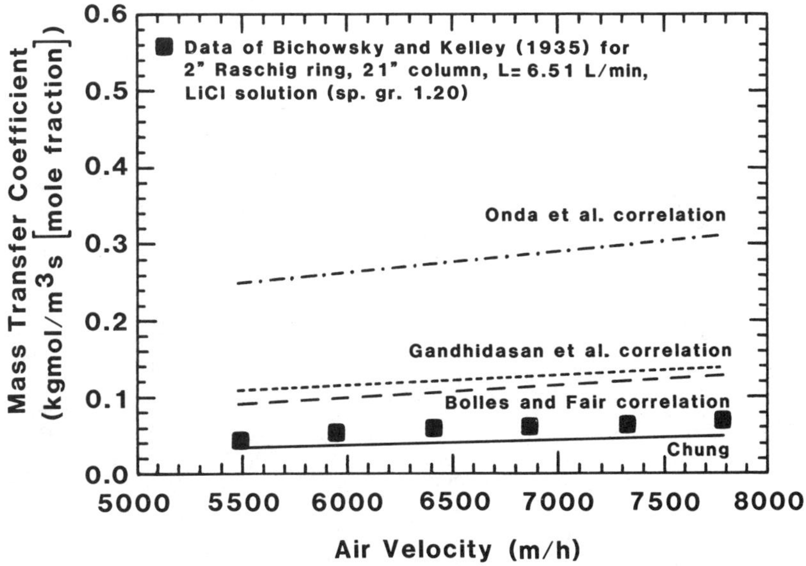

Figure 9.13 *Comparison of the experimental data of Bichowsky and Kelley (1935) with predictions by different correlations.*

used, including a column packed with an inert packing, a spray column, or a finned tube surface type of dehumidifier. Although all three types of systems are in use commercially and have certain advantages and disadvantages over others, only absorption in a packed tower will be considered here.

A schematic flow diagram of an absorber–stripper air–conditioning system in which the moisture is removed from the air by contact with a liquid desiccant in the absorber is shown in Figure 9.14. The dilute liquid solution is regenerated in the stripper and then cooled and fed back to the absorber, where it is sprayed on top of the packing to provide uniform wetting. The primary factors considered in the design of a packed–bed system include selection of the system configuration, the liquid desiccant, the type of packing and materials, the liquid flow rate, the packing height, and the column diameter. Although the most common configuration used by industry is shown in Figure 9.14, Esia et al. (1986) proposed the use of reflux to enhance system performance. It is important to note that for the water–lithium bromide solution studied by Esia et al., the use of reflux decreased both the solution temperature and the liquid concentration of the solution entering the absorber. The decrease in the solution temperature had a positive effect on the overall performance, while the decrease in liquid concentration had a negative effect. According to Esia et al., an optimum value of the reflux ratio can be obtained for the system that will improve its overall performance.

Various methods have been employed to regenerate the loaded liquid desiccant solution. One common method is to use steam or hot air as a stripping gas. The hot gas drives the moisture from the liquid, thereby reconcentrating it to its original value. A number of studies (Hollands, 1963; Peng and Howell, 1981 and 1984; and Gutkowski and Ryduchowski, 1986) have focused on the regeneration of solutions by using solar energy. Solar energy is used to generate steam from water or to heat air before introducing it into the stripping column. However, solar collectors generally require a relatively large space

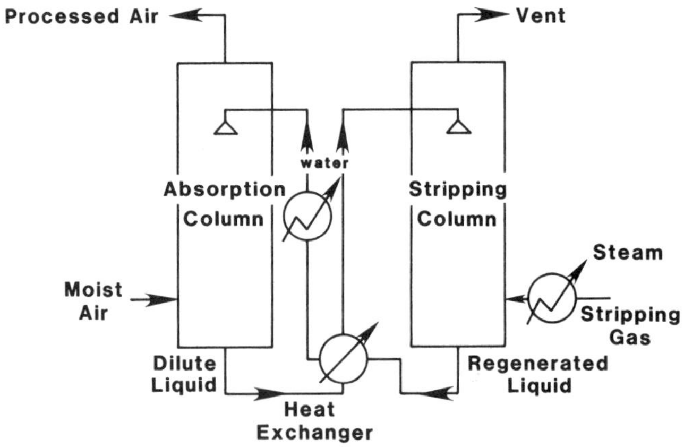

Figure 9.14 *Schematic flow diagram of an absorber-stripper air-conditioning system.*

and are expensive; also, their efficiency depends on weather.

The liquid flow rate, packing height, and column diameter for liquid desiccant dehumidification systems can be determined by using standard procedures, as described earlier and presented in the textbooks by McCabe et al. (1985), Hines and Maddox (1985), and Strigle (1987). A graphical method for designing packed columns was proposed by Czermann et al. (1958), who estimated the packing height by using the height of a transfer unit. Instead of using correlations, they determined the height of a transfer unit from the equilibrium diagram and the operating line. A summary of packed bed absorption studies for liquid desiccant systems is given in Table 9.4.

A significant reduction in pollutant concentrations can be achieved by controlling the humidity of the space to be air conditioned. Andersen et al. (1975) found that a decrease in relative humidity from 70% to 30% in an environmental chamber resulted in a 50% reduction in the formaldehyde concentration. They derived a linear correlation between formaldehyde concentration and the absolute humidity of a closed chamber. Berge et al. (1980) reported a 17% increase in formaldehyde concentration with a 10% increase in relative humidity and also derived a linear relationship between the formaldehyde concentration and the relative humidity. However, only a limited number of studies have been conducted in which liquid desiccant dehumidification systems were used for pollution control. In the study of Moschandreas and Relwani (1990), a lithium chloride solution was used for the removal of indoor air pollutants during the dehumidification process. They installed Kathabar humidity pumps on the roofs of several buildings for their study. The results showed that the absorption of total volatile organic compounds (TVOC), carbon monoxide, and carbon dioxide from the air was negligible.

Although liquid desiccant systems have been employed in large commercial buildings and in hospital operating rooms, they are not currently used in homes or small buildings. One reason could be that, because of the size, such a unit might not be cost effective for use on a small scale. Other studies that relate to the removal of pollutants by using a liquid desiccant include those of Sudakov and Sudakova (1986), who used triethylene glycol for removing ammonia, and Pedersen and Fisk (1984), who used an air washer to remove formaldehyde from indoor air. However, in the latter study, a refrigeration unit was needed to condense the extra water vapor from the air washer. Calculation procedures for the various design parameters of a dehumidification system that employs lithium chloride are discussed in the following two examples.

Example 9.3

An absorber packed with 5/8 in Flexi rings is to be used to absorb moisture from air by contacting it with a 40% LiCl–water solution. The air flow rate is 30 ft^3/min, and the absorption is to be carried out at an absolute pressure of 1 atm. The dry bulb temperature of the air is 75°F and the wet bulb temperature is 58°F, which corresponds to a relative humidity of 70%. The humidity of the air is to be reduced to 20%. Calculate the minimum liquid flow rate on the basis of equilibrium considerations only.

Solution: For inlet air, the moisture content corresponding to a dry bulb temperature of 75°F and a wet bulb temperature of 58°F obtained from the psychrometric chart is 0.007 lb of moisture/lb of dry air. The corresponding mole fraction of water is

TABLE 9.4 Selected Packed-bed Absorber Studies Using Liquid Desiccants

	Packing	*Desiccant*	*Gas flow rate ($kg/h{\cdot}m^2$) and Temperature (°C)*	*Liquid flow rate ($kg/h{\cdot}m^2$) and Temperature (°C)*	*Nature of study*	*Mass transfer coefficients ($kmol/s{\cdot}m^3$)*	*Heat transfer coefficients ($kw/K{\cdot}m^3$)*
Anderson (1967)	Berl saddles	15% LiCl	21–38 °C	29–32 °C	Efficiency of moisture removal		
Leboeuf and Lof (1980)	1.5″ Berl saddles	40% LiCl	4893 50–80 °C	978 30–50 °C	Heat and mass transfer analysis		
Factor and Grossman (1980)	Berl saddles and Raschig rings	58–60% LiBr	7340–9790 26–33 °C	1470–8810 25–30 °C	Computer simulation		
Gandhidasan et al. (1983)	1.5″ Berl saddles	41% LiCl	345.73 25–41 °C	933.5 24–40 °C	Tower performance	0.21–0.42	
Scalabrin and Scaltriti (1984)	4 × 6 mm Raschig rings and ¾″ Pall rings	40% LiCl	1648–9889 18.5–24.9 °C	2604–6617 16.7–34.8 °C	Mass transfer coefficients	0.0004–0.0011 0.0003–0.0009	
Lof et al. (1984)	1″ Raschig rings	26–35% LiCl	3355–5590 60–105 °C	2173–4274 34–39 °C	Heat transfer correlations	0.16–0.31 (regeneration)	6.0–10.7 (regeneration)
Gandhidasan et al. (1986)	1″ and 2″ Raschig rings; 1″ and 1.5″ Berl saddles	35–45% $CaCl_2$	1080–3960 30–70 °C	2520–24,120 30–60 °C	Heat and mass transfer correlations		
Gandhidasan et al. (1987)	2″ Raschig rings	40–45% $CaCl_2$	3006.7 30–32 °C	23,976 30–35.4 °C	Tower performance		
Chung et al. (1991)	⅝″ Flexi rings and ½″ Berl saddles	30–40% LiCl	3372–6744 2709–4335 23–28 °C	31,324–54,818 15,662–39,156 16–21 °C	Heat and mass transfer coefficients	0.1661–0.0981 0.1420–0.0791	11.98–3.16 9.59–4.17

$$y_{in} = \frac{0.007/18}{(0.007/18) + (1/28.8)} = 0.0111$$

and

$$\overline{Y}_{in} = \frac{0.007/18}{1/28.8} = 0.0112$$

The moisture content of the processed air (corresponding to 20% relative humidity) is 0.0034 lb of moisture/lb of dry air. The mole fraction is

$$y_{out} = 0.00541$$

$$\overline{Y}_{out} = 0.00544$$

Since the inlet solution contains 40% by weight of lithium chloride (whose molecular weight is 42.5) and 60% water by weight, the mole fraction of water is

$$x_{in} = \frac{0.6/18}{(0.6/18) + (0.4/42.5)} = 0.78$$

and

$$\overline{X}_{in} = \frac{0.6/18}{0.4/42.5} = 3.54$$

The minimum liquid to gas ratio is given by the slope of the operating line shown in Figure 9.15.

$$\left(\frac{L}{V}\right)_{min} = 0.043 \text{ lb of solution/lb of dry air}$$

For an air density of 0.08 lb/ft^3 at the column conditions, the dry air flow rate is 2.44 lb of dry air/min, which corresponds to a minimum liquid flow rate of 0.103 lb of solution/min.

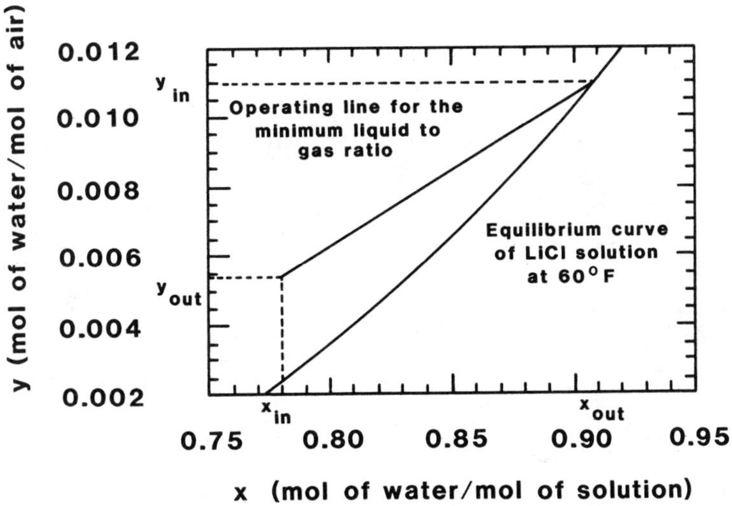

Figure 9.15 *Minimum liquid to gas ratio.*

Because the flow is too low to wet the packing adequately and provide good contact, the minimum liquid flow rate shown above cannot be used in an actual absorption process. Wetting of the packing is necessary for good column performance. We must therefore calculate the minimum flow rate on the basis of the packing that is to be used in the absorber and use this value for the design of the column. Wetting of the packing can be checked by using a simple expression based on the work of Morris and Jackson (1953), who found that the minimum wetting rate for most packing less than three inches in diameter is 0.85 $ft^3/hr \cdot ft^2$. This can be related to the specific surface area of the packing (F) and the liquid density of the absorbent (ρ_L) by the simple expression

$$MDR = 0.85\,(F)(\rho_L) \tag{9.25}$$

The difference between the minimum liquid rate calculated by this equation and that calculated from equilibrium considerations will be demonstrated next.

Example 9.4

If the absorption process described in Example 9.3 is carried out in a 6-inch ID column operated at 50% of flooding, (a) calculate the actual liquid flow rate, and (b) determine the minimum wetting rate by using the method proposed by Morris and Jackson (1953). (c) Using an overall mass transfer coefficient of 0.09 $kmol/m^3 \cdot s$, determine the height of packing necessary to dehumidify the air. The properties of the liquid absorbent, air, and packing are as follows:

Density of liquid (40% LiCl), $\rho_L = 78\ lb/ft^3$
Density of air, $\rho_V = 0.08\ lb/ft^3$
Packing factor, $F = 100\ ft^2/ft^3$
Surface area to volume ratio, $a = 104\ ft^2/ft^3$
Liquid viscosity, $\mu_L = 1.0$ cP
$\psi = \rho_{water}/\rho_L = 0.8$
Air flow rate, $V = 30\ ft^3/min = 0.204\ lb/ft^2 \cdot s$

Solution: (a) Given the above properties, the liquid flow rate at flooding conditions is calculated by using the relationship provided by Eckert (1970) in Figure 9.10. The ordinate for this system is

$$\frac{V^2 F \mu_L^{0.2} \psi}{g_c \rho_V \rho_L} = \frac{(0.204)^2(100)(1.0)^{0.2}(0.8)}{(32.2)(0.08)(78)} = 0.0166$$

From Figure 9.10, the value of the abscissa corresponding to the above value at flooding is given by

$$\left(\frac{L}{V}\right)\left(\frac{\rho_V}{\rho_L}\right)^{0.5} = 1.05$$

or

$$\left(\frac{L}{0.204}\right)\left(\frac{0.08}{78}\right)^{0.5} = 1.05$$

Therefore, the liquid flow rate at the flooding condition is equal to 6.68 $lb/ft^2 \cdot s$. If the system is operated at 50% of flooding, the actual liquid flow rate is

$$L = 0.5(6.68) = 3.34\ lb/ft^2 \cdot s = 3.7\ gal/min$$

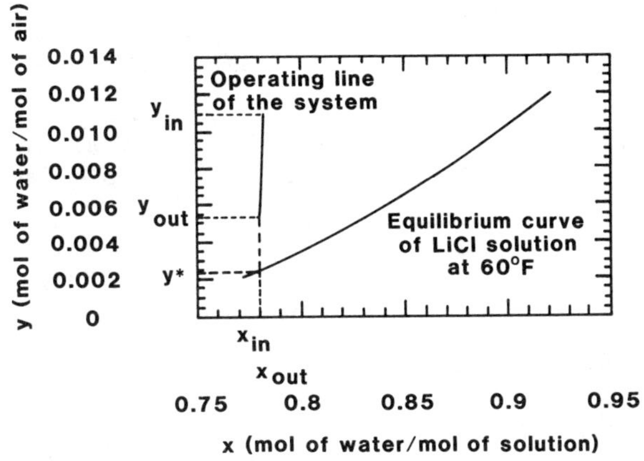

Figure 9.16 *Equilibrium curve of LiCl solution.*

(b) The minimum design flow rate (*MDR*) for the polyproplylene Flexi rings is calculated from Eq. (9.23). For polypropylene Flexi rings, the specific surface area is 104 ft^2/ft^3, and the density of the 40% lithium chloride solution is 78 lb/ft^3. Therefore,

$$MDR = 0.85(104)(78) = 6895.2 \text{ lb/ft}^2 \cdot \text{h}$$

The liquid flow rate at 50% of flooding is greater than that obtained from the minimum design rate necessary for wetting. Therefore, wetting of the packing will be achieved.

(c) The packing height can be calculated by using Eq. (9.13).

$$Z = \frac{V}{(K_y a)_{avg}} \int_{y_{A,a}}^{y_{A,b}} \frac{(1 - y_A)_{*M}\, dy_A}{(1 - y_A)(y_A - y_A^*)}$$

A numerical approach is needed to evaluate the integral portion of the above equation. A simple Simpson's rule technique can be used for this calculation.

As was shown in Figure 9.11, the equilibrium values, y_A^*, are obtained by using the operating line and equilibrium curve. Because the operating line in this system is nearly perpendicular to the x–axis, the equilibrium concentration, y_A^*, is assumed to be a constant throughout the column with a value of 0.0223. This is a particularly interesting case, as it is usually not true for most systems.

A table showing values of the integrand for several values of y_A from the top to the bottom of the column is shown next.

y_A	$f(y_A) = \frac{(1 - y_A)_{*M}}{(1 - y_A)(y_A - y_A^*)}$
0.0054	323
0.0063	250
0.0072	204
0.0082	170
0.0091	148
0.0101	129
0.0110	115

TABLE 9.5 Removal of Selected Indoor Air Pollutants by Liquid Desiccants

		Pollutant Removal Efficiency by Liquid Desiccants (%)[a]	
Pollutants	*Inlet Concentration (ppm)*	*Lithium Chloride (40% Solution)*	*Triethylene Glycol (95% Solution)*
Carbon dioxide	1000	20	56
Formaldehyde	0.02	3	30
Toluene	3	20	100
1,1,1-Trichloroethane	24	2.5	100
Formaldehyde and	0.02	20	23
carbon dioxide	1000	0	51
Formaldehyde and	0.02	13	29
toluene	3	11.3	100
Formaldehyde and	0.02	20	26
1,1,1-trichloroethane	24	0	100
Formaldehyde,	0.02	12	26
carbon dioxide,	1000	8	54
toluene, and	3	0	100
1,1,1-trichloroethane	24	0	100

[a]Removal efficiency $= \dfrac{\text{inlet concentration} - \text{outlet concentration}}{\text{inlet concentration}}$

Source: Hines and Ghosh, 1992.

Evaluating the integral numerically by Simpson's rule gives

$$\frac{9.33 \times 10^{-4}}{3}[323 + 4(250 + 170 + 129) + 2(204 + 148) + 115] = 1.04$$

The flow rate for air is 30 ft^3/min (0.0348 $kmol/m^3 \cdot s$), and the mass transfer coefficient is 0.09 $kmol/m^3 \cdot s$. Therefore, the packing height calculated by Eq. (9.13) is

$$Z = \frac{0.0348}{0.09}(1.04) = 0.4 \text{ m} = 1.32 \text{ ft}$$

The simultaneous removal of indoor pollutants during dehumidification of air by a liquid desiccant requires the solubility data of these pollutants. Although a liquid desiccant, such as 40% lithium chloride, contains a significant amount of water and the solubility of most of these pollutants in water is available, the presence of lithium chloride can decrease their solubility in water. This phenomenon is known as *salting out*. Therefore, the solubility of a pollutant in the liquid desiccant solution should be used to design the absorber, following the above procedures. Table 9.5 shows the simultaneous removal of water and some selected pollutants by 40% lithium chloride and 95% triethylene glycol solutions. Triethylene glycol is an organic solvent that is very good for most organic compounds. Consequently high removal efficiencies for organic pollutants are observed.

REFERENCES

ANDERSEN, I. B., LUNDQVIST, G. R., and MØLHAVE, L., *Atmos. Environ.*, **9**, 1121 (1975).

ANDERSON, G. W., *Actual Specifying Engineer*, **18**, G43 (1967).

BERGE, A., MELLEGAARD, B., HANETHO, P., and ORMSTAD, E. B., *Holz Roh–Werkst*, **37**(7), 251 (1980).

BICHOWSKY, F. R., and KELLEY, G. A., *Ind. Eng. Chem.*, **27**(8), 879 (1935).

BOLLES, W. L., and FAIR, J. R., *Int. Chem. Eng. Symp. Ser.*, No. 56, **2**, Paper No. 3.3/35 (1979).

CHUNG, T-W., "Design of a Packed Bed Dehumidification System," M. S. Thesis, University of Missouri-Columbia, Columbia, Mo., 1991.

CHUNG, T-W., GHOSH, T. K., and HINES, A. L., Seventh Symposium on Separation Science and Technology for Energy Applications, Knoxville, Tenn., October, 1991.

CORNELL, D., KNAPP, W. G., and FAIR, J. R., *Chem. Eng. Prog.*, **56**(7), 68 (1960a).

CORNELL, D., KNAPP, W. G., CLOSE, H. J., and FAIR, J. R., *Chem. Eng. Prog.*, **56**(8), 48 (1960b).

CZERMANN, J. J., GYOKHEGYI, S. L., and HAY, J. J., *Pet. Refin.*, **37**(4), 165 (1958).

ECKERT, J. S., *Chem. Eng. Prog.*, **66**(3), 39 (1970).

ECKERT, J. S., *Chem. Eng. Prog.*, **71**(8), 70 (1975).

ESIA, M. A. R., DIGGORY, P. J., and HOLLAND, F. A., *Energy Res.*, **10**, 333 (1986).

FACTOR, H. M., and GROSSMAN, G., *Solar Energy*, **24**, 541 (1980).

FELLINGER, L., "Absorption of Ammonia by Water and Acid in Various Standard Packings," Sc. D. Thesis, Massachusetts Institute of Technology, Cambridge, Mass., 1941.

GANDHIDASAN, P., and SATCUNANATHAN, S., *Proceedings Biennal Congress International Solar Energy Society 8th Meeting*, **3**, 1726 (1983).

GANDHIDASAN, P., KETTLEBOROUGH, C. F., and ULLAH, M. R., *J. Sol. Energy Eng.*, **108**, 123 (1986).

GANDHIDASAN, P., ULLAH, M. R., and KETTLEBOROUGH, C. F., *J. Sol. Energy Eng.*, **109**, 89 (1987).

GROVER, G. S., DEVOTTA, S., and HOLLAND, F. A., *Ind. Eng. Chem. Res.*, **28**(2), 250 (1989).

GUTKOWSKI, K. M., and RYDUCHOWSKI, K. W., *Int. J. Refrig.*, **9**, 39 (1986).

HINES, A. L., and MADDOX, R. N., *Mass Transfer: Fundamentals and Applications*, Prentice-Hall, Englewood Cliffs, N. J., 1985.

HINES, A. L., and GHOSH, T. K., "Air Dehumidification and Removal of Indoor Pollutants by Liquid Desiccants," Gas Research Institute, Chicago, Ill., Report No. GRI-92/0157.4, 1992.

HOLLAND, C. D., *Multicomponent Distillation*, Prentice-Hall, Englewood Cliffs, N. J., 1963.

HOLLANDS, K. G. T., *Solar Energy*, **7**(2), 39 (1963).

KREMSER, A., *Nat. Pet. News*, **22**(21), 48 (1930).

LEBOEUF, C. M., and LOF, G. O. G., *Proceedings of the American Section of the International Solar Energy Society*, **3**(1), 205 (1980).

LEVA, M., *Tower Packings and Packed Tower Design*, 2nd ed., The U. S. Stoneware Co., Akron, O., 1953.

LEWIS, W. K., *Trans. Am. Inst. Chem. Eng.*, **20**, 1 (1927).

LOF, G. O. G., LENZ, T. G., and RAO, S., *J. Sol. Energy Eng.*, **106**, 387 (1984).

MCCABE, W. L., SMITH, J. C., and HARRIOT, P., *Unit Operations of Chemical Engineering*, 4th ed., McGraw-Hill, New York, 1985.

MORRIS, G. A., and JACKSON, J., *Absorption Towers; with special reference to the design of packed towers for absorption and stripping*, Butterworths Publications, London, 1953.

MOSCHANDREAS, D. J., and RELWANI, S. M., "Impact of the Humidity Pump on Indoor Environments," Somerset Technology, Inc., New Brunswick, N. J. (1990).

ONDA, K., SADA, E., and MURASE, Y., *AIChE J.*, **5**, 235 (1959).

ONDA, K., TAKEUCHI, H., and OKUMOTO, Y., *J. Chem. Eng. Jpn.*, **1**, 6 (1968).

PEDERSEN, B. S., and FISK, W. J., "The Control of Formaldehyde in Indoor Air by Air Washing," Report LBL-17381, Lawrence Berkeley Laboratory, Berkeley, Calif., 1984.

PENG, C. S. P., and HOWELL, J. R., *J. Sol. Energy Eng.*, **103**, 67 (1981).

PENG, C. S. P., and HOWELL, J. R., *J. Sol. Energy Eng.*, **106**, 133 (1984).

PERRY, R. H., CHILTON, C. H., and KIRKPATRICK, S. D., (eds.), *Chemical Engineers' Handbook*, 4th ed. McGraw-Hill, New York, 1963.

SCALABRIN, G., and SCALTRITI, G., *Termotecnica*, **38**, 87 (1984).

SHERWOOD, T. K., and HOLLOWAY, F. A. L., *Trans. Am. Inst. Chem. Eng.*, **36**, 39 (1940).

STRIGLE, R. F., Jr., *Random Packings and Packed Towers: Design and Applications*, Gulf Publishing Co., Houston, Tex., (1987).

SUDAKOV, E. N., and SUDAKOVA, E. E., *Neft Gaz*, **29**(10), 46 (1986).

VAN KREVELEN, D. W., and HOFTIJZER, P. J., *Chem. Eng. Prog.*, **44**, 532, (1948).

10

Adsorption Methods

10.1 INTRODUCTION

For a number of years air cleaners have been used to remove air contaminants from both industrial and nonindustrial environments by employing adsorption processes. The oil and chemical industries make extensive use of adsorption in the cleanup and purification of wastewater streams, the dehydration of gases, the removal of sulfur dioxide from stack gases, and as a means of fractionating fluids that are difficult to separate by other separation methods. More recently adsorbents have been used to dehumidify the air in commercial buildings to reduce the latent energy load placed on vapor compression air-conditioning systems.

A number of porous solid materials are capable of adsorbing gases or vapors, but the majority of adsorbents used for air cleaning and/or dehydration of gas streams belong to one of four general types: silica gel, activated alumina, activated carbon, and zeolites (known as molecular sieves). Depending on the specific characteristics of their surface, some absorbents have a greater affinity for polar compounds, such as water, whereas other adsorbents, such as activated carbon preferentially adsorb nonpolar compounds. For example, silica- and alumina-based adsorbents are used primarily for dehydration, while activated carbon is best suited for the adsorption of organic vapors. Commercially available silica-based adsorbents include silica gel and sorbead (silica based beads). The best known alumina-based materials are activated alumina and activated bauxite. Activated carbons can be produced from a number of base materials, but the two most readily available commercially are coal based and coconut based. Molecular sieves, however, have been employed for both dehydration and selective adsorption of other components. A variety of molecular sieves is available commercially; they are identified as 3A, 4A, 5A, 10X, and 13X, depending on the average pore diameter and which cation is present in the adsorbent. Also, several naturally occurring porous clays are available that have excellent adsorption properties; these include chabazite, mordenite, chinoptilolite, and phillipsite and are collectively referred to as zeolites. All these possess different adsorption characteristics, making the adsorption process the most versatile of any of the indoor air treatment methods, provided that the proper adsorbent or combination of adsorbents is used.

10.2 FUNDAMENTALS

Adsorption involves the transfer of a material from one phase to a surface where it is bound by intermolecular forces. Although adsorption is usually associated with the transfer from a gas or liquid to a solid surface, transfer from a gas to a liquid surface also occurs. The substance being concentrated on the surface is defined as the *adsorbate*, and the material on which the adsorbate accumulates is defined as the *adsorbent*.

The design of an effective adsorption process requires information about the amount of gases or vapors adsorbed as a function of their concentration in the gas phase at different temperatures. This type of information is generally referred to as an *adsorption isotherm* or *equilibrium adsorption data* and is a function of the surface characteristics of the adsorbent and the physical and chemical properties of the gas and/or vapor being adsorbed. Various shapes of curves representing the isotherm data can be obtained. The great majority of the isotherms observed to date can be classified into the five types shown in Figure 10.1. The different shapes are an indication of the actual adsorption mechanism, which is related to the properties of the adsorbate and adsorbent. This is explained in more detail later in the chapter.

Adsorption isotherm data are frequently obtained using a static system in which the amount of gas or vapor adsorbed on the solid surface is measured as a function of pressure or concentration when the solid adsorbent is exposed to an atmosphere containing the adsorbate at constant temperature. Experimental data can also be obtained by using a flow system. When employing a flow system, a stream containing the adsorbate (contaminants) is passed through a fixed bed of solid adsorbent. As the bed becomes saturated, the adsorbate concentration in the outlet stream becomes equal to that of the inlet stream. The data obtained from such systems are expressed in terms of the concentration profile of the adsorbate in the effluent gas stream as a function of time and is typically referred to as a *breakthrough curve*. The properties of several adsorbents are given in Table 10.1.

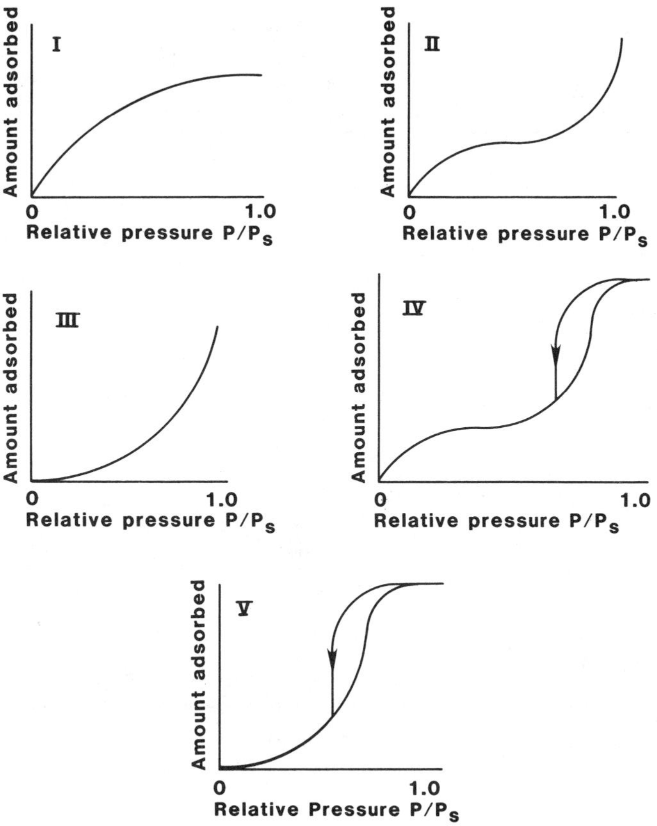

Figure 10.1 *Brunauer's classification of adsorption isotherms, showing amount adsorbed vs. final concentration in the fluid.*

Equilibrium Considerations

The selection of an adsorbent includes a consideration of surface area as well as the type of adsorbate involved in the adsorption process, since these relate to the types of bonds formed. Depending on the types of bonds, adsorption is described as either *physical adsorption* or *chemical adsorption* (*chemisorption*). Physical adsorption results when the adsorbate adheres to the surface by van der Waals forces (that is, by dispersion and Coulombic forces). Although a displacement of electrons may occur, electrons are not shared between the adsorbent and adsorbate. During adsorption, a quantity of heat, described as the *heat of adsorption*, is released. The quantity of heat released during physical adsorption is approximately equal to the heat of condensation, resulting in adsorption being frequently described as a condensation process. As expected, the quantity of material physically absorbed increases as the adsorption temperature decreases. The nature of the forces for phys-

TABLE 10.1 Adsorbent Properties [a]

Adsorbents	Pore Volume (cm^3/g)	Particle Density (g/cm^3)	Bulk Density (g/cm^3)	Surface Area (m^2/g)	Physical Form
Silica gel					
Regular density [b]	0.43	1.13	0.72–0.77	750–800	Granular
Intermediate density [b]	1.15	0.62	0.40	340	Granular
Molecular sieve					
5A [b]	0.32–0.33	1.14–1.16	0.71	—	Pellets
13X [b,c]	0.41	1.13	0.72	395	Spherical beads,
5A [d]	0.32	1.07	(bead)	—	pellets
			0.72		Pellets
Activated alumina					
F1[e]	0.20	1.40	—	250	Granular
A-201[f]	0.50	1.60	0.77	325	Spherical beads
Activated carbon					
Type BPL (coal based)[g]	0.80	0.80	0.48	1050–1150	Granular
Type PCB (coconut based)[g]	0.72	0.85	0.44	1150–1250	Granular

[a]Manufacturer's data unless stated otherwise. [b] Davison Chemical Division of W.R. Grace & Co.
[c]Measured in our laboratory. [d] Linde Division of Union Carbide Co.
[e]Aluminum Co. of America (Alcoa). [f] LaRoche Chemicals. [g] Calgon Carbon Corporation.

ical adsorption is such that multiple layers of the adsorbate can accumulate on the surface of the adsorbent.

The primary difference between physical adsorption and chemisorption is the nature of the bond formed between the adsorbed molecule and the adsorbent surface. Chemisorption is characterized by a sharing of electrons between the adsorbent and adsorbate, which results in the liberation of a quantity of heat that is approximately equal to the heat of reaction. As a result of the sharing of electrons with the surface, chemisorbed materials are restricted to the formation of a monolayer. Although chemical and physical adsorption are characterized by different thermal effects, a clear line between the two adsorption mechanisms does not exist. In some cases it has been observed that the quantity of adsorbate chemisorbed on a surface increases with increased temperature. Considering the nature of the bond, this effect might be expected.

The equilibrium isotherms shown in Figure 10.1 for the adsorption of vapors were classified into five principle forms by Brunauer (1945) based on the adsorption mechanism. Type I is classified as the *Langmuir* type and is characterized by a monotonic approach to a limiting adsorption capacity that corresponds to the formation of a complete monolayer. This type is found for systems in which the adsorbate is chemisorbed. Type I isotherms have been observed for microporous adsorbents such as charcoal, silica gel, and molecular sieves in which the capillaries have a width of only a few molecular diameters. A type II isotherm is characteristic of the formation of multiple layers of adsorbate molecules on the solid surface. This type, which is known as the *BET* after Brunauer, Emmett, and Teller (1938), has been found to exist for nonporous solids. Type III isotherms, although similar to type II because they have been observed for nonporous solids, are relatively rare. The shape of type III isotherms also suggests the formation of multilayers. Types IV and V are considered to reflect capillary condensation since they level off when the saturation pressure of the adsorbate vapor is reached. Both types of isotherms exhibit a hysteresis loop during desorption; porous adsorbents provide isotherms of this shape.

Single-Component Monolayer Models

Adsorption equilibrium studies, which correspond to no net mass transfer between phases, are used to determine the distribution of an adsorbate between the bulk fluid phase and the phase adsorbed on the surface of a solid adsorbent. The equilibrium distribution (equilibrium isotherm) is generally measured at a constant temperature. A number of mathematical models have been proposed to describe the adsorption process. In addition to monolayer and multilayer adsorption equations, models have been developed to describe situations in which the adsorbate occurs either locally on specific sites or is mobile over the surface of the adsorbent. Consideration has also been given to cases in which the adsorbed molecules interact both with the surface and with each other. Rather than attempting to discuss all the isotherm equations that have been developed over the years, only selected equations will be presented in this section. For a comprehensive discussion of the various isotherms and the mechanisms of adsorption, the reader is referred to the monographs of Young and Crowell (1962), Ross and Olivier (1964), Ruthven (1984), and Yang (1987).

One of the oldest and most frequently used isotherm equations was developed by Langmuir (1918). The Langmuir equation applies to cases in which there is no interaction among molecules on the surface, and the surface is homogenous. It provides a reasonable description of type I systems and is often justified based on its ability to fit equilibrium data. The Langmuir equation can be expressed as

$$q_A = \frac{QKC_A}{1 + KC_A} \tag{10.1}$$

where q_A = equilibrium uptake of adsorbate A by the adsorbent corresponding to concentration C_A

Q = weight of adsorbate contained in the monolayer on the surface

C_A = concentration of adsorbate A in the fluid phase in equilibrium with the concentration q_A on the solid

K = constant

Another frequently used isotherm is the semiempirical Freundlich equation given as

$$q_A = K(C_A)^{1/n} \tag{10.2}$$

where q_A = equilibrium uptake of adsorbate A by the adsorbent corresponding to concentration C_A

C_A = concentration of A in the fluid phase that is in equilibrium with the concentration q_A on the solid

K, n = constants

The Freundlich equation was derived by Zeldowitsch (1934) from a consideration of the decrease in the number of adsorption sites.

It is important to note that the Langmuir equation reduces to a linear form, frequently identified as the Henry's law equation, as the fluid phase concentration approaches zero; whereas, the Freundlich equation does not. The Henry's law equation is given as

$$q_A = K'C_A \tag{10.3}$$

When fitting equilibrium adsorption data for indoor air pollutants, which are usually present at low concentrations, one should be cautious about using the Freundlich equation.

Jovanovic (1969) proposed another simple equation that applies to type I isotherms, which can be expressed in terms of the volume adsorbed as

$$V = V_m\,(1 - \exp(-a\,P/P_s)) \tag{10.4}$$

where V_m is the volume adsorbed in the monolayer, P_s is the saturation pressure, and a is a constant that describes adsorption in the monolayer. The term a is expressed as

$$a = \sigma\tau P_s\,/\sqrt{2\pi mkT} \tag{10.5}$$

where σ is the area occupied by one molecule on the surface, τ is the average settling time of a molecule adsorbed in the first layer, m is the mass of one molecule, T is the absolute temperature, and k is the Boltzmann constant. Although not as widely used as the isotherms discussed earlier, the Jovanovic equation has been used extensively in the development of several more complicated and accurate isotherm models that apply to heterogeneous surfaces.

All the above models apply to type I isotherms but frequently fail to fit the equilibrium data with suitable accuracy. This is often attributed to the heterogeneous nature of the absorbent surface. As a result, other models have been developed for type I systems based on a consideration of the energy of an adsorption site. It is well known that for a large class of solid-vapor systems, the surface heterogeneity plays an important role in determining adsorption characteristics, and its effects require adequate treatment. The general approach used to describe adsorbent heterogeneity is to postulate that the heterogeneous surface exhibits a distribution of adsorptive potentials that are either grouped in patches or distributed randomly on the surface. Even small variations in the adsorption potential have been shown to influence the adsorption behavior.

As shown by Ross and Olivier (1964), the overall isotherm is obtained by integrating the contribution of each patch over the energy distribution range. The adsorption isotherm is thus given as

$$Q(P,T) = \int_0^\infty Q_1(P,T,e)E(e)\,de \tag{10.6}$$

where $E(e)$ is a probability distribution function and $Q(P, T)$ is the overall adsorption isotherm on the heterogeneous adsorbent. The term $Q_1(P, T, e)$ describes a specific adsorption isotherm for homotactic sites of adsorptive energy e. Both Q and Q_1 are amounts adsorbed per unit mass of adsorbent. Ross and Olivier (1964) used a number of probability distribution functions for $E(e)$, including a Gaussian function, and employed the two-dimensional van der Waals equation of state for the adsorption isotherm $Q(P, T)$. They solved the model equations numerically and found that the experimental isotherms of argon and nitrogen on various carbon blacks and a synthetic zeolite agreed well with the predicted values obtained when using a Gaussian distribution function. A detailed review of the studies that employ this approach is given by Jaroniec et al. (1981).

As shown by Ross and Morrison (1975) and House and Jaycock (1978), most of the published studies have focused on estimating $E(e)$ numerically from experimental $Q(P, T)$ data by assuming some well-known expression for $Q_1(P, T, e)$, such as the Langmuir model. However, analytical expressions for $E(e)$ were obtained by Misra (1969) and Jaroniec (1975) by assuming that the overall adsorption isotherm, $Q(P, T)$, could be represented by some well-known isotherm model, and the local isotherm, $Q_1(P, T, e)$, could be described by either the Langmuir model or the step isotherm.

Several studies have been carried out to develop analytical functions for $Q(P, T)$ since they are useful for data extrapolation and for the modeling of separation processes by adsorption. Among these, Misra (1970, 1973) derived analytical expressions that employed exponential and constant distribution functions in conjunction with either the Langmuir or the Jovanovic isotherm equation to represent the specific isotherm for homotactic sites. The overall isotherm equations he obtained by using different local isotherm equations were found to be comparable, provided that the distribution functions were the same. Cerofolini et al. (1978) used the Langmuir local isotherm equation but incorporated a condensation approximation as the adsorption energy distribution to obtain an expression for the overall isotherm equation. In contrast to the models developed by Misra, Cerofolini et al.'s equation reduced to the Henry's isotherm in the very low pressure limit. However, it was not tested using actual data. Sircar (1984a) developed an expression for $Q(P, T)$ by assuming that the Langmuir model could be used to represent the local isotherm and that the gamma probability density function described the energetic heterogeneity of the adsorbent surface. Sircar's model successfully described the adsorption of various gases on activated carbons and zeolites over large ranges of pressure and temperature. The resulting equation, however, is somewhat cumbersome to use. Sircar (1984b) later developed a simpler model that had the same degree of versatility by using the Jovanovic local isotherm model in conjunction with the gamma distribution function. Sircar analyzed the effect of the choice of the local isotherm on the adsorbent heterogeneity and reported that both models were capable of accurately describing the overall adsorption data. His work and the work of others suggest that the choice of the probability distribution function has a greater impact on the overall outcome than does the selection of the local isotherm equation.

Hines et al. (1990) developed an adsorption equation to describe the adsorption of gases on energetically heterogeneous surfaces, using the Jovanovic model as the local isotherm in conjunction with a modified Morse-type distribution function to describe the energy distribution of the heterogeneous surface. The Hines et al. equation is

$$Q(P,T) = Q_m [1 - \frac{K_1K_3}{K_3 - K_1K_2}(\frac{1}{P + K_1} - \frac{K_2}{P + K_3})] \tag{10.7}$$

where Q_m is the saturation adsorption capacity and the K values are constants.

The accuracy of their equation was evaluated by using published experimental data for several systems. Twelve adsorbate-adsorbent pairs were correlated over a wide range of pressures, and the majority of the pairs were analyzed at three different temperatures. The adjustable model parameters were determined for each data set by using a nonlinear regression analysis. Their model successfully represents experimental adsorption isotherm data for several gases and vapors on different heterogeneous adsorbents over a wide range of temperatures and pressures. Furthermore, it is easy to use, and it provides a useful tool for accurately correlating isotherm data on heterogeneous surfaces.

Single Component Multilayer Models

The formation of multiple adsorbed layers has been observed for a large number of adsorbate-adsorbent systems. Although multilayer adsorption is a physical process, only a limited number of models have contributed significantly to its understanding, and these models often fail to fit equilibrium adsorption data over a wide range of pressures.

The earliest multilayer model was proposed by Brunauer et al. (1938) and includes the basic assumptions of the Langmuir equation, except that multilayers are formed, and the heats of adsorption for these layers are different from the first layer. Their model, which is typically referred to as the BET equation, is given as

$$\frac{P}{V(P_s - P)} = \frac{1}{V_mC} + \frac{(C - 1)P}{(CV_mP_s)} \tag{10.8}$$

where V is the volume adsorbed at pressure P, V_m is the volume occupied in a monolayer, P_s is the saturation pressure, and C is a constant that is related to the energy of adsorption. Although the BET equation typically is valid over a relative pressure range, P/P_s, from 0.05 to 0.35, and it would consequently not be used to fit adsorption data, it is widely used as a method for determining surface areas of adsorbents. A plot of $P/(V(P_s - P))$ versus P/P_s should give a straight line with a slope

$$S = \frac{C - 1}{V_mC}$$

and an intercept

$$I = \frac{1}{V_mC}$$

The volume of the adsorbed gas that corresponds to a monolayer can be obtained by solving the above equations. Thus

$$V_m = \frac{1}{S + I}$$

A more recent multilayer adsorption model proposed by Jovanovic (1969) provides a method of correlating data over a wider relative pressure range, $0.2 < P/P_s < 0.7$, than does the BET equation. This model is given as

$$V = V_m(1 - \exp(-a\, P/P_s)) \exp(b\, P/P_s) \tag{10.9}$$

where a was defined earlier for the Jovanovic monolayer model, and b is related to uptake in the second and higher layers.

Ghosh (1989) developed a multilayer adsorption model for heterogeneous surfaces by using the same method described earlier for the development of the model of Hines et al., except that the Jovanovic multilayer adsorption isotherm equation was employed as the local isotherm. Ghosh's equation is

$$Q(T,P) = Q_m\left[1 - \frac{K_1K_3}{K_3 - K_1K_2}\left(\frac{1}{K_1 + P} - \frac{K_2}{K_3 + P}\right)\right]\frac{\exp(-b_0P)}{\varepsilon_M P}\left[\exp(\varepsilon_M P - 1)\right] \tag{10.10}$$

where K_1, K_2, and K_3 are energy parameters in a Morse-type probability distribution function, Q_m is the adsorption capacity at saturation, b_0 is the limiting value of the constant b given in the Jovanovic equation, and ε_M is the energy parameter for second and higher layers in the adsorbed phase. Ghosh's model was tested for several systems. The adsorbents for the systems tested were microporous, and the data for these systems could not be accurately described with existing homogeneous isotherm equations. Equation (10.10) provided excellent correlation of the data for all systems tested. This model is recommended for type II isotherms.

Multicomponent Models

The design of an adsorber to remove two or more pollutants from an indoor air stream requires multicomponent adsorption equilibrium data. While multicomponent data may not be available and are difficult to obtain experimentally, for some cases it is possible to calculate multicomponent equilibrium from pure component isotherm data by means of a suitable model. A number of such models have been developed, but it is not our intent here to provide an exhaustive discussion of all the existing multicomponent models. We will, therefore, limit our discussion to only selected models that have found wide use in adsorber design. The application of many multicomponent models is beyond the scope of this book. The interested reader is referred to the work of Ruthven (1984) and Yang (1987).

One of the earliest multicomponent models was developed by Markham and Benton (1931), who extended the single-component Langmuir equation to describe adsorption in a multicomponent system. Their equation for an n-component system is written in terms of the amount adsorbed as

$$q_i = \frac{Q_iK_iP_i}{1 + \sum_{j=1}^{n} K_jP_j} \tag{10.11}$$

where P_i is the partial pressure of component i in the gas phase, Q_i is the amount adsorbed in the monolayer for component i, and K_i is a constant for component i. The total quantity adsorbed for all components is thus given as

$$q_T = \sum_{i=1}^{n} q_i = \frac{\sum_{i=1}^{n} Q_i K_i P_i}{1 + \sum_{j=1}^{n} K_j P_j} \tag{10.12}$$

In this model the assumption is made that the amount of one component that can be adsorbed in the monolayer does not influence the amount of other components that are adsorbed on the surface. Broughton (1948) and Kemball et al. (1948) showed that the above equation was thermodynamically consistent if the amounts of each component adsorbed in the monolayer are equal. Because this is not typically the case, Innes and Rowley (1947) proposed that an average value be used for the monolayer coverage. For a binary system, the average value is obtained from the expression

$$\frac{1}{Q_a} = \frac{x_1}{Q_1} + \frac{x_2}{Q_2} \tag{10.13}$$

where Q_a is the average amount adsorbed in the monolayer, and x_1 and x_2 are mole fractions of the two components in the adsorbed phase. Although the above modification provides significant improvement in terms of its ability to fit the model to experimental data, the model is limited to pure component data that can be fit with the Langmuir equation. In addition, the adsorbate molecules must be of comparable size.

An improvement in the ability to fit multicomponent equilibrium data is obtained by using a combined Langmuir-Freundlich equation, as demonstrated by Sips (1948), Koble and Corrigan (1952), and Yon and Turnock (1971). The Langmuir-Freundlich equation is given as

$$\frac{q_i}{Q'_i} = \frac{K_i P_i^{1/n_i}}{[1 + \sum_{j=1}^{n} (K_j P_j)^{1/n_j}]} \tag{10.14}$$

where Q'_i was defined by Yon and Turnock as the maximum obtainable loading instead of the monolayer capacity. Since q_i/Q'_i for a pure component represents the loading ratio for the adsorbent, Eq. (10.14) has been referred to as the *loading ratio correlation* (LRC). The LRC is not thermodynamically consistent and does not have a theoretical foundation. It does, however, provide a good fit to experimental data and, consequently, is used widely in adsorber design.

It is important to note that the parameters in both the multicomponent Langmuir equation and the LRC can be obtained from pure component isotherm data. In principle, it is thus possible to predict multicomponent equilibrium. For adsorbate-adsorbent systems in which there is strong interaction between the adsorbates and between the adsorbates and the adsorption sites, pure component data cannot be relied on to provide an adequate fit to multicomponent data. Yon and Turnock (1971) followed the method of Schay (1956) to modify the LRC to account for such interactions.

A number of researchers (Arnold, 1949; Lewis et al., 1950; Myers and Prausnitz, 1965; Cook and Basmadjian, 1965) developed models for multicomponent adsorption by assuming that the adsorbed phase formed an ideal solution on the solid surface. The basic assumption of the ideal adsorbed solution (IAS) theory is that equilibrium exists between the gas phase and adsorbed phase for each component of the mixture; it is given by

$$Py_i = P_i^0 x_i \tag{10.15}$$

where P_i^0 is a function of the spreading pressure (π) and is interpreted as the equilibrium pressure that pure component i should have in the gas phase to produce the same spreading pressure as that of the mixture (π_m) at the same temperature when adsorbed on the solid surface. The term P is the total pressure, and y_i and x_i are mole fractions in the gas phase and on the solid surface, respectively.

Sircar and Myers (1973) showed that the models developed by the previous researchers differed from each other primarily in their choice of standard states. Unfortunately, all the models performed poorly for multilayer adsorption. Sircar and Myers developed a model that was based on a surface potential theory of multilayer adsorption. They also considered the adsorbed phase to be ideal, and they were successful in predicting the data for O_2-N_2 on anatase and for benzene-2, 4-dimethylpentane on silica gel.

O'Brien and Myers (1984) developed an algorithm based on the IAS theory to predict the adsorbed phase mole fractions from pure component isotherm data. The algorithm required an isotherm equation to represent the pure component data accurately that, at the same time, could be integrated analytically to obtain the spreading pressure. Otherwise, their model required repeated numerical integrations that increased the computation time. Later Moon and Tien (1987) tried to improve the algorithm by reformulating the equations of O'Brien and Myers. Substantial improvement in computing time was reported for mixtures having more than five components. Moon and Tien compared the predictive capability of their model with the algorithm of O'Brien and Myers (1984).

Many researchers (Costa et al., 1981; Hyun and Danner, 1982; Talu and Zwiebel, 1986) have pointed out that only a few systems behave ideally in the adsorbed phase. However, the nonideality of the adsorbed phase can be taken into account by introducing the activity coefficient in Eq. (10.15). Assuming that the gas phase is ideal, Eq. (10.15) can be written as

$$Py_i = \gamma_i P_i^0 x_i \tag{10.16}$$

where γ_i is the activity coefficient. The activity coefficient for each component is calculated from the binary mixture data. Costa et al. (1981) used the Wilson and the UNIQUAC-type activity coefficient models to estimate the activity coefficient of each component in the mixture, while the binary interaction parameters for these models were obtained from the binary mixture data. Talu and Zwiebel (1986) developed an expression from the superlattice model and followed the approach of Maurer and Prausnitz (1978) to calculate the activity coefficients. Paludetto et al. (1987) used an approach similar to the one followed by

Costa et al. (1981) to predict the adsorption equilibrium data of ternary mixtures of xylenes on a zeolite but used a different approach to calculate the spreading pressure of the mixture. However, one of their assumptions, that there was no area change upon mixing at constant π and T, might not be valid for other systems. Paludetto et al. employed the Wilson and the Hilderband equations to estimate the activity coefficients and obtained good agreement between the experimental data and the predicted values for ternary mixtures.

Jaroniec (1975, 1977, 1978) and Jaroniec and Toth (1976, 1978) developed partial isotherms for individual components of a binary gas mixture that was adsorbed on a heterogeneous solid surface. Various pure component isotherm equations, such as the Jovanovic, Freundlich, Langmuir, and Toth models, were used along with a combination of different distribution functions. The predictive capabilities of these isotherms are rather limited and were tested for only a few systems. Jaroniec and Toth (1976) successfully predicted the partial isotherms of C_2H_4 from the C_2H_4-C_2H_6 mixture and C_2H_6 from the C_2H_6-C_2H_4 mixture, both on Nuxit-AL at 293 K. However, a large deviation was reported for C_3H_6 from the C_3H_6-C_3H_8 mixture on the same adsorbent at the same temperature. Their treatment was also limited to binary systems and was not extended to ternary or higher-order mixtures.

A review of the different methods used to predict multicomponent adsorption has been provided by Sircar and Myers (1973). Jaroniec (1980) reviewed the kinetics and the equilibrium state of adsorption for multicomponent systems. Still other methods that have been used to predict binary and multicomponent adsorption data are discussed in detail by Yang (1987).

10.3 FUNDAMENTALS OF DYNAMIC ADSORPTION

The analysis of adsorption in packed beds is based on the development of effluent concentration-time curves, which are a function of adsorber geometry and operating conditions and equilibrium adsorption data. These breakthrough curves are obtained by flowing a fluid that contains an adsorbable solute with an initial concentration C_0 through a packed bed that contains a clean or regenerated adsorbent. As the flow of the fluid continues, the bed becomes saturated with the adsorbate at a given position, and a concentration distribution is established within the bed as shown in Figure 10.2.

At time t_i, the solute first appears in the effluent stream. Time t_b is defined as the time required to reach the breakpoint concentration, indicated as C_b. This corresponds to the maximum concentration allowable in the effluent stream. Time t_e is the time at which the bed becomes totally saturated with the adsorbate. At this time the bed is exhausted and must be regenerated prior to reuse. The time period from t_i to t_e corresponds to the thickness of the adsorption or mass transfer zone in the bed and is related to the mechanism of the adsorption process. It is readily seen that the area behind the breakthrough curve represents the quantity of adsorbate retained by the adsorbent contained in the column. This corresponds to a point on the equilibrium isotherm. If the isotherm can be represented by the Langmuir equation, this point is expressed as

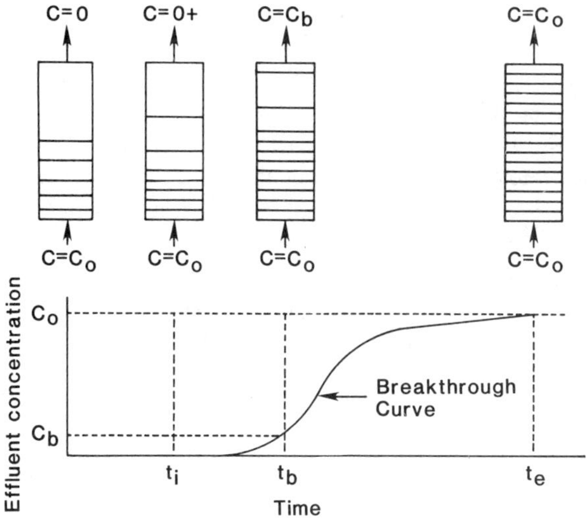

Figure 10.2 *Adsorption wavefront.*

$$q^{\infty} = \frac{QKC_0}{1 + KC_0} \quad or \quad C_s^{\infty} = \frac{Q'KC_0}{1 + KC_0} \tag{10.17}$$

where C_0 is the concentration of solute in the influent solution, and q^{∞} and C_s^{∞} are the saturation capacities of the adsorbate in the adsorbent bed. Consistent concentration units must be used in these equations.

For the purpose of discussing dynamic adsorption, we will classify the equilibrium isotherms as either (a) favorable, (b) linear, or (c) unfavorable. These are shown in Figure 10.3. If the isotherm is concave in the direction of the fluid phase concentration, as shown by curve (a), layers of high concentration in the bed move faster than layers of lower concentration. This results in the adsorption zone becoming thinner as the wavefront moves through the bed and gives a breakthrough curve that is self-sharpening. For a favorable isotherm, the breakthrough curve develops and moves through the packed column in a constant pattern. The unfavorable isotherm results in a breakthrough curve that becomes more diffuse as it traverses the bed length. For nonequilibrium adsorption, a favorable isotherm will yield a constant pattern breakthrough curve after a period of time. In most industrial adsorption systems, equilibrium between the adsorbate and adsorbent is not reached.

The velocity of the gas stream flowing through the bed also changes the shape of the

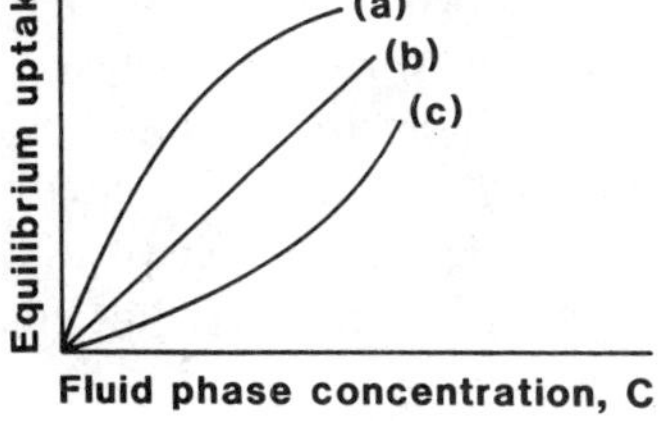

Figure 10.3 *Shape of equilibrium isotherms.*

breakthrough curve and influences the rate of adsorption. At low flow rates, equilibrium conditions are approached, but axial dispersion of the adsorbate can be significant and must be considered in the bed design. At lower velocities, the rate of adsorption may be limited by the rate of mass transfer from the fluid phase to the solid surface. For higher flow rates, axial dispersion is typically insignificant, relative to its impact on the rate of mass transfer and its influence on the shape of the breakthrough curve, but adsorbate-adsorbent equilibrium is not attained. The overall rate of adsorption is controlled by a number of factors, including external mass transfer to the surface, internal mass transfer through the fluid that fills the pores of the adsorbent, internal mass transfer across the solid surface, and the actual rate of adsorbate uptake. Discussions of the various mechanisms that impact the rate of adsorption in a bed are given by Hines and Maddox (1985) and Yang (1987).

LUB-Equilibrium Method

The region of the bed over which the adsorbate concentration changes from the specified breakthrough concentration, C_b, to the inlet or fully loaded concentration, C_0, is defined as the height or thickness of the mass transfer zone. The thickness of the zone is a function of the adsorbate-adsorbent system and depends specifically on the rate of mass transfer in the solid, the transfer rate to the solid surface, and the concentration difference. Ideally, we would like to have an effluent-concentration curve in addition to equilibrium data prior to calculating the amount of adsorbent necessary to make a certain separation. If neither are available, however, it may be more expedient to obtain a breakthrough curve than to measure equilibrium isotherm data. A breakthrough curve obtained at the expected design temperature and pressure provides uptake data under dynamic conditions in addition to the height of the mass transfer zone, which gives the length of the unused bed (LUB).

The LUB-equilibrium method is frequently used in adsorber design. In this method the packed-bed adsorber is viewed as consisting of two sections, the equilibrium section and the LUB (length of unused bed) section. The size of the equilibrium section is found from equilibrium adsorption data at the bed design temperature. The length of the equilibrium section represents the shortest bed length possible and can be described as the stoichiometric length, since the adsorbent in the equilibrium section of the bed is assumed to be in equilibrium with the adsorbate in the fluid. The stoichiometric wavefront moves through the bed as a step function.

Because of the presence of the mass transfer zone, all the adsorbent behind the actual wavefront will not be at its maximum capacity. Therefore, we must add an additional quantity of adsorbent to the bed to compensate for the presence of the mass transfer zone, which is described as the LUB. The stoichiometric wavefront, relative to the actual stable wavefront, is shown in Figure 10.4. In this figure, t_b is defined as the time when the leading edge of the breakthrough curve leaves the bed, t_e is the time when the trailing edge of the wavefront leaves the bed, and t_s is the time when the stoichiometric wavefront would leave. The stoichiometric front is found by equating the unused bed capacity behind the front to the used capacity ahead of the front. The stoichiometric time is found by adjusting t_s until the areas indicated by A' and B' are equal. At breakthrough, the leading edge of the actual wavefront, relative to the stoichiometric front, is

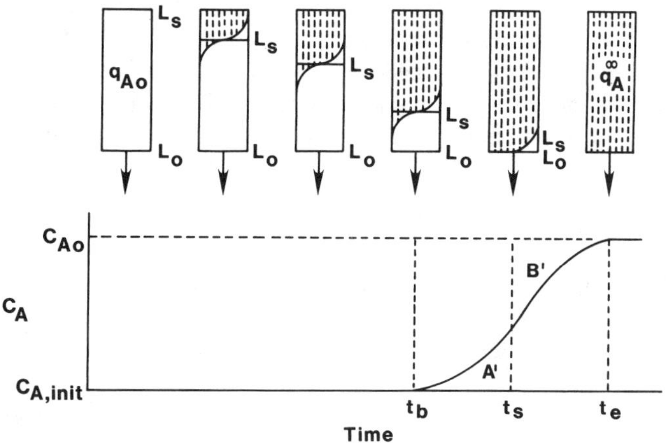

Figure 10.4 *Stoichiometric front relative to the mass transfer zone.*

$$LUB = L_0 - L_s \tag{10.18}$$

where L_0 is the total bed length and L_s is the distance the stoichiometric front has moved through the bed. The important assumptions on which the LUB-equilibrium section concept is based have been summarized by Collins (1967).

From Figure 10.4, the length of the unused bed can be shown to be

$$LUB = L_0 \frac{(t_s - t_b)}{t_s} \tag{10.19}$$

As shown by Collins (1967), the length of the stoichiometric wavefront can be expressed as

$$L_s = \frac{U\,(C_{A0} - C^*_{A,init})t_b M_A}{\rho_b(q_A^\infty - q_{A0})} \tag{10.20}$$

where

q_A^∞ = saturation capacity of the adsorbent, g solute/g solid

q_{A0} = initial concentration of solute on the adsorbent, g solute/g solid

C_{A0} = concentration of solute in the influent, gmol/cm^3

$C^*_{A,\text{init}}$ = concentration of solute in equilibrium with q_{A0}, gmol/cm^3

M_A = molecular weight of the adsorbate

u = superficial velocity

e_b = density of the adsorbent in the bed, g solid/cm^3

From this equation, the velocity of the stoichiometric wavefront is

$$u' = \frac{L_s}{t_b} = \frac{U(C_{A0} - C^*_{A,\text{init}})M_A}{\rho_b\,(q_A^\infty - q_{A0})} \tag{10.21}$$

Examination of the material balance used by Collins (1967) shows that the maximum driving force for mass transfer was used for both phases, and the driving force is constant. The approximate nature of the material balance above should be recognized since the concentration differences vary with time and position. Equation (10.21) does provide a method of obtaining the length of an adsorber bed if the adsorption cycle time is specified.

Fluid Velocity and Bed Diameter

Generally, the thickness of the mass transfer zone is a function of gas velocity and decreases as the velocity is increased. Because of this, a longer adsorption bed operated at a higher gas velocity will usually result in a higher uptake capacity than will a shorter bed. Obviously, the treatment of a fixed quantity of gas per day using a higher gas velocity would result in a smaller bed diameter, but would cause a greater pressure drop through the bed. In some cases, the magnitude of the pressure drop may be the limiting factor in the design.

The maximum velocity in a bed is related to the direction of gas flow. If the flow is upward, the maximum velocity is the velocity required to fluidize the bed. Due to the difficulties in trying to operate an adsorption bed in a fluidized state, a velocity no greater than about 80% of that value is recommended in a preliminary design. For downward flow, manufacturers frequently recommend a velocity equal to about 1.8 times the fluidization velocity. The superficial fluidization velocity can be estimated from any of several equations, as shown by Couderc (1985). For our purposes, an equation proposed by Richardson and Jeronimo (1979) is recommended:

$$\mathrm{Re}_{mf} = [(25.7)^2 + 0.0365\ \mathrm{Ga\ Mv})\ ]^{0.5} - 25.7 \qquad (10.22)$$

where Ga is the Galileo number and is equal to

$$\mathrm{Ga} = \frac{d_p^3 \rho_f^2\, g}{\mu^2} \qquad (10.23)$$

and Mv is the density ratio given as

$$\mathrm{Mv} = \frac{\rho_p - \rho_f}{\rho_f} \qquad (10.24)$$

where d_p is the particle diameter, ρ_f and ρ_p are the densities of the fluid phase and solid particle, respectively, μ is the viscosity of the fluid, and g is the gravitational constant.

For cases in which the rate of internal mass transfer is very low and limits the rate of adsorption, it may become necessary to use a lower gas velocity to increase the contact time. Operating a bed at a low velocity, however, will increase the bed size and also may result in a poor bed performance. To prevent this, a good starting velocity in a preliminary design corresponds to conditions at which the flow becomes turbulent.

According to Zenz and Othmer (1960), the Reynolds number at which flow changes from laminar to turbulent flow depends on the percent of void volume in the bed. For bed void fractions of 0.3, 0.4, 0.5, 0.6, and 0.7, the Reynolds numbers at which this occurs are approximately 100, 55, 30, 15, and 10, respectively; these values are shown in Figure 10.5. The Reynolds number is defined as

$$\mathrm{Re} = \frac{\rho_f U d_p}{\mu_f} \tag{10.25}$$

where U is the superficial velocity and is based on the cross-sectional area of an empty bed. For adsorbent particles other than spheres, Zenz and Othmer (1960) recommend calculating an equivalent diameter as

$$d_p = \frac{6V_p}{S_p} \tag{10.26}$$

where V_p is the particle volume and S_p is the external surface area. Most packed adsorbent beds will have void fractions that range from 0.30 to 0.70.

Example 10.1

Calculate the fluidization velocity and the velocity at which flow becomes turbulent for adsorption from an air stream flowing in a packed bed of regular-density silica gel particles. Assume that the particles have an equivalent diameter of 3 mm, and the air temperature is 20°C. Assume the bed has a void fraction of 0.35. The properties of the air and silica gel are

$$\mu_f = 1.8 \times 10^{-5}\ \mathrm{kg/m \cdot s}$$
$$\rho_p = 1130\ \mathrm{kg/m^3}$$
$$\rho_f = 1.2\ \mathrm{kg/m^3}$$

Solution: From Figure 10.5 for a void fraction of 0.35, the Reynolds number at which the air flow becomes turbulent is about 75. Therefore, the superficial velocity is

$$U = \frac{\mu_f \mathrm{Re}}{\rho_f d_p} = \frac{(1.8 \times 10^{-5})(75)}{(1.2)(3 \times 10^{-3})} = 0.375\ \mathrm{m/s} = 73.8\ \mathrm{ft/min}$$

The fluidization velocity is calculated from Eq. (10.22) proposed by Richardson and Jeronimo (1979).

$$U_f = \frac{1.8 \times 10^{-5}}{(1.2)(3 \times 10^{-3})}\{[(25.7)^2 + (0.0365)\frac{(3 \times 10^{-3})^3(1.2)^2(9.81)(1130 - 1.2)}{(1.8 \times 10^{-5})^2(1.2)}]^{0.5} - 25.7\}$$

$$= 0.88\ \mathrm{m/s} = 173\ \mathrm{ft/min}$$

The complete design of an adsorption system requires equilibrium adsorption data, along with an effluent concentration-time curve, or the ability to predict one using mass transfer data and an appropriate model. Even though a complete design is beyond the scope of this work, the amount of adsorbent necessary to make a certain separation can be estimated by using isotherm data only. In reality, this method underestimates the amount of adsorbent needed for a specific separation, because when the maximum breakthrough concentration, given as the breakthrough point t_b in Figure 10.2, is reached, the operation of the bed must be discontinued. This results in the amount of adsorbent contained in the bed between points t_b and t_e not being available for adsorption. To account for this loss of ca-

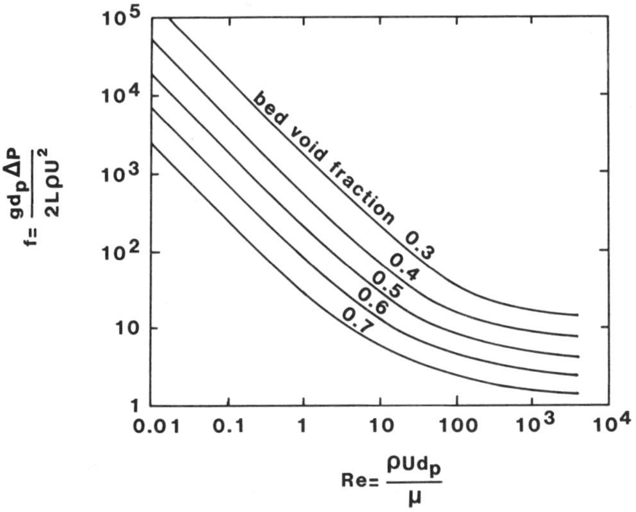

Figure 10.5 *Correlation of Reynolds number with respect to bed void fraction.*

pacity, a design based on equilibrium data only must be increased by some value, say 20% to 25%. This method is readily demonstrated by considering the following example.

Example 10.2

Use the equilibrium adsorption isotherm data of Yeh (1991) to estimate the amount of activated carbon required to remove 120 ppm of 1,1,1-trichloroethane from the air in a manufacturing facility. A total of 96,000 ft^3/day of air is to be processed.

Solution: The volume of 1,1,1-trichloroethane to be removed in one day is

$$\left(96{,}000\frac{\text{ft}^3}{\text{day}}\right)\left(120 \times 10^{-6}\text{ft}^3 \frac{\text{ft}^3 \text{ of adsorbate}}{\text{ft}^3 \text{ of air mixture}}\right) = 11.5\frac{\text{ft}^3 \text{ of adsorbate}}{\text{day}}$$

Assuming that the adsorption process will be carried out at 1 atm pressure and 25°C, the ideal gas law can be used to determine the mass adsorbed.

$$n = \frac{PV}{RT} = \frac{(1 \text{ atm})(11.5 \text{ ft}^3)(30.48)^3 \text{ cm}^3/\text{ft}^3}{(82.05 \text{ cm}^3 \text{ atm/gmol} \cdot \text{K})(298 \text{ K})} = 13.34 \text{ gmol}$$

Multiplying by the molecular weight of 133.42 g/gmol gives a mass of 1780 g of 1,1,1-trichloroethane to be removed in one day. The amount of 1,1,1-trichloroethane that can be adsorbed by activated carbon is determined by using the isotherm data shown in Table 10.2. To use these data, we must find the partial pressure of 1,1,1-trichloroethane in air at 1 atm pressure, which is

$$P = 760 \text{ mmHg } (120 \times 10^{-6}) = 0.0912 \text{ mmHg}$$

Using the data in Table 10.2, the uptake for a pressure 0.0912 mmHg is 300 mg of 1,1,1-trichloroethane per gram of activated carbon. If the bed is to operate for one day without being regenerated, the amount of activated carbon required is 5933 g or 13.1 lb.

The above value represents the minimum amount of activated carbon that would be required since the data shown in Table 10.2 are equilibrium data. To account for the length of the mass transfer zone, which represents the amount of the bed that is not available for use, the amount of activated carbon should be increased by some amount. As an estimate, we will assume that the bed should be increased by 25%. Therefore, 16.4 lb of activated carbon would be recommended.

TABLE 10.2 Equilibrium Adsorption Data for 1,1,1-Trichloroethane on Activated Carbon at 298 K

Pressure (mmHg)	*Uptake (mg/g)*
0.1	319
1.5	420
5.0	449
9.0	468
9.9	470
12.8	479
13.9	482
16.5	487
18.0	492

Source: Yeh, 1991.

10.4 APPLICATION OF SELECTED ADSORBENTS TO INDOOR AIR POLLUTION PROBLEMS

Silica gel, activated alumina, and molecular sieves are used primarily as solid desiccants to dehumidify gas streams. These adsorbents have been employed extensively for natural gas dehydration and the adsorption of H_2S and CO_2 from natural gas, but their use in indoor air dehumidification systems is more recent. Research work has been reported on the development of solid-desiccant-based dehumidification systems by several authors. The desiccant system converts the latent heat of vaporization to sensible heat by adsorbing moisture from the air. The dry air can then be further conditioned to meet specifications. The regeneration of the desiccant bed is usually accomplished by flowing a stream of return air heated by either a gas-fired or electric heater through the bed. Other forms of energy, such as waste heat and solar energy, have also been used in the regeneration step. These adsorbents have also been shown to have significant adsorption capacities for organic pollutants such as aldehydes, alcohols, aromatics, and chlorinated hydrocarbons.

Silica Gel

Inorganic pollutants. Cohen et al. (1985) installed a silica gel-based dehumidification system in a supermarket as an adjunct to the conventional vapor compression air conditioning system. Load models were used to obtain seasonal cooling cost projections for both systems operating under different conditions. An initial estimate indicated that the payback period of the hybrid system could be under one year.

An economic analysis by Marciniak (1985) showed that an integrated desiccant dehumidifier-electric air conditioner was economically superior to conventional systems in a climate like Miami, which generally has a high relative humidity. Marciniak also tested the unit in two other cities, Atlanta and Chicago, and came to the conclusion that such systems can provide enhanced comfort with lower operating cost.

Several studies have been conducted that are related to the development of desiccant-based dehumidification systems, with one of the objectives being to evaluate the effectiveness of these systems in removing pollutants from indoor air. The gas-fired desiccant system developed by Novosel et al. (1987) with Cargocaire Engineering Corporation is currently being used by 14 supermarket chains throughout the United States. Furthermore, the system has been specified as one of several alternative systems applicable for military commissaries. A systematic study was not conducted to monitor the enhanced comfort in the places where these units were installed, but random interviews of the occupants showed a positive response. In a later study, Novosel et al. (1988) provided a conceptual framework of a desiccant-based air conditioner that would actively control air quality in residences. From the published report, the authors expected that the silica gel could remove formaldehyde and other volatile organics during the dehumidification process.

One of the earliest investigations of silica gel to dehumidify air was made by Taylor (1945), who measured the water adsorption capacity of a commercially available silica gel in the temperature range of 70° to 450°F. A few years later, Hubard (1954) studied the adsorption by a similar type of the silica gel over the range of 45° to 340°F. Almost 18 years later, Jury and Edwards (1971) compared their data with that of Hubard and observed only a 3% to 5% difference in the adsorption capacity. They also noticed that silica gel produced a hysteresis loop during desorption runs, which has been attributed to capillary forces and the physical structure of the pores. A water molecule can penetrate the type of pore that causes hysteresis rather easily, but it requires a greater amount of energy to leave the pore. This is an important consideration when selecting an adsorbent for a particular application. Although adsorption data for water vapor on silica gel are available from the manufacturer, the capacity reported by them may be different from values obtained in other laboratory tests. Pedram and Hines (1983) obtained water adsorption data on Mobil Sorbead *R* silica gel at three temperatures, 301, 310, and 326 K. They found that their data were about 20% lower than that reported by Mobil Corporation. Yeh et al. (1992) also observed a similar result from their study with Davison silica gel grade 40. A significant difference in capacity was noted among the literature data and from one batch of adsorbent to another. Yeh et al. found that the regeneration temperature and time had a significant impact on the adsorption capacity. The manufacturer often measures the data at conditions that are very difficult to maintain during field operations. Therefore, care must be taken when calculating the amount of an adsorbent required for a specific application.

Organic pollutants. Bartell and Bower (1952) were one of the first to measure the adsorption capacities of organic vapors on silica gel. They studied the mechanisms by which silica gel attracts molecules from the gas phase and retains them on the surface.

Three different types of silica gel were used to measure the adsorption capacities of methyl alcohol, ethyl alcohol, *n*-propyl alcohol, benzene, hexane, and carbon tetrachloride. Their study showed that silica gel would adsorb significant amounts of these chemicals. In the 1970s, Mikhail and Shebl (1970) and Mikhail and El-Akkad (1975) studied in more detail the adsorption characteristics of silica gel for several organic compounds. A commercially available silica gel, Davison grade 03, was used in most of their work to study the adsorption of carbon tetrachloride, benzene, cyclohexane, isopropanol, and trichlorofluoromethane. They reported that these chemicals were adsorbed from their gaseous state on the solid surface, but condensed and remained as liquids in the pores. They observed that a significant amount of organic vapors can be removed by silica gel. Shen and Smith (1968) measured the amount of benzene and hexane adsorbed by silica gel in the temperature range of 70° to 130°C as pure components and then in a mixture. The amounts of benzene and hexane retained by silica gel in this temperature range were extremely low, suggesting that a part of the circulating air from homes or buildings can be used for regeneration of silica gel.

A systematic evaluation of the adsorption capabilities of a commercial grade silica gel for various indoor pollutants was started by Hines and co-workers in the late 1980s. The silica gel used in their studies was Davison grade 40. This particular grade is used extensively in desiccant humidity control systems. Kuo and Hines (1988) measured the adsorption capacities of six chlorinated hydrocarbons: methylchloride, dichloromethane, chloroform, carbon tetrachloride, 1,1,1-trichloroethane, and tetrachloroethylene. They found that silica gel can adsorb approximately 40% of its weight of these pollutants. However, the amount it can adsorb at very low concentrations—in the parts per million level—is of interest for applications in homes and commercial buildings, particularly in the presence of water vapor. The amounts of these chlorinated hydrocarbons adsorbed by silica gel at room temperature (25°C) and at concentrations of 1, 10, and 100 ppm are given in Table 10.3. Methylchloride, dichloroethane, and 1,1,1-trichloroethane have a greater affinity for silica gel than the other chlorinated hydrocarbons. Also, none of the six hydrocarbons exhibited any hysteresis during the desorption experiments. The adsorption capacities of three aldehydes, acetaldehyde, propionaldehyde, and butyraldehyde, on the same silica gel were found by Ghosh and Hines (1990) to be lower than the above chlorinated hydocarbons. The adsorbed amounts of aldehydes under similar conditions are also given in Table 10.3. Neither of the above studies addressed the impact of water vapor on the adsorption capacity.

Hines and his co-workers measured the adsorption capacities of six representative indoor pollutants on silica gel, molecular sieve 13X, and activated carbon (BPL). The pollutants studied included radon, CO_2, toluene, 1,1,1-trichloroethane, formaldehyde, and water vapor. Since the removal of any indoor pollutant has to be accomplished in the presence of water vapor, one objective of their studies was to investigate simultaneous adsorption of water vapor with each of the indoor pollutants. Silica gel has considerable capacity for toluene in the low concentration range and is comparable with that of activated carbon. A similar result was found when formaldehyde was adsorbed on silica gel. In an experiment carried out in a packed column that contained 5 g of silica gel, 100% of the formaldehyde was removed from nitrogen during the initial 150 minutes of operation. The gas stream

TABLE 10.3 Adsorption Capacities of Selected Indoor Pollutants (Chlorinated Hydrocarbons and Aldehydes) on Silica Gel at Room Temperature (25°C)

Pollutant	*Concentration in Air (ppm)*	*Adsorption Capacity*[a] *(mg/g of silica gel)*
Methyl chloride	1	0.064
	10	0.607
	100	4.090
Methylene chloride	1	0.206
	10	1.975
	100	14.630
Chloroform	1	0.555
	10	5.322
	100	39.635
Carbon tetrachloride	1	1.030
	10	9.806
	100	68.809
1,1,1-Trichloroethane	1	1.337
	10	12.709
	100	87.969
Tetrachloroethylene	1	9.358
	10	77.659
	100	297.930
Acetaldehyde	1	0.834
	10	7.807
	100	48.066
Propanal (propionaldehyde)	1	4.383
	10	33.802
	100	103.911
Butanal (butyraldehyde)	1	4.967
	10	39.227
	100	127.495

[a]Adsorption capacities are calculated using experimental data of Kuo and Hines (1988), Ghosh and Hines (1990), and Equation 10.7.

contained 9 to 10 ppm of formaldehyde, and its flow rate through the bed was 0.54 L/min. The results of this study suggest that desiccant humidity control devices can be effective in removing formaldehyde and other indoor pollutants from air.

When more than one adsorbable component is present in air, the adsorption process becomes more complex, and the relative affinity of each pollutant for the solid surface plays a significant role. If two pollutants have the same affinity, they will compete with each other for an adsorption site. When the attraction is unequal, the component with the stronger affinity can displace the weakly adsorbed one from the surface. In any event, pure component adsorption behavior must be known before predicting mixture behavior.

Activated Carbon

Organic pollutants. Activated carbon has been used extensively for industrial gas cleaning operations, although it has been used sparingly for indoor air cleaning applications. Portable air cleaners that contain an activated carbon are available commercially for cleaning the air in a single room. However, systematic studies that deal with the removal efficiency, humidity effects on the adsorption process, and the competitive adsorption of various contaminants are not reported in the literature. Also, research studies on the use of carbon filter units for indoor environments, such as homes, offices, schools, and hospitals, have not been reported. Since concentration levels of air contaminants in industrial effluents are much higher than indoor levels, care must be taken when interpreting, analyzing, and subsequently applying these data to indoor environments. Data obtained for industrial applications are often incomplete or are not reported in the open literature because of company restrictions regarding proprietary information. Hobbs et al. (1983) conducted an extensive survey of the industrial applications of vapor phase activated carbon adsorption and attempted to develop a database on the basis of information attained from manufacturers of organic compounds. Although information is available about the types of organic compounds that can be adsorbed, a generalization cannot be made about the effectiveness or economic viability of a carbon adsorption process for indoor applications.

Activated carbon based adsorption systems have been used effectively to control organic vapors in the workplace. The design of indoor carbon-based air cleaning systems is frequently not based on adsorption data but rather on the myth that carbon can adsorb any organic compound under any circumstance. Activated carbon has the greatest affinity for organic molecules because of the nonpolar nature of its surface as compared to other solid adsorbents, such as silica gel and molecular sieve. It is assumed that water vapor does not interfere with the adsorption process of organics and other compounds on activated carbon. This may be true if the relative humidity of the air is below 40%. However, when the relative humidity exceeds about 40%, activated carbon can adsorb a significant amount of water vapor, which can severely reduce the adsorption capacity for organic molecules. Therefore, the designer of a carbon adsorption system must also be familar with the water adsorption characteristics of carbons. It is important to note the amount of water adsorbed can vary from one batch of activated carbon to another (Hassan et al., 1991a).

Freeman and Reucroft (1979) found that ASC wheterlite carbon (an activated carbon impregnated with an inorganic salt) has almost 10 times greater water adsorption capacity than type BPL in the relative pressure range 0.001 to 0.01. Both carbons showed a strong hysteresis effect upon desorption. Some of the most interesting work on the adsorption of water vapor by activated carbon was carried out by Dubinin (1980) and Dubinin and Serpinsky (1981). Adsorption isotherms were determined on four types of carbons, and all showed hysteresis effects during desorption. Okazaki et al. (1978) obtained adsorption isotherms for water vapor and several organic vapors, including acetone, methanol, benzene, and toluene, on two activated carbons and found that the presence of water vapor in the air stream decreased the adsorption capacities for organic molecules.

For many adsorption processes, volatile organic compounds are perceived as being a single entity. As a result, design criteria are often established based on the total VOC content of an effluent stream. Sansone and Jonas (1981) obtained equilibrium adsorption data for six carcinogenic vapors at 23°C on a Barnebey-Cheney granular carbon. They proposed a method in which the carbon was first characterized by a reference vapor. The data for the reference vapor were then used to predict the performance of that carbon for other vapors. Carbon tetrachloride was used as the reference vapor to predict the adsorption capacity and adsorption rate constant for benzene, chloroform, *p*-dioxane, acrylonitrile, and 1,2-dichloroethane. Furthermore, the authors predicted the adsorption parameters for 31 carcinogen vapors. The average error was expected to be in the range of ± 15%. Urano et al. (1982) also provided a method to predict the adsorption capacities of organic vapors on activated carbons by using the molecular weight, vapor pressure, density, and parachors. The proposed method was tested with the adsorption data of 13 organic vapors on seven commercial granular activated carbons. Both of these methods are semiempirical and were tested basically at one data point (that is, the adsorption data were taken at one temperature and one concentration). A reliable method is not available for predicting adsorption data *a priori*. Generally, the data are taken at one or more temperatures and then are correlated with one of the classical models, such as the Langmuir and/or BET equations.

Nelson and his co-workers (Nelson and Correia, 1976) conducted extensive studies to evaluate the service life of several respirator cartridges that contained activated carbons, by measuring the breakthrough point for a number of organic vapors at 22°C. The effects of various factors were determined from the experiments, including concentration, humidity, temperature, and the cartridge size and configuration. The 10% breakthrough times, which was defined as the time required for the outlet concentration to reach 10% of that of the inlet stream, are presented for several pollutants in Table 10.4. A pulsating breathing simulator was also used to obtain more accurate data, but there was little difference from the data taken under steady-state conditions. It appears that activated carbon has a good retention capability and can be effective for the removal of certain pollutants that are emitted intermittently.

The behavior of several organic compounds in their binary mixtures was studied by Reucroft et al. (1986). The binary mixtures employed in their study included chloroform-carbon tetrachloride, benzene-*n*-hexane, chloroform-methylene chloride, and *n*-hexane-methylene chloride. It was noted that the higher-molecular-weight organics had the greatest adsorption capacity. The adsorption uptake was highest for carbon tetrachloride, followed by chloroform, methylene chloride, *n*-hexane, and benzene. The total amount adsorbed from the binary mixtures was always between the pure component adsorption capacities of the two chemicals. As an example, for chloroform-carbon tetrachloride mixtures, the amount adsorbed from their mixtures was less than that of carbon tetrachloride alone, but greater than that of chloroform.

McNulty et al. (1977) examined several techniques that can be used to regenerate a carbon bed that contains contaminants found in a spacecraft cabin. The regeneration systems that were tested included thermal desorption with a vacuum, thermal desorption with a nitrogen purge, *in-situ* catalytic oxidation of adsorbed contaminants, and *in-situ* noncatalytic oxidation of adsorbed contaminants. Bench scale studies of three organic vapors [diisobutyl ketone (DIBK), caprylic acid, and acrolein] were conducted. On the basis of

TABLE 10.4 Breakthrough Times of Various Organic Pollutants from Respiratory Cartridges

Solvent	*Concentration (ppm)*	*Flow Rate (L/min)*	*Test Relative Humidity (%)*	*Breakthrough Time When Outlet Concentration Was 10% of the Inlet Concentration (min)*
Benzene [a]	125	53.3	50	355
	500	53.3	50	134
	2000	53.3	50	41.9
Benzene [b]	1000	53.3	20	101
	1000	53.3	50	101
	1000	53.3	80	87.4
Toluene [a]	1000	20.6	50	288
	1000	36.7	50	164
	1000	53.3	50	114
Methanol [c]	1000	53.3	50	3.2
Isopropanol [c]	500	53.3	50	126
	2000	53.3	50	54.7
Butanol [c]	1000	53.3	50	141
Pentanol [c]	1000	53.3	50	130
Vinyl chloride [b]	50	40.0	50	77
	250	40.0	50	52.5
	1000	40.0	50	22.7
Ethyl chloride [a]	1000	53.3	50	10.7
1-Chlorobutane [a]	1000	53.3	20	86.3
	1000	53.3	50	87.3
	1000	53.3	90	68
Chlorobenzene	1000	53.3	50	131
Dichlorobenzene [a]	500	53.3	50	30
	2000	53.3	50	17.3
o-Dichlorobenzene [a]	1000	53.3	50	132
Chloroform [a]	1000	53.3	50	52.4
Methyl chloroform [a]	250	53.3	50	207
	2000	53.3	50	56.1
Trichloroethylene [a]	1000	53.3	50	83
Carbon tetrachloride [a]	1000	53.3	20	84.9
	1000	53.3	50	68.8
	1000	53.3	90	66
Perchloroethylene [a]	1000	53.3	50	129
Methyl acetate [a]	100	53.3	50	146
	1000	53.3	50	45.9
Ethyl acetate [b]	1000	53.3	20	88.4
	1000	53.3	65	90.4
	1000	53.3	90	68.4
Propyl acetate [a]	1000	53.3	50	99
Butyl acetate [a]	1000	53.3	50	96.9

(*continued*)

TABLE 10.4 (continued)

Solvent	*Concentration (ppm)*	*Flow Rate (L/min)*	*Test Relative Humidity (%)*	*Breakthrough Time When Outlet Concentration Was 10% of the Inlet Concentration (min)*
Acetone [b]	100	53.3	50	245
	500	53.3	50	96.7
	1000	53.3	50	66.3
Acetone[b]	1000	53.3	20	61.1
	1000	53.3	80	54.5
	1000	53.3	90	53.1
Pentane [d]	1000	53.3	50	71.3
Hexane [b]	100	53.3	50	565
	500	53.3	50	143
	2000	53.3	50	37.9
Hexane [a]	1000	53.3	0	76.7
	1000	53.3	65	68.1
	1000	53.3	90	64
Cyclohexane [d]	1000	53.3	50	82.3
Heptane [d]	1000	53.3	50	80.5
Methylamine [d]	1000	53.3	50	17.9
Ethylamine [d]	1000	53.3	50	49.7
Diethylamine [a]	250	53.3	50	92.5
	1000	53.3	50	35.6
	2000	53.3	50	20.6
Dipropylamine [d]	1000	53.3	50	105

[a]MSA cartridges, coconut base, 52.2 g/pair.
[b]AO cartridges, petroleum base, 70.5 g/pair.
[c]Corrected for humidity.
[d]AO cartridges, coconut base, 62.2 g/pair.
Source: Nelson and Correia, 1976.

repetitive adsorption and desorption cycles, the authors concluded that nitrogen purge with thermal desorption might be the best regeneration method.

Alumina

Activated aluminas have been in use for more than 25 years in a variety of drying operations for both gaseous and liquid streams, but they have not been employed in desiccant-based air-conditioning systems. The initial granular form of activated alumina did not prove satisfactory because of excessive particle breakdown. In the last few years, with the introduction of spherical products, activated alumina is finding increasing application as a desiccant.

The earlier grade (Grade F-1, manufactured by the Aluminum Company of America, Alcoa) had a very low capacity for water vapor. The amount of water adsorbed was only

6% to 7% of the weight of the adsorbent at room temperature and at a dew point of 55°F. However, another Alcoa activated alumina, which was a modification of the earlier grade, performed somewhat better. A comparison of water capacities of different types of activated aluminas, along with activated bauxite, can be found in the ASHRAE Fundamentals (1988). Typical water vapor adsorption data can also be obtained from the manufacturer. LaRoche Chemicals is presently marketing an activated alumina, A-201, for use as a desiccant material. The manufacturer's data show that its water adsorption capacity is lower than that of molecular sieve at relative humidities below 60%, but its capacity is about 10% higher than that of molecular sieve at the saturation pressure. The adsorption characteristics are basically the same as that of silica gel.

Chou (1987) conducted a laboratory scale experiment using a packed bed containing 22.6 kg of the Alcoa activated alumina under adiabatic conditions. The air inlet temperature and relative humidity were 23°C and 84%, respectively. Water vapor appeared in the outlet gas stream after about 37 hours under these conditions. The particle size, relative humidity of the inlet air, temperature, linear velocity, and bed length had a significant influence on the time required to saturate the bed and on the temperature of the outlet air stream.

The face velocity of air flowing through commercial solid-desiccant dehumidifiers is in the range of 150 to 200 ft/min. At this flow rate, the desiccant bed operates essentially at isothermal conditions. No significant difference in the breakthrough time was observed by Carter (1968) in his experiment with an adiabatic and an isothermal bed containing activated alumina. But he reported an interesting phenomenon: The surface area of alumina decreased from 275 to 153 m^2/g after 160 adsorption and regeneration cycles. A particle size analysis of the granules showed a reduction in the mean particle diameter with use, which was attributed to the thermal stresses induced by the heating and cooling cycles.

Activated aluminas are also used for the removal of various contaminants from gases, such as trace fluorides, chlorides, H_2S, COS, alcohols, and ethers. However, a major application of activated aluminas is in the removal of fluorides from alkylate streams and H_2S and other sulfur compounds from natural gas by the Claus process. Fluorides, H_2S, and other sulfur compounds are removed primarily via chemical oxidation reactions rather than by adsorption. Because of their surface properties, activated aluminas are now being investigated more extensively for use in removing indoor air pollutants.

Zeolites and Molecular Sieves

Inorganic pollutants. Natural zeolites were known nearly two centuries ago for their ability to adsorb water vapor. The word zeolite comes from the Greek word zeo, meaning to boil, and lithos, meaning stone, because zeolites give off water vapor when heated gently. About 40 types of zeolites occur naturally and have the ability to adsorb water vapor and other gases. Around 1920, chemists discovered the molecular sieving property of zeolites; that is, zeolites can preferentially adsorb some gases over others, because of the shape and size of their crystalline cavities. However, the composition of natural zeolites varied too much from one deposit to another. In 1948 Linde Company began exploring the development of synthetic zeolites that had uniform pore openings in the crystal structure. Approximately 10 years later they succeeded in producing a synthetic zeolite that had a controlled pore size. The first family of synthetic zeolites was known as type A. Later a type X zeolite was developed with a larger pore size than the type A. Types 3A, 4A, and 5A synthetic zeolites (also called molecular sieves) rapidly found commercial applications in processes other than the adsorption of water vapor. Their sieving behavior was exploited in separating gas mixtures. The average molecular diameters of several inorganic gases along with the average pore diameters of molecular sieves are given in Figure 10.6.

The 4A and 5A molecular sieves are best suited for water vapor adsorption, because their average pore diameters are such that they allow water to diffuse into the pores while blocking other molecules. If only the adsorption capacities of water vapor are compared for the various molecular sieves, 13X has a slightly greater capacity at a certain temperature and pressure. However, if other gases are present, they may be co-adsorbed and reduce the water adsorption capacity for 13X.

The pore diameter is not the only factor associated with the adsorption characteristic of gas molecules. Various adsorption forces play a significant role in determining the amount that can be adsorbed. For example, a significant amount of CO_2 can be adsorbed on molecular sieves 4A and 5A, in spite of its having an average molecular diameter of 3.914Å, which is larger than the average pore diameter of either 4A or 5A. However, if the objective is to remove both water vapor and other pollutants simultaneously from indoor air, either 10X or 13X may be a better choice, because of their larger pore diameters. Typical equilibrium data of water vapor and air contaminants, such as SO_2, CO_2, and H_2S, are available from the manufacturer, but care must be taken when using the data in a design. Hassan et al. (1991b) noted that rather stringent experimental conditions were required to match the manufacturer's data. The difference in the adsorption capacity depends on how the material is treated prior to an adsorption run. When the 13X molecular sieve was heated at 200°C under a vacuum of 1×10^{-4} mmHg for 16 hours before beginning the experiment, the laboratory data matched the manufacturer's data (see Figure 10.7).

The adsorption characteristics of natural and synthetic zeolites are basically the same; the adsorption isotherms, which represent the amount of water uptake, rise sharply in the low humidity range and then level off quickly. However, the presence of impurities and the crystalline structure of the natural zeolites seem to affect the capacity for water vapor adsorption. Yamanka et al. (1989) evaluated 10 zeolite samples from the western part of the United States and found their water adsorption capacities to be comparable to those of synthetic molecular sieves. The adsorption capacities of these zeolites at 20°C and three relative humidities are given in Table 10.5. However, the shape of the equilibrium curve, when plotted as the amount of water adsorbed versus the relative humidity, is not particularly favorable for dehumidification applications. All these zeolites exhibited a rather large hysteresis loop during desorption, suggesting that a large energy input may be required for complete regeneration. Several attempts have been made to alter the crystalline structure and the chemical nature of the zeolite by changing the cations of the surface. Although the adsorption behavior in the high relative pressure range changed, it remained basically the same at low pressures.

Organic pollutants. The adsorption capacities of 4A and 5A molecular sieves drop dramatically as the size of the gas molecule increases. Thomas and Mays (1961) showed that 13X can adsorb substantial amounts of aromatics, particularly those with side

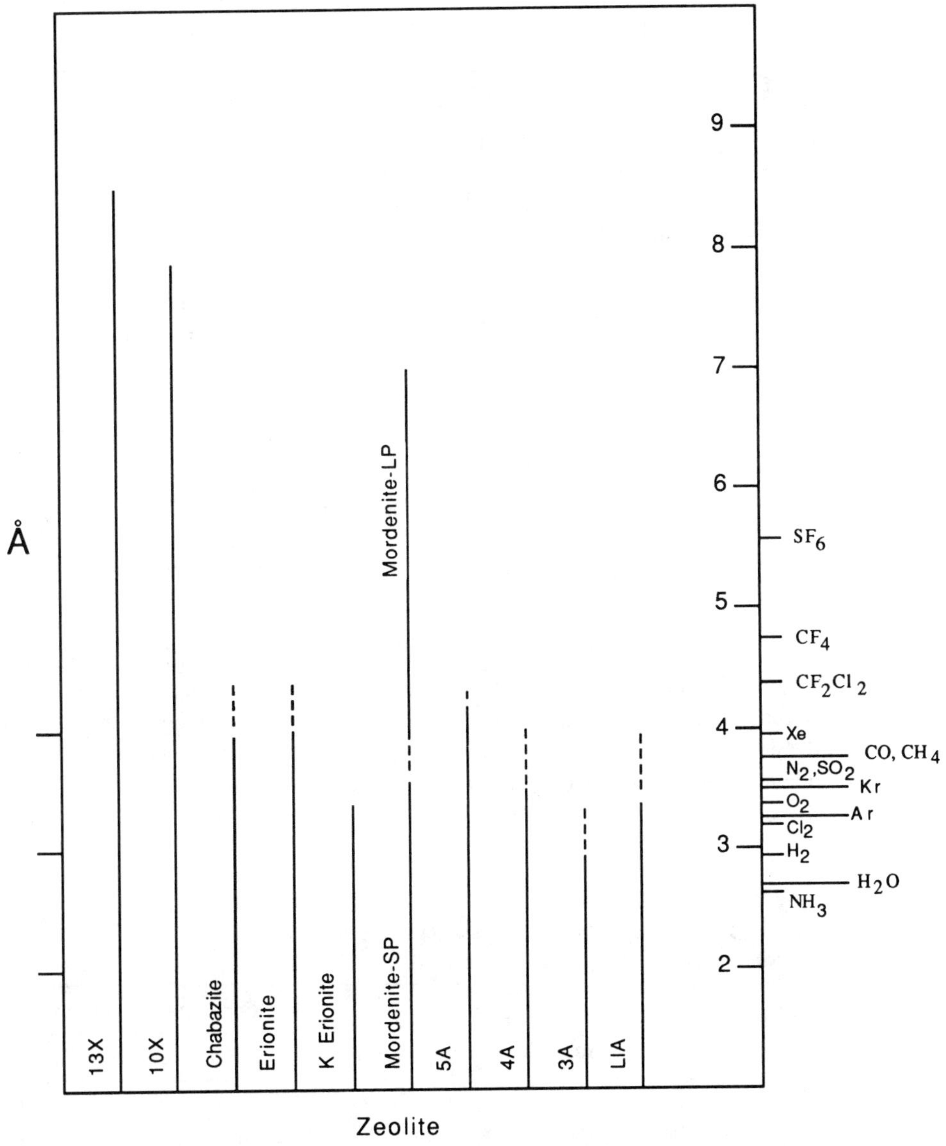

Figure 10.6 *Effective pore sizes of zeolites and molecular diameters (Lennard-Jones Kinetic Diameter) of pollutants.* (Source: *Breck, 1984.)*

chains, such as 1,3,5-triethylbenzene. They also noted that adsorption by molecular sieves is not based solely on the molecular sieving effect but also on the polarity and aromaticity of the molecules. The more polar and unsaturated molecules are more strongly adsorbed on the surface.

The use of molecular sieves for the removal of indoor pollutants has not been fully

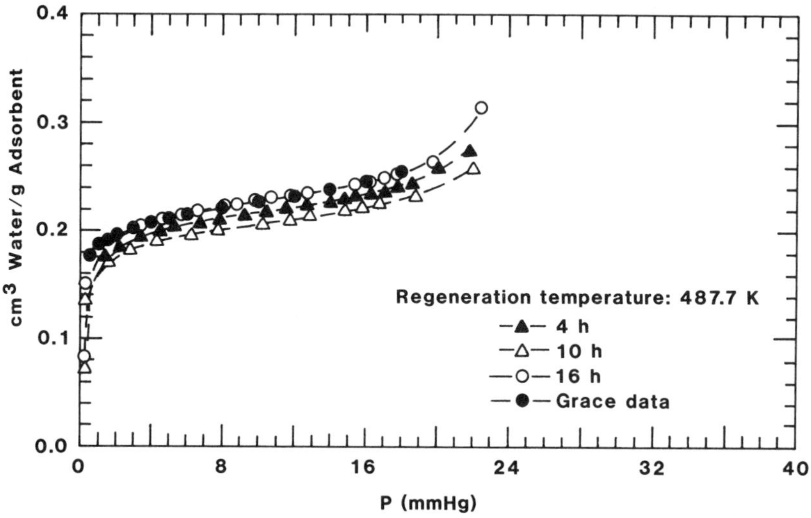

Figure 10.7 *Effect of regeneration times on water vapor isotherms on molecular sieve 13X at 298 K.* (Source: *Hassan et al., 1991b.)*

explored. Most organic compounds found indoors have molecular diameters larger than the pore diameters of 4A and 5A. As a result, either 10X or 13X might be expected to be a better adsorbent for the simultaneous removal of water vapor and indoor pollutants. Ghosh and Hines (1991) investigated the adsorption characteristics of acetaldehyde, propionaldehyde, and butyraldehyde on 13X as pure components and in their mixtures. The 13X can adsorb considerable amounts of these aldehydes in the low concentration range, but the

TABLE 10.5 Water Adsorption Capacity of Natural Zeolites at 20°C and Various Relative Humidities

	Adsorption Capacities (g/g) at:		
Zeolite and Molecular Sieves	*30% RH*	*70% RH*	*Saturation Pressure*
Chabazite (Bowie, Arizona)	0.193	0.228	0.312
Chabazite (Christmas, Arizona)	0.170	0.210	0.334
Clinoptilolites (Mountain Green, Utah)	0.128	0.122	0.189
Clinoptilolites (Castle Greek, Idaho)	0.130	0.158	0.216
Clinoptilolites (Hector, California)	0.115	0.140	0.244
Erionites (Pine Valley, Nevada)	0.150	0.170	0.213
Erionites (Shoshone, California)	0.130	0.140	0.147
Mordenites (Union Pass, Arizona)	0.098	0.110	0.139
Mordenites (Lovelock, Nevada)	0.070	0.081	0.173
Phillipsite (Pine Valley, Nevada)	0.160	0.190	0.389

Source: Yamanka et al., 1989.

13X reached its saturation capacity rather quickly. All three aldehydes appeared to have an equal affinity for the molecular sieve surface. When a mixture of these aldehydes in nitrogen was passed through a stainless steel column that contained 13X, all three aldehydes broke through the bed at about the same time. During desorption, hysteresis loops were present. A relatively high temperature, 300°C, was required to regenerate the 13X.

The ability of 13X to adsorb and retain aldehydes was used by Matthews and Howell (1982) to measure formaldehyde concentrations in air. Formaldehyde was adsorbed from air on 13X and then extracted with water for subsequent analysis. Our own studies also suggest that 13X can be effective in removing formaldehyde from a nitrogen stream. In a 100-hour laboratory test, 100% of the formaldehyde was removed from a nitrogen stream by an adsorption bed that contained 5 g of the 13X. The nitrogen stream contained 9 ppm of formaldehyde and flowed through the bed at a rate of 540 cm^3/min. As expected, the bed had to be heated to 300°C for more than four hours to remove all the adsorbed formaldehyde.

In addition to the formaldehyde, the representative chemicals mentioned earlier (radon, CO_2, toluene, 1,1,1-trichloroethane, and water) were also used to evaluate the indoor pollutant removal capability of the 13X. The 13X had capacities comparable to silica gel in the low concentration range, but the presence of hysteresis loops during desorption may make it unattractive.

Molecular sieves are generally very good catalysts and can convert some volatile organics to other compounds. Our studies with 1,1,1-trichloroethane in a molecular sieve 13X bed showed that 1,1,1-trichloroethane was converted to the more toxic vinylidene chloride and hydrochloric acid at room temperature. Both vinylidene chloride and the unconverted 1,1,1-trichloroethane were retained on the surface initially; but as the adsorption progressed, both of them started to break through the column. During regeneration, the high temperature facilitated the further conversion of 1,1,1-trichloroethane to vinylidene chloride and hydrochloric acid. One should exercise caution when designing a molecular sieve removal system to avoid the production of toxic organics like vinylidene chloride.

Some carbon-based materials are also known to have molecular sieving properties similar to those of the synthetic zeolites. These are called carbon molecular sieves (CMS). There are two major sources of CMSs. They can be prepared from some coals and the carbonization of polymers. Compared to activated carbons, the CMSs have a large number of micropores that are accessible only to smaller molecules. Depending on the polymer used and the process variables, the average pore diameter of these materials can be controlled. Like activated carbon, the CMSs are also nonpolar and have a low affinity for water, they can be used to adsorb organic molecules with little interference. However, carbon molecular sieves have found very limited commercial success and, accordingly, are not readily available.

Polymers

Polymers are versatile materials and can be synthesized for use as desiccants, adsorbents, membranes, and thin films. Several studies have noted the development of supersorbent polymers that are capable of holding up to 10,000 times their own weight of water. Unfortunately, they quickly form a gelatinous mass, making them unsuitable for dehumidification applications. However, some polymers can adsorb 50% of their weight of water vapor and still remain in their granule-free flowing state. Organic polymers often exhibit an interesting phenomenon: The adsorption data pass through a maximum at high relative humidities. As a polymer adsorbs water vapor, its structure becomes highly plasticized by the vapor, and the polymer matrix goes through a structural rearrangement that releases some of the water.

Tock (1983) evaluated the performance of a polymer [hydrolyzed-starch polyacrylonitrile (H-SPAN)] for controlling the humidity of an enclosed space. The adsorption capacity of the polymer was two times greater than that of silica gel or alumina over the 60% to 90% relative humidity range. Below 60%, all three sorbents had essentially the same capacity. Probably the greatest advantage of this polymer is its low regeneration temperature (120°F) compared to the typical range of 250° to 400°F for inorganic desiccants. However, the H-SPAN polymer has an unfavorably shaped adsorption curve, making it unsuitable for applications in a humidity control system. Polymers synthesized from polyvinyl alcohol and polyvinylene glycol showed a behavior similar to that of molecular sieve during water vapor adsorption. Polyacrylic acid/poly-4-vinylpyridine polymer changed from a glassy to a rubbery state at a relative humidity of about 50% (Hirai and Nakajima, 1989). The interaction of water with specific polymer sites played an important role in the adsorption process. The water tended to cluster inside the polymer matrix.

Lunde and Kester (1975) developed a regenerable polymeric sorbent material that was capable of adsorbing water vapor and CO_2 simultaneously from an air stream. The polymer can be regenerated by vacuum to its full capacity within a very short period of time. A potential application for this polymer is in the humidity control system of a space shuttle cabin. Although the polymer's capacities for CO_2 and water vapor were only 2% and 4% of its weight, respectively, the polymer performed satisfactorily in a 2000-hour cyclic operation without degradation. One problem with the polymer was its degradation at temperatures higher than 180°F. Since a small amount of NH_3 was produced, the polymer may not be suitable for indoor applications in which thermal regeneration is employed. A number of other polymers can also be used for water adsorption. A list of leading polymer candidates is given in Table 10.6. Except for CO_2, their ability to adsorb indoor pollutants has not been investigated.

Table 10.7 provides a summary of pure component adsorption data for several indoor air pollutants on selected adsorbents.

10.5 CO-ADSORPTION IN FIXED BEDS

The role of water vapor during the co-adsorption of various air contaminants is somewhat controversial. For example, activated carbon has a very low affinity for water vapor, and numerous gases and vapors can readily displace the adsorbed water from the carbon and progressively reduce its sorptive capacity for water. This was demonstrated by Turk and Van Doren (1953) in the study of an apple storage facility; they showed that at the beginning of the experiment the adsorption rate of water vapor was very high, but the ad-

TABLE 10.6 Polymer Candidates for Use in Desiccant Cooling Systems

Material	*Sorption Capacity at 80% Relative Humidity (wt.%)*	*Other Factors (x = wt. % H_2O)*
Polystyrene sulphocationates	30–55	Swelling of 1.3 to 5 times
Cationites based on methacrylic acid	18–55	Swelling of 1.2 to 3 times
Aminoalkylalkoxysilanes	10–90	Supported by glass fibers; most uptake is between x = 0.6 and 0.9
Copolyoxamides	6–68	Made for membrane applications
Wool plus 8% Calgon Polymer 261	28	Other modified wools also studied
Viscose rayon	18	Sorption studies at 35°C, so uptake may be greater at 25°C; cyclic studies done
Triethylbenzl ammonium cationites	18–22	Swelling of 1.2 to 5 times
Perfluorosulfonic acid membranes (Du Pont Nafion)	13	Influence of processing on isotherm shape and capacity studies; heat of adsorption determined
Cellophane	15	Capacity increases to over 20% at x = 0.9
Vinylon	12	Capacity ranges from 4% to 16% between x = 0.5 and 0.9
Polyurethanes, poly(ethylene oxide)	13	Diffusivity is 4×10^{-7} cm^2/s and the sorption kinetics are fast; capacity range is 3% to 23% from x = 0.6 to 0.9
Styrene divinylbenzene copolymer	9–18	Influence of amino groups on sorption capacity was studied
Epoxy resins	10	Sorption kinetics measured slow uptakes
Poly (cross-lined) methacrylates	10	Sorption at 35°C and larger uptakes from x = 0.8 to 0.9
Polystyrene sulfonates	Over 20	Ion exchange resins known to be good water sorbers
Nylon 6,6	9.5	Good diffusivity
Polyurethanes	5–7	Diffusivity and sorption isotherms
Polycarbonate	4	Compositional and block length studies on capacity
Polyester	8	Measured at 30°C
Cellulose acetate	7	Measured at 30°C
Polyacrylonitrile	4	Sorption kinetics measured
Chromia gels	32	Water content from outgassing
Carbon blacks	20–40	Depends on preparation method

Source: Czanderna and Thomas, 1986.

sorbed water was gradually displaced by the organic vapors as the adsorption continued. Adsorption of a mixture of methanol and water vapor on Calgon BPL carbon was examined by Nemeth et al. (1984) in a dynamic system. Breakthrough curves of pure water, pure methanol, and their mixtures were obtained at various conditions at 20°C. A small amount of adsorbed water was displaced in the initial period of the experiment but was readsorbed

TABLE 10.7 Summary of the Literature Search for Pure Component Adsorption Data of Indoor Pollutants

	Number of Studies with Adsorbents					
Pollutants	*Activated Carbon*	*Silica Gel*	*Molecular Sieve*	*Activated Alumina*	*Polymers*	*Temperature (K)*
Chloromethane	1	1				288–298
Dichloromethane	2	1	1			288–298
Chloroform	6	1				277–318
Carbon tetrachloride	10	8	1	1	1	282–318
Trichloroethane	2	1	1			288–298
Tetrachloroethylene	1	1	1		1	288–298
p-Dichlorobenzene	1					296
Ozone	1					
Acetone	6					223–303
Toluene	9		2		1	298–503
Ammonia	7	1	4			298–473
CO	6		11	1		193–348
H_2O	16	17	12	6	5	273–373
CO_2	22	1	22	5		193–423
NO	2					298–348
NO_2	2			3		
SO_2	3		2			273–1073
H_2S	5		2			273–303
p-Nitrobenzene		1				296–303
Nitrotoluene		1				296–303
Trichlorobenzene		1				296–303
Tetrachlorobenzene		1				296–303

Source: Hines et al, 1991.

on the carbon as the experiment progressed. Finally, the bed became saturated with water. The authors concluded that under dynamic conditions, methanol and water vapor enhanced each other's adsorption.

This cooperative behavior was not observed by Werner (1985) and Werner et al. (1985) in their experiments with trichloroethylene (TCE) and water vapor in a glass column packed with activated carbon. Werner (1985) observed that the breakthrough for both water vapor and TCE (inlet concentration: 65% relative humidity and 1000 mg/m^3 of TCE) from the column occurred at the same time. However, the bed became saturated with water vapor at a much faster rate than it did with TCE. Water was not displaced by TCE in his study. Werner et al. (1985) measured the equilibrium uptake of TCE at inlet concentrations ranging from 300 to 1350 mg/m^3 of TCE in air. The relative humidity of the air varied from 0% to 85%. The adverse effect caused by the presence of water vapor was more significant at lower TCE concentrations. The presence of water vapor decreased not only the adsorption capacity of TCE on the carbon but also the bed efficiency by increasing the width of the breakthrough curve.

Jonas et al. (1985) studied the effect of moisture on the adsorption of chloroform on activated carbon at 23°C and noted that, if the carbon bed was maintained at or below 45% relative humidity, the uptake of chloroform was not affected. However, for higher relative humidities the uptake of chloroform was reduced. Considering the shape of the isotherm for water on activated carbon and the fact that chloroform is not soluble in water, this result might have been expected; the uptake of water by activated carbon is primarily due to pore filling. At 23°C, capillary condensation and pore filling begin at about 45% relative humidity. Because the filling of the pores by water reduces the access of chloroform to certain adsorption sites, and chloroform is not soluble in water, its overall uptake would decrease at higher relative humidities. In a study of carbonyl chloride adsorption on activated carbon (Noyes, 1946), equilibrium uptake increased at higher relative humidities. This can be attributed to its higher water solubility.

In a number of applications, the stream to be processed could have more than one component. Therefore, depending on the type of the adsorbent employed and the chemical nature of the adsorbates, the overall uptake capacity could be either enhanced or reduced. If a second adsorbate is present in the process stream and if it competes with the primary adsorbate for the same adsorption site, the uptake of the primary adsorbate would be reduced. If, however, the second adsorbate does not compete for the same adsorption site and the primary adsorbate is soluble in the second adsorbate, the uptake of the primary component would be enhanced. A situation such as this could occur when certain chlorinated hydrocarbons are adsorbed on activated carbon from an air stream that has a high relative humidity. Thus, when carrying out design calculations, competitive adsorption must be considered, and the size of the bed must be adjusted accordingly. A summary of adsorption studies for several indoor pollutant mixtures on a number of adsorbents is given in Table 10.8.

The impact of adsorbing more than one component in a packed bed from a mixture of gases is shown in Figure 10.8. For adsorption of a single component, the breakthrough curve approaches the normalized concentration of $C/C_0 = 1$ as the bed is exhausted. When there is more than one adsorbate, however, the concentration of one or more adsorbates can actually exceed the inlet concentration at positions within the bed. This is the result of one component displacing the other, which raises the concentration of the displaced solute in the fluid phase. This effect is typically referred to as *roll-up* or *roll-over*. This phenomenon could be expected in treating indoor air by adsorption methods. A roll-over or chromotographic effect would occur in an adsorption system that is used to treat indoor air during the start-up phase of the adsorber as a result of the inlet concentration being higher initially. As the air is repeatedly cleaned, the inlet concentration of the pollutants would become equal to the emanation rate of the source. As shown in Figure 10.8, the concentration of the displaced adsorbate in the fluid phase is a function of the inlet concentration of both adsorbates.

TABLE 10.8 Summary of Mixture Data for Indoor Pollutants

	Number of Studies with Adsorbents			
Pollutants	*Activated Carbon*	*Silica Gel*	*Molecular Sieve*	*Temperature (K)*
i - $C_4H_{10} + C_2H_4$	1			273–373
$CO_2 + C_2H_4$			2	273–373
$CO + O_2$		1		144–373
$CO + CH_4$	1		1	298
$CO + N_2 + O_2$			1	143
$CO + CO_2$	1	1		273–298
$CO + N_2$			1	145–366
$CO_2 + CH_4$	5		1	260–304
$CO_2 + H_2S$	2		2	303
$CO_2 + C_3H_8$			1	303
$CO_2 + He$	1			304
$CO_2 + C_2H_6$	2		1	213–323
$CO_2 + N_2 + C_2H_6$			1	298
$CO_2 + H_2O + H_2S + C_2H_4$			1	298–323
$C_6H_{14} + C_6C_{16}$	1	1		343–403
$CHCl_3 + C_6H_6$	1			298
$CCl_4 + C_6H_6$	1			298
C_6H_6 + Toluene	1			423
Acetone + H_2O	1			303
$CH_3OH + H_2O$	2			293–303
Toluene + H_2O	1			293–303
$C_6H_6 + H_2O$	2			303
$CO_2 + H_2O$			4	273–498
$H_2S + C_3H_8$			1	303
$H_2S + CO_2 + C_3H_8$			1	303
$CO_2 + H_2S + CO + CH_4 + H_2$	1			293
$CO_2 + CO + CH_4 + H_2$	1			298

Source: Hines et al., 1991.

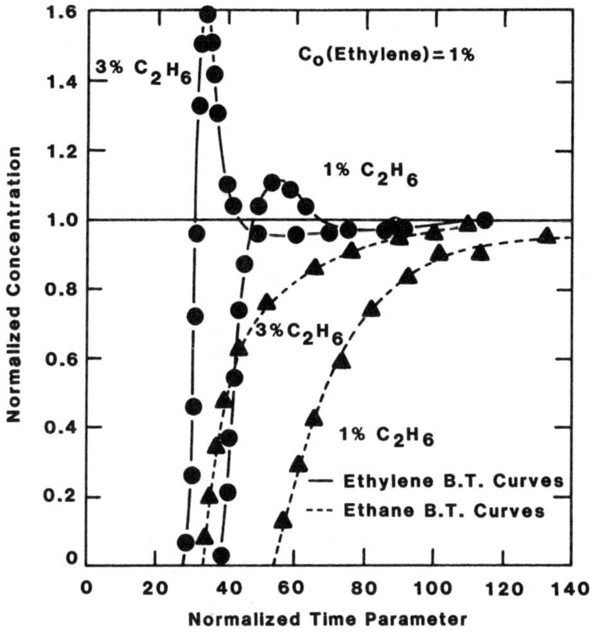

Figure 10.8 *Displacement of lighter components by heavier components in a dynamic bed.* (Source: *Gariepy and Zwiebel, 1971.*)

REFERENCES

ARNOLD, J. R., *J. Am. Chem. Soc.*, **71**, 104 (1949).

ASHRAE (American Society of Heating, Refrigerating and Air-Conditioning Engineers), *ASHRAE Handb. Fundamentals*, The Society, Atlanta, Ga., 1988.

BARTELL, F. E., and BOWER, J. E., *J. Colloid. Sci.*, **7**, 80 (1952).

BRECK, D. W., *Molecular Sieve Zeolites*, Wiley, N. Y., 1984.

BROUGHTON, D. B., *Ind. Eng. Chem.*, **40**, 1506 (1948).

BRUNAUER, S., EMMETT, P. H., and TELLER, E., *J. Am. Chem. Soc.*, **60**, 309 (1938).

BRUNAUER, S., *The Adsorption of Gases and Vapors*, Princeton University Press, Princeton, N. J., 1945.

CARTER, J. W., *Trans. Inst. Chem. Eng.*, **46**, T213 (1968).

CEROFOLINI, G. F., JARONIEC, M., and SOKOLOWSKI, S., *Colloid Polym. Sci.*, **256**, 471 (1978).

CHOU, C. L., *Chem. Eng. Commun.*, **56**(1-6), 211 (1987).

COHEN, B. M., LEVINE, A. H., and ARORA, R., "Field Development of a Desiccant Based Space Conditioning System for Supermarket Applications," GRI-83/0013 (1983), Gas Research Institute, Chicago, Ill., NTIS Document No. PB85-101921, 1985.

COLLINS, J. J., *Chem. Eng. Prog. Symp. Ser.*, Vol. 63, No. 74, 31 (1967).

COOK, W. H., and BASMADJIAN, D., *Can. J. Chem. Eng.*, **43**, 78 (1965).

COSTA, E., SOTELO, J. L., CALLEJA, G., and MARRON, C., *AIChE J.*, **27**(1), 5 (1981).

COUDERC, J. P., Chapter 1, *Fluidization*, 2nd ed., (eds. J. F. Davidson, R. Clift, and D. Harrison), Academic Press, London, 1985.

CZANDERNA, A. W., and THOMAS, T. M., "Advanced Desiccant Materials Research," Report No. SERI/PR-255-2887, Solar Energy Research Institute, Golden, Colo., 1986.

DUBININ, M. M., *Carbon*, **18**, 355 (1980).

DUBININ, M. M., and SERPINSKY, V. V., *Carbon*, **19**, 402 (1981).

FREEMAN, G. B., and REUCROFT, J. P., *Carbon*, **17**, 313 (1979).

GARIEPY, R. L., and ZWIEBEL, I., *AIChE Symp. Ser.*, **67**, No. 117, 17 (1971).

GHOSH, T. K., "Adsorption of Acetaldehyde, Propionaldehyde, and Butyraldehyde on Silica Gel and Molecular Sieve-13X," Ph. D. Dissertation, Oklahoma State University, Stillwater, Okla., 1989.

GHOSH, T. K., and HINES, A. L., *Sep. Sci. Technol.*, **25**(11 & 12), 1101 (1990).

GHOSH, T. K., and HINES, A. L., *Sep. Sci. Technol.*, **26**(7), 931 (1991).

HASSAN, N. M., GHOSH, T. K., HINES, A. L., and LOYALKA, S. K., *Carbon*, **29**, 681 (1991a).

HASSAN, N. M, GHOSH, T. K., HINES, A. L., and LOYALKA, S. K., *Chem. Eng. Commun.*, **105**, 1 (1991b).

HINES, A. L., and MADDOX, R. N., *Mass Transfer: Fundamentals and Applications*, Prentice Hall, N. J., 1985.

HINES, A. L., KUO, S. L., and DURAL, N., *Sep. Sci. Technol.*, **25**(7), 869 (1990).

HINES, A. L., GHOSH, T. K., LOYALKA, S. K., and WARDER, R. C., Jr., "Investigation of Co-Sorption of Gases and Vapors as a Means to Enhance Indoor Air Quality," Report No. GRI-90/0194, Gas Research Institute, Chicago, Ill., NTIS Document No. PB91-178806, 1991.

HIRAI, Y., and NAKAJIMA, T., *J. Appl. Polym. Sci.*, **37**, 2275 (1989).

HOBBS, F. D., PARMELE, C. S., and BARTON, D. A., "Survey of Industrial Applications of Vapor-Phase Activated-Carbon Adsorption for Control of Pollutant Compounds from Manufacture of Organic Compounds," Report No. EPA-600/Z-83-035, Envirnomental Protection Agency, Washington, D. C., 1983.

HOUSE, W. A., and JAYCOCK, M. J., *Colloid Polym. Sci.*, **256**, 52 (1978).

HUBARD, S. S., *Ind. Eng. Chem.*, **46**, 356 (1954).

HYUN, S. H., and DANNER, R. P., *J. Chem. Eng. Data*, **27**, 196 (1982).

INNES, W. B., and ROWLEY, H. H., *J. Phys. Chem.*, **51**, 1154 (1947).

JARONIEC, M., *J. Colloid Interface Sci.*, **53**(3), 422 (1975).

JARONIEC, M., and TOTH, J., *Colloid Polym. Sci.*, **254**, 643 (1976).

JARONIEC, M., *Colloid Polym. Sci.*, **255**, 32 (1977).

JARONIEC, M., and TOTH, J., *Colloid Polym. Sci.*, **256**, 690 (1978).

JARONIEC, M., *Colloid Polymer Sci.*, **256**, 1089 (1978).

JARONIEC, M., *Thin Solid Films*, **71**, 273 (1980).

JARONIEC, M., PATRYKIEJEW, A., and BOROWKO, M., *Progress in Surface and Membrame Science*, Vol. 14, Academic Press, New York, 1981.

JONAS, L. A., SANSONE, E. B., and FARRIS, T. S., *Am. Ind. Hyg. Assoc. J.*, **46**, 20 (1985).

JOVANOVIC, D. S., *Kolloid Z. Z. Polym.*, **235**, 1203 (1969).

JURY, S. H., and EDWARDS, H. R., *Can. J. Chem. Eng.*, **49**, 663 (1971).

KEMBALL, C., RIDEAL, E. K., and GUGGENHEIM, E. A., *Trans. Faraday Soc.*, **44**, 948 (1948).

KOBLE, R. A., and CORRIGAN, T. E., *Ind. Eng. Chem.*, **44**, 383 (1952).

KUO, S. L., and HINES, A. L., *Sep. Sci. Technol.*, **23**(4 & 5), 273 (1988).

LANGMUIR, I., *J. Am. Chem. Soc.*, **40**, 1361 (1918).

LEWIS, W. K., GILLILAND, E. R., CHERTOW, B., and CADOGAN, W. P., *Ind. Eng. Chem.*, **42**, 1319 (1950).

LUNDE, P. J., and KESTER, F. L., "Desiccant Humidity Control System," NASA-CR-115568, Hamilton Standard Div., United Technologies, Windsor Locks, Conn., 1975.

MARCINIAK, T. J., "Solid Desiccant Dehumidification Systems for Residential Applications," Gas Research Institute, Chicago, Ill., NTIS Document No. PB85-198489, 1985.

MARKHAM, E. D., and BENTON, A. F., *J. Am. Chem. Soc.*, **53**, 497 (1931).

MATTHEWS, T. G., and HOWELL, T.C., *Anal. Chem.*, **54,** 1495 (1982).

MAURER, G., and PRAUSNITZ, J. M., *Fluid Phase Equilib.*, **2**, 91 (1978).

MCNULTY, K. J., GOLDSMITH, R. L., GOLDSMITH, G. A., HOOVER, P. R., NWANKWO, J., and TUSK, A., "Evaluation of Techniques for Removal of Spacecraft Contaminants from Activated Carbon", NTIS Document No. NASA-8742, 1977.

MIKHAIL, R. S., and SHEBL, F. A., *J. Colloid Interface Sci.*, **26** (5), 505 (1970).

MIKHAIL, R. S., and EL-AKKAD, T., *J. Colloid Interface Sci.*, **32**(2), 260 (1975).

MISRA, D. N., *Surf. Sci.*, **18**, 367 (1969).

MISRA, D. N., *J. Chem. Phys.*, **52**, 5499 (1970).

MISRA, D. N., *J. Colloid Interface Sci.*, **43**, 85 (1973).

MOON, H., and TIEN, C., *Ind. Eng. Chem. Res.*, **26**, 2042 (1987).

MYERS, A. L., and PRAUSNITZ, J. M., *AIChE J.*, **11**, 121 (1965).

NELSON, G. P., and CORREIA, A. N., *Am. Ind. Hyg. Assoc. J.*, **37**, 514 (1976).

NEMETH, J., BATICZ, S., and HORANYI, M. P., *Hung. J. Ind. Chem.*, **12**, 199 (1984).

NOVOSEL, D., RELWANI, S. M., and MOSCHANDREAS, D. J., Proceedings IAQ '87: Practical Control of Indoor Air Problems, American Society of Heating, Refrigerating and Air-Conditioning Engineers, Atlanta, Ga., 261, 1987.

NOVOSEL, D., MCFADDEN, D. H., and RELWANI, S. M., Proceedings IAQ'88: Engineering Solution to Indoor Air Problems, American Society of Heating, Refrigerating and Air-Conditioning, Engineers, Atlanta, Ga., 148, 1988.

NOYES, W. A., Jr., Ed., Summary Technical Report of Division 10 (Absorbents and Aerosols), Vol. 1, "Military Problems with Aerosols and Nonpersistent Gases", NTIS Document No. AD 221 596, 1946.

O'BRIEN, J. A., and MYERS, A. L., *J. Chem. Soc., Faraday Trans.* I, **80**, 1467 (1984).

OKAZAKI, M., TAMON, H., and TOEI, R., *J. Chem. Eng. Jpn.*, **11**(3), 209 (1978).

PALUDETTO, R., STORTI, G., GAMBA, G., CARRA, S., and MORBIDELLI, M., *Ind. Eng. Chem. Res.*, **26**, 2250 (1987).

PEDRAM, E. O., and HINES, A. L., *J. Chem. Eng. Data.*, **28**, 11 (1983).

REUCROFT, P. J., PATEL, H. K., RUSSELL, W. C., and KIM, W. M., "Modeling of Equilibrium Gas Adsorption for Multicomponent Vapor Mixtures, Part II", NTIS Document No. AD-174 058, 1986.

RICHARDSON, J. F., and JERONIMO, M. A., *Chem. Eng. Sci.*, **34**, 1419 (1979).

ROSS, S., and MORRISON, I. D., *Surf. Sci.*, **52**, 103 (1975).

ROSS, S., and OLIVIER, J. P., *On Physical Adsorption*, Wiley-Interscience, New York, 1964.

RUTHVEN, D. M., *Principles of Adsorption and Adsorption Processes*, Wiley, New York, 1984.

SANSONE, E. B., and JONAS, L. A., *Am. Ind. Hyg. Assoc. J.*, **42**, 688 (1981).

SCHAY, G. J., *Chem. Phys. Hung.*, **53**, 691 (1956).

SHEN, J., and SMITH, J. M., *Ind. Eng. Chem Fundam.*, **7**(1), 100 (1968).

SIPS, R. J., *Chem. Phys.*, **16**, 490 (1948).

SIRCAR, S., and MYERS, A. L., *Chem. Eng. Sci.*, **28**, 489 (1973).

SIRCAR, S., *J. Colloid Interface Sci.*, **98**, 306 (1984a).

SIRCAR, S., *J. Colloid Interface Sci.*, **101**, 452 (1984b).

TALU, O., and ZWIEBEL, I., *AIChE J.*, **32**(8), 1263 (1986).

TAYLOR, R. K., *Ind. Eng. Chem.*, **37**, 649 (1945).

THOMAS, T. L., and MAYS, R. L., in *Physical Methods in Chemical Analysis* (W. G. Berl ed.), Academic Press, New York, 1961.

TOCK, R. W., "Characterization of Supersorbent Polymers for Dehumidification", Texas Energy and Natural Resources Advisory Council, Austin, Tex., NTIS Document No. DE83-900812, 1983.

TURK, A., and VAN DOREN, A., *Agr. Food Chem.*, **1**, 145 (1953).

URANO, K., OMORI, S., and YAMAMOTO, E., *Environ. Sci. Technol.*, **16** (1), 10 (1982).

WERNER, M. D., *Am. Ind. Hyg. Assoc. J.*, **46**, 585 (1985).

WERNER, M. D., GROSS, R. L., and HEYSE, E. C., "Predictive Models for Gaseous-Phase Carbon Adsorption and Humidity Effect", NTIS Document No. AD-159 167, 1985.

YAMANKA, S., MALLA, P. B., and KOMARNENI, S., *Zeolites*, **9**, 18 (1989).

YANG, R. T., *Gas Separation by Adsorption Processes,* Butterworths, Boston, Mass., 1987.

YEH, R-L., "Adsorption of Water Vapor, Toluene, 1, 1, 1-Trichloroethane, and Carbon Dioxide on Silica Gel, Molecular Sieve-13X, and Activated Carbon", M. S. Thesis, University of Missouri-Columbia, Columbia, Mo., 1991.

YEH, R-L., GHOSH, T. K., and HINES, A. L., *J. Chem. Eng. Data*, **37**, 259 (1992).

YON, C. M., and TURNOCK, P. H., *AIChE Symp. Ser.*, Vol. 67, No. 117, 3 (1971).

YOUNG, D. M., and CROWELL, A. D., *Physical Adsorption of Gases*, Butterworths, London, 1962.

ZELDOWITSCH, J., *Acta Physicochim. U. R. S. S.*, **1**, 961 (1934).

ZENZ, F. A., and OTHMER, D. F., *Fluidization and Fluid-Particle Systems*, Reinhold, New York, 1960.

Appendix A

Unit Conversion Factors and Constants

TABLE A.1 Conversion Factors

To Convert From:	*To:*	*Multiply by:*
Length		
Å (Angstrom)	m (meter)	1.000×10^{-10}
cm	m	1.000×10^{-2}
ft	m	3.048×10^{-1}
in	m	2.540×10^{-2}
μm (micrometer or micron)	m	1.000×10^{-6}
ft	cm	3.048×10^{1}
Area		
cm^2	m^2	1.000×10^{-4}
ft^2	m^2	9.290×10^{-2}
in^2	m^2	6.452×10^{-4}
ft^2	cm^2	9.290×10^{2}
Volume		
bbl (42 gallon)	m^3	1.590×10^{-1}
cm^3	m^3	1.000×10^{-6}
ft^3	m^3	2.832×10^{-2}
gal (U.S.)	m^3	3.785×10^{-3}
ft^3	cm^3	2.832×10^{4}

(*continued*)

TABLE A.1 (continued)

To Convert From:	*To:*	*Multiply by:*
Mass		
lb	kg (kilogram)	4.536×10^{-1}
g	kg	1.000×10^{-3}
Density		
g/cm^3	kg/m^3	1.000×10^{3}
lb/ft^3	kg/m^3	1.602×10^{1}
lb/ft^3	g/cm^3	1.602×10^{-2}
Force		
lb_f	N (Newton)	4.448
Pressure		
inHg (60°F)	N/m^2 (Pa, Pascal)	3.377×10^{3}
inH_2O	N/m^2 (Pa)	2.488×10^{2}
lb_f/ft^2	N/m^2 (Pa)	4.788×10^{1}
lb_f/in^2 (psi)	N/m^2 (Pa)	6.895×10^{3}
torr (mmHg, 32°F)	N/m^2 (Pa)	1.333×10^{2}
lb_f/in^2 (psi)	mmHg (32°F)	5.171×10^{1}
Viscosity		
P (poise, g/cm · s)	Pa · s	1.000×10^{-1}
cP (centipoise)	Pa · s	1.000×10^{-3}
St (Stokes)	m^2/s	1.000×10^{-4}
Energy Terms		
Btu	J (Joule)	1.055×10^{3}
Btu/h	W (Watt)	4.187×10^{3}
cal	J	4.187
ft · lb_f	J	1.356
Temperature		
°F	K (Kelvin)	$T_K = (T_F + 459.67)/1.8$
°R	K	$T_K = T_R/1.8$
°C	K	$T_K = T_C + 273.15$
°F	°C	$T_C = (T_F - 32)/1.8$

TABLE A.2 Values of Universal Gas Constant (R) in Various Units

$$
\begin{aligned}
R &= 82.06 \text{ atm} \cdot \text{cm}^3/\text{gmol} \cdot \text{K} \\
&= 0.7302 \text{ atm} \cdot \text{ft}^3/\text{lbmol} \cdot {}^\circ\text{R} \\
&= 1.987 \text{ Btu/lbmol} \cdot {}^\circ\text{R} \\
&= 1.987 \text{ cal/gmol} \cdot \text{K} \\
&= 1545 \text{ ft} \cdot \text{lb}_f/\text{lbmol} \cdot {}^\circ\text{R} \\
&= 8314 \text{ N} \cdot \text{m/kgmol} \cdot \text{K}
\end{aligned}
$$

Appendix B

Conversion of Parts per Million (ppm) to mg/m³

Assuming that the ideal gas law applies to a mixture of air and a pollutant,

$$ppm = \frac{R\,T\,C}{P_i\,M}$$

where

ppm = pollutant concentration in parts per million
R = universal gas constant
T = temperature
C = pollutant concentration in mg/m^3
P_i = partial pressure of pollutant in the mixture
M = molecular weight of pollutant

Then,

$$C = \frac{ppm\,P_i\,M}{R\,T}$$

The conversion factors for selected indoor pollutants, from ppm to mg/m^3 at room temperature (25°C or 77°F) and atmospheric pressure (760 mmHg or 14.7 psi), are given in Table B.1. The following values and units are used in the calculation.

$$R = 82.06 \text{ cm}^3 \cdot \text{atm/gmol} \cdot \text{K} = 82.06 \times 10^{-9} \text{ m}^3 \cdot \text{atm/mgmol} \cdot \text{K}$$
$$P_i = P_T \times 10^{-6} \ (P_T = \text{total pressure} = 1 \text{ atm})$$
$$T = 25°\text{C} = 298.15 \text{ K}$$

TABLE B.1 Conversion Factors for ppm to mg/m^3 for Selected Indoor Air Pollutants at Atmospheric Pressure and 25°C

Pollutants	*Molecular Weight*	*Multiply By*
Acetaldehyde	44	1.80
Acetone	58	2.37
Ammonia	17	0.70
Benzene	78	3.19
Butane	58	2.37
Butanol	74	3.02
Carbon dioxide	44	1.80
Carbon monoxide	28	1.14
Carbon tetrachloride	152	6.21
Chlorine	35	1.43
Chlorobenzene	112	4.58
Chloroform	119	4.86
Cresol	108	4.41
Cyclohexane	84	3.43
Dichloroethane	98	4.00
Diethylbenzene	134	5.48
Dioxane	88	3.60
Ethanol	46	1.88
Ethylene glycol	62	2.53
Formaldehyde	30	1.23
Hexane	86	3.52
Hydrogen chloride	36	1.47
Hydrogen sulfide	36	1.47
Methanol	32	1.31
Methyl chloride	50	2.04
Methyl ethyl ketone	72	2.93
Naphthalene	128	5.23
Nitrogen dioxide	46	1.88
Ozone	48	1.96
Phenol	94	3.84
Styrene	104	4.25
Sulfur dioxide	64	2.62
Tetrachloroethane	166	6.78
Toluene	92	3.76
Trichloroethane	133	5.44
Vinyl chloride	62	2.53
Xylene	106	4.33

Appendix C

Pollutants Found in Various Products

TABLE C.1 Organic Compounds Found in Household Products and Building Materials

Material/Product	*Organic Compounds Identified*
Latex caulk	Methyl ethyl ketone, butyl propionate, 2-butoxyethanol, butanol, benzene, toluene
Floor adhesive (water-based)	Nonane, decane, undecane, dimethyloctane, 2-methylnonane, dimethylbenzene
Particleboard	Formaldehyde, acetone, hexanal, propanol, butanone, benzaldehyde, benzene
Moth crystals	*p*-Dichlorobenzene
Floor wax	Nonane, decane, undecane, dimethyloctane, trimethylcyclohexane, ethylmethylbenzene
Wood stain	Nonane, decane, undecane, methyloctane, dimethylnonane, trimethylbenzene
Latex paint	2-Propanol, butanone, ethylbenzene, propylbenzene, 1,1-oxbisbutane, butylpropionate, toluene
Furniture polish	Trimethylpentane, dimethylhexane, trimethylhexane, trimethylheptane, ethylbenzene, limonene
Polyurethane floor finish	Nonane, decane, undecane, butanone, ethylbenzene, dimethylbenzene
Room freshener	Nonane, decane, undecane, ethylheptane, limonene, substituted aromatics (fragrances)
Tape	1,2-Dichloroethane, benzene, carbon tetrachloride, chloroform, ethyl benzene, methyl chloroform, styrene, tetrachloroethylene, trichloroethylene

(*continued*)

TABLE C.1 (Continued)

Material/Product	*Organic Compounds Identified*
Cosmetics	Methyl chloroform, styrene, tetrachloroethylene, trichloroethylene
Deodorants	Limonene, styrene
Health and beauty aids	Benzene, limonene, styrene, methyl chloroform, trichloroethylene
Electrical equipment	1,2-Dichloroethane, benzene, carbon tetrachloride, chloroform, ethyl benzene, methyl chloroform, styrene, tetrachloroethylene, trichloroethylene
Ink pen	Benzene, carbon tetrachloride, chloroform, methyl chloroform, styrene, tetrachloroethylene, trichloroethylene
Paper	Benzene, chloroform, methyl chloroform, tetrachloroethylene, trichloroethylene, formaldehyde
Photo equipment	Benzene, carbon tetrachloride, chloroform, ethyl benzene, methyl chloroform, styrene, tetrachloroethylene
Photofilm	Benzene, chloroform, ethylbenzene, methyl chloroform, styrene, trichloroethylene

Appendix D

Radon Concentrations in U.S. Counties

TABLE D.1 Radon Concentrations in U.S. Counties

	Living Areas				Basements			
Counties	*Number of Measurements*	*Mean*[a] *(pCi/L)*	*Average*[b] *(pCi/L)*	*Number of Measurements Above 4 pCi/L*	*Number of Measurements*	*Mean (pCi/L)*	*Average (pCi/L)*	*Number of Measurements Above 4 pCi/L*
Alabama								
Jefferson	94	1.03	1.62	8	80	1.78	2.47	14
Madison	61	2.83	5.57	19	27	4.61	7.96	14
Mobile	15	0.53	0.55	0				
Montgomery	17	0.82	1.21	2	1	0.90	0.90	0
Morgan	14	1.28	2.41	3	4	1.20	1.43	0
Shelby	13	1.13	1.63	1	16	1.68	2.73	4
Alaska								
Anchorage	19	0.56	0.61	0	7	0.94	1.10	0
Arizona								
Coconino	15	1.56	3.17	3	2	8.88	15.95	1
Maricopa	99	1.45	1.91	7	7	1.37	1.99	2
Pima	27	1.61	2.01	3				
Yavapai	11	1.17	3.26	1				
Arkansas								
Benton	13	3.03	9.12	5	8	4.72	8.29	4
Boone					17	2.81	6.23	6
Garland	15	1.35	1.87	1	6	2.14	2.55	1
Pulaski	34	0.69	0.83	0	11	0.98	1.35	0
Sebastien +	34	0.62	0.80	0				
Washington	17	1.21	2.16	1				
California								
Alameda	38	0.69	0.86	1				
Contra Costa	13	0.98	1.55	1				
Contra Costa +	42	0.80	1.01	0				
Fresno	30	1.67	2.18	4	5	1.65	1.78	0
Humboldt +	54	0.90	1.30	6				
Kern	12	1.32	1.53	0				
Los Angeles	255	0.76	1.26	7	28	0.91	2.33	3
Monterey	12	0.57	0.87	1	3	1.66	1.77	0
Orange	80	0.68	0.82	0	6	1.00	1.35	0

Placer	10	0.57	0.57	0				
Riverside	28	0.85	1.07	0	4	1.43	4.17	1
Sacramento	13	0.87	1.02	0				
San Benito	16	1.06	1.34	0	1	6.80	6.80	1
San Bernardino	32	0.84	1.32	3	5	1.38	2.78	1
San Diego	65	0.73	0.94	1	2	0.30	0.15	0
San Francisco	17	0.62	0.80	1	2	0.67	0.80	0
San Luis Obispo	12	1.06	1.60	2	1	0.30	0.30	0
San Mateo +	16	0.70	0.93	6				
Santa Barbara	16	0.99	1.43	2	3	0.38	0.37	0
Santa Clara	12	0.69	0.76	0	2	2.22	2.30	0
Solano +	18	0.82	1.00	0				
Sonoma	17	0.72	0.86	0				
Ventura	39	0.89	1.32	1	7	1.09	1.80	1
Ventura +	30	1.43	1. 72	7				
Colorado								
Adams	30	2.51	3.15	8	74	5.41	15.34	51
Arapahoe	144	3.54	5.26	65	383	6.08	8.02	285
Boulder	206	3.50	5.49	85	162	5.69	8.90	104
Boulder +	50	2.60	3.71	3				
Chaffee	14	4.48	6.26	8	8	5.34	10.57	9
Clear Creek	14	3.24	4.23	7	5	4.59	5.18	2
Denver	130	3.14	4.23	52	229	4.67	6.03	149
Douglas	40	3.99	7.59	21	94	6.77	9.46	69
Eagle	12	3.82	6.18	7	9	5.78	8.97	4
El Paso	280	2.73	4.81	86	249	4.18	7.59	119
Fremont	15	2.59	3.39	4	4	3.10	5.72	2
Garfield	12	3.39	4.21	4	6	6.65	9.70	5
Jefferson	367	3.50	8.40	157	476	5.48	8.11	333
La Plata	14	1.91	3.19	2	8	7.99	10.34	7
Larimer	138	3.86	5.97	67	117	7.21	10.35	92
Mesa	18	2.02	3.89	2	6	3.84	4.97	3
Park	13	5.54	16.96	7	6	13.60	21.82	5
Pitkin	17	3.51	8.97	8	8	3.21	4.42	5
Teller	33	3.42	6.34	11	10	12.27	8.40	9
Weld	7	2.60	3.37	2	12	5.26	6.70	6
Connecticut								
Fairfield	694	1.24	2.01	73	1183	2.65	4.52	378

(continued)

TABLE D.1 (continued)

Counties	Living Areas: Number of Measurements	Living Areas: Mean[a] (pCi/L)	Living Areas: Average[b] (pCi/L)	Living Areas: Number of Measurements Above 4 pCi/L	Basements: Number of Measurements	Basements: Mean (pCi/L)	Basements: Average (pCi/L)	Basements: Number of Measurements Above 4 pCi/L
Hartford	290	0.76	1.16	13	1079	1.36	2.65	150
Litchfield	113	1.18	1.92	8	261	1.87	3.25	49
Middlesex	84	1.20	1.71	8	191	2.14	3.25	44
New Haven	286	1.00	1.75	25	857	1.91	3.48	197
New London	102	1.18	1.91	5	257	2.09	3.86	70
Tolland	88	1.02	1.53	6	271	1.60	2.46	37
Windham	30	1.41	2.01	5	111	2.08	3.26	26
Delaware								
Delaware +	62	0.63	0.83	3				
Kent					10	1.31	1.46	0
New Castle	46	0.89	1.19	2	204	1.37	2.26	20
New Castle +	33	0.70	0.90	3				
Sussex	22	0.47	0.45	0	9	1.18	1.43	0
Sussex +	17	0.60	0.65	0				
District of Columbia	483	0.88	1.31	21	1181	1.48	2.32	142
Florida								
Alachua	21	1.93	3.64	4				
Brevard	24	0.60	0.67	0				
Broward	59	0.70	0.93	1	4	0.84	4.75	1
Charlotte	17	1.25	1.83	1				
Citrus	26	1.62	2.64	4				
Clay	12	0.61	0.73	0				
Dade	49	1.29	2.00	6				
Duval	46	0.53	0.55	0				
Duval +	13	0.60	0.70					
Escambia	18	0.81	0.94	0				
Hillsborough	174	1.12	1.88	19	6	2.08	4.57	3
Lake	19	0.71	0.99	1				
Lee	26	1.14	1.59	2				
Leon	30	1.78	2.45	7				
Manatee	36	0.83	1.25	3				

Marion	43	3.05	15.43	17	1	17.50	17.50	1
Okaloosa	13	0.66	0.78	0				
Orange	53	0.59	0.80	2	4	1.08	1.18	0
Osceola	10	0.53	0.57	0				
Palm Beach	42	0.76	1.10	1				
Pasco	32	0.87	1.27	2				
Pinellas	88	0.67	1.38	4				
Polk	120	1.33	2.85	17				
Sarasota	96	1.14	1.60	5				
Sarasota +	59	0.90	1.20	3				
Seminole	37	0.57	0.64	0				
St. Johns	11	0.75	1.43	1	3	1.10	1.83	1
Volusia	24	0.47	0.50	0	4	0.62	0.73	0
Georgia								
Bibb	16	0.71	0.80	0	8	0.84	1.10	0
Chatham	29	0.79	1.36	1	2	0.65	0.65	0
Chatham +	29	0.85	1.24	3				
Cherokee	11	0.79	0.92	0	13	2.08	2.63	2
Clayton	10	1.05	1.28	0	9	1.82	2.68	3
Cobb	81	0.90	1.08	0	55	1.70	2.22	7
De Kalb	80	1.37	2.00	7	55	2.29	3.34	14
Fayette	16	1.06	1.81	3	7	2.65	2.91	1
Fulton	83	1.17	1.63	6	70	2.39	3.23	17
Gwinnett	45	1.43	1.90	4	43	2.97	3.42	12
Henry	12	1.36	1.58	1	6	2.76	3.13	2
Richmond	10	1.01	1.25	0	4	0.86	0.98	0
Walton	412	1.16	1.63	29	311	2.31	3.21	77
Idaho								
Ada	30	1.64	2.22	4	3	2.56	3.53	2
Adams	126	2.04	2.56	23	31	2.75	3.63	8
Bannock	47	2.29	3.07	13	13	2.82	4.12	4
Bear Lake	19	4.05	5.87	10				
Bingham	20	2.23	2.94	5	6	3.16	3.65	2
Blaine	124	5.66	8.92	76	34	12.67	19.14	29
Boise	23	2.13	2.70	6				
Bonner	38	1.66	2.46	4	14	3.89	5.89	6
Bonneville	64	1.84	2.46	11	26	2.60	3.29	9
Boundary	10	2.14	4.54	2				

(*continued*)

TABLE D.1 (continued)

	Living Areas				Basements			
Counties	*Number of Measurements*	*Mean[a] (pCi/L)*	*Average[b] (pCi/L)*	*Number of Measurements Above 4 pCi/L*	*Number of Measurements*	*Mean (pCi/L)*	*Average (pCi/L)*	*Number of Measurements Above 4 pCi/L*
Butte	17	5.71	9.89	12				
Canyon	53	1.34	1.74	1	7	1.71	1.90	0
Caribou	15	1.90	2.50	2				
Cassia	20	2.38	4.13	6				
Clearwater	11	2.44	3.11	4				
Custer	17	3.27	4.31	6				
Elmore	19	3.68	6.46	10				
Franklin	15	2.35	5.07	5				
Fremont	26	2.56	3.30	6				
Gem	12	2.08	2.15	0				
Gooding	17	1.92	2.41	3				
Idaho	16	1.90	3.09	2				
Jefferson	21	2.95	3.89	7				
Jerome	15	1.87	2.29	1				
Kootenai	88	3.47	7.90	39	45	6.78	11.63	33
Latah	19	1.78	2.69	3				
Latah +	32	1.10	2.20	9				
Lemhi	28	2.96	4.30	9				
Madison	13	3.60	5.32	6				
Minidoka	14	2.41	2.85	2				
Nez Perce	30	1.86	2.73	4	5	1.64	2.20	1
Oneida	12	3.15	3.49	5				
Owyhee	14	2.00	2.85	4				
Payette	14	1.76	2.27	2				
Power	17	1.49	1.86	1				
Teton	30	6.78	13.56	18				
Twin Falls	29	1.93	2.54	4	9	2.01	2.29	1
Valley	48	1.16	1.57	4	8	2.89	3.83	4
Washington	11	1.87	1.95	0				
Illinois								
Adams	6	1.73	2.43	1	15	7.84	10.41	13
Boone	4	1.29	1.50	0	14	4.21	5.16	6

Bureau	8	2.75	3.48	4	18	5.82	8.09	11
Champaign	27	1.75	2.92	7	47	3.55	4.76	22
Cook	1056	1.06	1.47	60	1053	1.72	2.67	160
De Kalb	13	2.25	2.78	2	62	3.46	6.90	26
Du Page	433	1.78	2.56	72	574	2.97	4.34	207
Effingham	7	1.77	2.76	2	13	2.78	3.55	4
Grundy	13	1.34	1.59	1	15	2.38	3.53	4
Henry	7	2.68	4.80	7	46	4.34	8.54	23
Jo Daviess	6	3.40	5.60	2	10	5.54	11.30	5
Kane	83	2.16	2.89	10	142	3.67	5.57	62
Kankakee	12	1.55	2.58	3	20	3.82	6.20	11
Kendall	12	2.34	3.67	2	32	4.79	7.34	20
Knox	6	3.87	6.78	3	20	8.93	12.16	16
Lake	223	1.34	1.84	19	220	2.21	3.02	45
La Salle	21	1.89	3.65	3	46	3.37	7.92	19
Macon	17	2.44	2.96	3	37	3.59	11.69	17
Madison	9	0.99	1.31	0	15	2.99	4.06	5
McHenry	60	1.60	2.69	7	99	3.22	4.92	34
McLean	14	3.77	5.37	8	51	5.71	8.68	35
Ogle	5	2.76	2.86	0	23	4.02	6.72	12
Peoria	13	2.34	3.17	5	28	5.00	6.62	21
Rock Island	29	2.68	4.73	14	227	4.55	6.69	142
Sangamon	28	2.38	3.48	11	93	3.20	4.11	33
St. Clair	11	2.20	2.42	2	21	3.29	4.03	6
Tazewell	15	3.73	5.63	5	26	5.30	8.27	17
Vermilion	16	1.69	2.48	4	22	2.91	4.37	9
Whiteside	6	1.87	1.92	0	11	2.18	3.84	4
Will	83	2.21	4.34	21	157	3.49	6.59	68
Winnebago	18	1.27	2.29	4	88	2.49	3.27	23
Indiana								
Allen	41	2.15	4.49	10	46	3.50	6.07	16
Bartholomew	6	2.32	2.83	2	4	4.14	6.08	2
Clark	59	2.25	4.19	12	62	5.08	9.13	29
Clinton					10	7.34	9.34	7
Elkhart	5	1.23	1.42	0	17	3.04	6.04	7
Floyd	34	1.64	2.47	3	72	2.37	3.13	21
Hamilton	14	2.25	3.66	4	21	5.93	7.40	16
Jefferson	18	1.02	1.33	1	5	1.31	1.60	0

(continued)

TABLE D.1 (continued)

	Living Areas				Basements			
Counties	Number of Measurements	Mean[a] (pCi/L)	Average[b] (pCi/L)	Number of Measurements Above 4 pCi/L	Number of Measurements	Mean (pCi/L)	Average (pCi/L)	Number of Measurements Above 4 pCi/L
Johnson	12	2.89	3.90	4	15	4.50	6.54	9
Lake	38	1.15	2.01	2	67	0.91	1.29	4
La Porte	20	1.62	2.61	5	24	1.77	3.41	8
Madison	16	2.97	6.10	6	6	5.17	8.75	3
Madison +	36	2.00	2.73	14				
Marion	96	1.86	3.00	17	96	4.61	6.54	59
Marshall	3	1.66	1.70	0	10	2.26	2.55	2
Monroe	5	2.15	3.42	2	13	5.24	10.85	7
Porter	20	1.26	1.77	2	26	2 38	4.25	10
Scott	21	1.09	1.58	2	4	1.98	4.88	1
St. Joseph					28	2.42	3.92	7
Tippecanoe	12	1.69	2.02	0	19	5.10	6.21	11
Vigo	21	2.66	3.27	6	7	3.07	4.37	4
Iowa								
Adams	4	0.99	1.35	0	18	4.89	6.89	8
Allamakee	10	2.96	4.87	5	17	4.19	5.24	10
Appanoose	10	1.32	1.90	1	21	3.32	4.23	9
Audubon	4	6.21	7.05	3	12	9.21	10.33	10
Benton	8	3.89	4.70	5	28	6.95	9.70	21
Black Hawk	48	3.50	5.02	20	305	5.79	8.04	215
Boone	17	3.54	5.09	8	103	4.52	6.20	58
Bremer	20	2.00	4.18	4	53	4.57	6.24	31
Buchanan	9	2.31	3.01	3	18	2.85	4.21	4
Buena Vista	11	2.80	4.38	5	47	7.16	9.03	39
Butler	12	3.64	4.29	6	25	5.22	6.18	16
Calhoun	11	3.87	5.01	6	46	4.62	5.97	26
Carroll	9	2.35	2.62	0	42	5.81	7.65	32
Cass	17	3.96	6.38	9	22	8.09	9.61	18
Cedar	13	2.77	3.82	5	31	4.44	6.41	19
Cerro Gordo	41	2.55	3.43	12	149	4.21	6.10	79
Cherokee	4	6.03	7.12	3	20	6.90	11.30	16

Chickasaw	11	2.43	3.28	2	19	4.10	6.98	7
Clarke	6	1.63	1.90	0	14	3.57	8.74	5
Clay	8	3.51	6.18	3	40	8.16	10.48	33
Clayton	27	2.54	3.77	7	71	6.28	9.61	49
Clinton	13	1.66	2.17	2	79	3.27	4.44	35
Dallas	27	3.15	4.53	10	165	6.10	8.85	111
Davis	8	1.84	4.01	1	12	2.69	4.17	4
Decatur	15	1.60	2.19	2	20	2.17	3.02	6
Delaware	5	1.71	2.08	0	11	1.88	2.51	2
Des Moines	17	3.33	4.52	7	83	5.06	7.90	47
Dickinson	12	2.65	3.97	5	42	7.55	12.34	32
Dubuque	43	2.03	3.00	10	135	3.72	5.82	63
Emmet					15	6.77	7.38	13
Fayette	17	2.24	3.20	4	38	4.60	6.50	22
Floyd	10	2.14	3.05	2	44	4.00	5.12	24
Franklin	5	4.46	4.72	2	21	5.02	6.29	11
Greene	12	4.16	4.38	6	43	7.10	10.61	32
Grundy	10	3.83	4.50	4	33	5.91	8.05	24
Guthrie	12	3.35	4.31	6	32	5.34	6.70	20
Hamilton	19	3.48	4.19	8	77	5.27	6.78	48
Hancock	8	2.79	3.15	2	33	4.19	5.26	15
Hardin	6	1.63	1.90	0	14	3.57	8.74	5
Henry	9	6.17	7.98	7	34	3.90	5.66	15
Humbolt	5	3.08	4.54	4	32	6.35	10.04	22
Iowa	6	4.34	9.47	4	28	7.41	10.12	23
Jackson	10	2.92	3.97	4	24	4.27	5.22	14
Jasper	30	4.50	14.65	18	156	8.15	12.47	130
Jefferson	9	3.06	3.18	1	36	5.51	8.33	24
Johnson	109	3.19	4.78	40	397	5.30	7.65	263
Jones	11	2.25	2.85	3	31	4.83	8.28	20
Keokuk	6	2.38	2.75	1	17	8.98	12.67	15
Kossuth	9	3.13	3.49	4	43	4.72	6.11	25
Lee	16	2.19	2.61	3	50	3.56	4.79	22
Linn	65	2.53	4.18	19	404	3.00	5.13	155
Linn +	52	1.21	1.62	8				
Louisa	13	2.14	2.70	3	16	4.23	5.03	9
Lucas	12	1.83	2.18	1	31	3.62	5.41	16

(continued)

TABLE D.1 (continued)

Counties	Living Areas				Basements			
	Number of Measurements	Mean[a] (pCi/L)	Average[b] (pCi/L)	Number of Measurements Above 4 pCi/L	Number of Measurements	Mean (pCi/L)	Average (pCi/L)	Number of Measurements Above 4 pCi/L
Madison	13	3.50	4.86	6	34	8.36	12.36	30
Mahaska	12	2.74	3.53	5	58	4.75	24.05	31
Marion	73	3.32	5.46	37	271	6.04	8.79	191
Mitchell	15	3.86	6.07	8	29	5.17	7.49	16
Montgomery	1	4.30	4.30	1	13	7.74	9.73	11
Muscatine	19	1.94	2.36	2	44	3.92	7.00	20
O'Brien	5	1.45	2.08	1	24	8.58	10.71	21
Page	9	5.19	6.21	6	11	7.89	9.34	9
Palo Alto	8	3.89	5.64	4	14	6.46	10.33	11
Plymouth	9	9.38	11.41	8	44	7.70	9.32	39
Pocahontas	6	3.42	4.48	3	29	4.30	5.66	15
Polk	358	4.24	6.09	199	1826	6.88	9.20	1421
Pottawattamie	12	3.24	6.53	6	72	5.58	7.58	50
Poweshiek	25	4.28	5.57	11	104	7.13	9.55	78
Sac					13	6.57	7.51	10
Scott	53	3.11	4.00	21	261	4.53	6.47	163
Shelby	5	8.41	9.34	4	8	4.67	5.58	5
Sioux					41	7.26	9.78	33
Story	136	3.24	4.72	54	604	5.08	7.22	394
Tama	7	3.38	3.91	3	44	7.02	9.28	33
Taylor					11	5.38	6.86	7
Union					35	5.05	7.63	22
Van Buren	4	1.75	1.88	0	10	5.37	9.66	6
Wapello	17	1.85	2.58	3	56	3.37	4.71	22
Warren	48	4.03	5.96	25	155	5.95	9.00	104
Washington	12	4.41	10.62	8	34	5.42	8.26	21
Webster	34	3.29	4.09	15	204	5.60	7.48	143
Winnebago	10	2.32	3.52	3	42	4.67	6.00	26
Winneshiek	39	3.06	4.07	14	30	4.47	5.58	17
Woodbury	23	3.86	7.24	12	84	4.95	6.89	48
Worth	10	3.28	5.65	3	11	6.08	7.25	9
Wright	11	3.17	4.82	7	54	4.41	5.72	32

Kansas								
Johnson	61	2.46	3.61	17	135	3.67	7.23	59
Sedgwick	9	2.66	4.01	4	20	2.43	3.18	6
Shawnee	7	2.08	2.36	1	11	4.52	7.07	5
Kentucky								
Boyle	11	3.97	5.87	4	8	3.75	5.46	4
Bullitt	7	4.84	6.84	5	11	6.47	9.89	8
Campbell +	19	0.72	0.94	0				
Christian	4	1.65	1.67	0	13	2.38	2.94	4
Fayette	44	4.07	5.71	25	42	9.15	15.37	34
Fayette +	35	2.80	6.90	40				
Franklin	15	2.37	4.52	5	14	3.79	6.66	6
Hart	3	1.33	2.97	1	10	5.28	10.27	5
Jefferson	94	2.23	3.50	27	299	4.43	8.85	163
Kenton	8	1.51	3.55	1	14	1.18	1.52	1
Madison	19	2.78	4.15	9	16	3.17	8.56	6
Oldham	9	2.18	3.31	2	18	3.74	6.63	7
Warren	17	4.99	13.19	10	16	21.08	39.91	14
Louisiana								
Bossier	17	0.58	0.72	0				
Caddo	38	0.46	0.45	0				
Jefferson	12	0.48	0.53	0				
Lafayette	12	0.76	0.92	0	1	0.60	0.60	0
Ouachita	12	0.87	1.07	0				
Maine								
Androscoggin	29	1.27	1.97	3	29	1.74	3.16	6
Aroostook	15	1.21	2.69	3	41	3.73	7.32	23
Cumberland	106	2.04	3.44	27	155	4.55	11.25	77
Franklin	9	0.83	1.72	2	18	1.27	2.06	2
Hancock	54	1.39	3.77	9	23	3.03	5.97	11
Kennebec	33	1.36	2.70	2	64	2.26	3.52	19
Knox	37	1.01	1.39	1	30	2.22	3.26	6
Lincoln	34	1.05	1.65	3	29	2.20	2.97	9
Oxford	29	1.77	2.44	4	38	3.67	6.24	20
Penobscot	35	0.93	1.47	3	56	1.74	2.54	10
Piscataquis					12	2.67	4.31	5
Sagadahoc	14	0.80	1.42	1	12	2.29	3.90	4
Somerset	13	0.68	0.86	0	19	2.56	6.16	4

(continued)

TABLE D.1 (continued)

Counties	*Living Areas* Number of Measurements	Mean[a] (pCi/L)	Average[b] (pCi/L)	Number of Measurements Above 4 pCi/L	*Basements* Number of Measurements	Mean (pCi/L)	Average (pCi/L)	Number of Measurements Above 4 pCi/L
Waldo	21	1.08	1.29	0	22	2.34	3.77	5
Washington	13	0.78	1.26	1	31	2.00	3.09	3
York	73	1.18	2.16	6	131	3.06	4.60	53
Maryland								
Allegany	22	1.60	2.32	3	92	2.80	6.01	28
Anne Arundel	579	1.16	2.00	62	1246	2.07	3.56	293
Baltimore	879	1.42	2.84	60	3387	2.09	4.13	896
Calvert	74	2.19	3.50	18	98	4.45	7.10	52
Caroline	35	0.65	0.82	0	13	1.39	2.03	3
Carroll	281	4.24	7.72	143	695	7.45	16.57	465
Cecil	25	0.75	1.00	0	53	1.60	2.22	7
Charles	172	0.89	1.35	8	110	1.70	6.92	14
Dorchester	16	0.47	0.49	0	5	0.55	0.72	0
Frederick	294	3.02	7.12	114	575	5.70	12.39	339
Garrett	16	2.19	3.88	4	33	3.68	6.70	12
Harford	154	1.43	2.82	27	608	2.95	6.50	219
Howard	764	2.66	4.52	263	1749	4.92	7.94	1066
Kent	31	0.63	1.49	1	17	2.21	3.02	5
Montgomery	2664	1.63	2.56	413	5385	2.68	4.54	1750
Prince Georges	1221	1.11	1.75	102	537	1.72	2.90	288
Prince Georges +	47	1.03	1.31	2				
Queen Annes	38	0.74	0.94	0	22	1.66	2.30	3
Somerset	41	0.45	0.43	0	9	1.21	1.52	0
St. Marys	38	1.14	1.47	2	41	2.03	2.98	9
Talbot	30	0.72	0.98	1	4	0.95	1.02	0
Washington	193	4.12	8.43	99	332	8.49	14.88	262
Wicomico	40	0.68	22.16	2	23	1.10	1.87	2
Worchester	42	0.52	0.55	0	7	0.78	0.93	0
Massachusetts								
Barnstable	30	0.73	1.04	1	88	1.85	3.17	15
Berkshire	180	1.08	1.76	14	370	1.99	4.73	88

Bristol	68	1.24	2.43	7	268	1.87	2.84	53
Essex	338	1.25	2.16	40	946	2.75	4.85	311
Franklin	38	0.88	1.19	1	65	1.63	2.47	10
Hampden	131	0.77	1.01	3	519	1.41	2.21	64
Middlesex	854	1.14	1.81	70	2304	2.20	3.70	584
Norfolk	293	1.19	2.14	32	921	2.17	3.66	227
Plymouth	112	0.77	1.19	10	344	1.57	2.44	55
Suffolk	79	0.76	1.23	6	196	1.41	2.08	26
Worcester	353	1.34	2.50	47	1119	2.87	5.46	408
Michigan								
Allegan	24	0.76	1.00	0	19	1.87	2.53	3
Barry	12	1.67	2.37	1	3	2.97	4.57	1
Bay	8	0.39	0.33	0	18	1.79	3.01	3
Berrien	11	0.93	1.05	0	18	2.12	2.39	1
Berrien +	45	1.00	1.20	0				
Calhoun	15	2.20	2.73	2	15	3.00	4.54	7
Delta					15	1.22	1.87	2
Eaton	8	1.30	1.75	1	13	2.55	3.24	4
Genesee	31	1.06	1.45	1	94	1.81	2.49	16
Hillsdale					10	4.24	4.91	5
Houghton	6	0.58	0.72	0	27	1.32	4.56	5
Ingham	37	1.23	1.64	3	75	2.16	2.81	14
Iron					10	4.53	5.67	5
Jackson	23	1.55	2.00	2	57	4.67	6.61	31
Kalamazoo	81	1.83	2.67	15	98	2.69	3.70	28
Kalamazoo +	26	1.33	1.90	8				
Kent	128	1.30	1.86	8	116	1.85	2.91	19
Lapeer	9	0.92	1.44	1	15	4.05	11.09	6
Lenawee	15	2.38	4.52	7	59	4.81	7.97	33
Livingston	37	2.49	3.64	12	77	4.20	8.18	39
Macomb	99	0.85	1.11	1	360	1.45	2.12	24
Marquette	6	0.95	1.13	0	13	1.41	2.07	1
Midland	4	0.75	0.78	0	12	1.33	1.79	1
Monroe	23	1.00	1.32	1	53	2.10	3.11	11
Muskegon	9	0.75	0.83	0	13	1.28	1.46	1
Oakland	351	1.38	2.10	40	809	2.00	3.82	154
Ottawa	36	1.40	2.09	4	35	1.56	2.04	5
Saginaw	8	0.80	0.94	0	21	1.22	1.42	1

(continued)

TABLE D.1 (continued)

	Living Areas				Basements			
Counties	Number of Measurements	Mean[a] (pCi/L)	Average[b] (pCi/L)	Number of Measurements Above 4 pCi/L	Number of Measurements	Mean (pCi/L)	Average (pCi/L)	Number of Measurements Above 4 pCi/L
Shiawassee	6	1.19	1.78	1	21	3.61	4.02	1
St. Clair	24	0.99	2.22	3	39	1.35	1.81	3
St. Clair +	48	0.72	0.91	2				
Van Buren	7	0.88	1.06	0	18	1.13	1.57	1
Washtenaw	100	2.34	3.99	26	238	3.40	5.36	92
Washtenaw +	61	1.71	3.10	18				
Wayne	243	1.03	1.39	9	726	1.53	2.15	61
Minnesota								
Anoka	32	1.20	1.54	2	241	2.33	2.98	55
Beltrami	4	0.93	0.97	0	20	2.77	3.92	5
Blue Earth	10	3.02	4.74	5	15	2.91	5.12	9
Brown					14	5.54	9.69	10
Carlton	2	0.71	0.85	0	12	2.56	2.95	2
Carver	14	1.91	2.99	4	30	5.65	8.32	23
Cass					12	3.90	6.46	6
Chisago	2	2.08	2.15	0	28	1.47	2.03	4
Clay	7	2.78	3.86	3	17	6.26	14.62	14
Crowwing	10	1.09	1.60	1	22	1.99	3.23	5
Dakota	50	1.85	2.58	12	305	3.46	4.74	127
Douglas	5	2.02	3.20	1	20	4.73	5.80	14
Freeborn					12	4.27	5.04	6
Goodhue	5	4.42	5.34	2	22	3.46	4.64	8
Hennepin	258	1.79	2.44	42	1238	3.70	4.89	571
Isanti	4	1.88	1.90	0	13	2.78	4.63	5
Itasca	6	1.23	1.45	0	16	2.55	3.25	4
Kandiyohi	1	14.00	14.00	1	32	5.15	7.61	20
Le Sueur					11	7.46	9.25	8
McLeod	4	4.10	4.25	2	14	6.74	8.26	11
Morrison	11	1.85	2.19	1	16	2.42	3.04	4
Mower					19	2.54	3.97	5
Nicollet					21	6.24	8.04	18
Olmstead	8	3.63	4.38	3	48	3.76	6.10	24

Otter Tail	10	1.72	3.36	2	24	4.38	5.94	13
Pine	7	0.82	0.96	0	16	2.52	3.49	7
Polk	5	1.34	2.10	1	16	6.79	9.17	12
Ramsey	76	1.70	2.37	10	427	2.81	3.57	135
Rice	10	2.27	2.54	1	50	5.14	6.71	31
Scott	9	2.16	2.42	1	46	4.22	5.35	23
Sherburne	8	3.28	4.11	3	14	4.17	5.60	6
Steele	5	5.42	5.80	5	22	5.12	6.76	14
Stevens	9	2.30	2.98	3	34	3.50	4.76	15
St. Louis	51	1.51	2.42	6	117	2.08	2.84	16
Waseca	2	3.15	3.15	0	12	4.77	6.20	9
Washington	26	1.54	2.18	4	178	3.03	3.93	67
Winona	5	1.52	2.36	1	12	3.71	4.73	6
Wright	6	2.20	2.93	2	28	5.48	6.79	24
Mississippi								
Harrison	14	0.55	0.59	0				
Harrison +	15	1.00	1.10	0				
Hinds	13	0.72	0.92	0				
Missouri								
Clay	9	1.69	1.83	0	34	3.36	5.99	13
Greene	8	4.26	4.81	6	15	3.00	4.05	7
Jackson	42	1.91	3.17	9	96	3.85	9.56	49
Jasper	18	0.83	1.02	0				
Jefferson	7	1.50	1.67	0	11	1.56	2.45	2
Jefferson +	25	1.04	1.44	0				
Platte	5	1.14	2.50	1	15	3.33	4.99	5
St. Charles	12	2.22	2.62	17	26	1.78	2.47	3
St. Louis	104	1.60	2.10	8	110	1.97	2.67	17
Montana								
Flathead	13	1.98	7.75	2	23	3.26	4.30	8
Ravalli	18	2.75	6.26	6	4	6.80	8.83	3
Yellowstone	16	1.54	1.83	2	10	2.54	2.98	3
Nebraska								
Douglas	10	2.83	3.52	4	37	4.79	6.40	24
Nevada								
Clark	27	0.61	0.71	0	5	0.48	0.50	0

(continued)

TABLE D.1 (continued)

	Living Areas				Basements			
Counties	Number of Measurements	Mean[a] (pCi/L)	Average[b] (pCi/L)	Number of Measurements Above 4 pCi/L	Number of Measurements	Mean (pCi/L)	Average (pCi/L)	Number of Measurements Above 4 pCi/L
Clark +	33	1.10	1.41	3				
Washoe	21	2.71	6.47	7	4	4.63	6.15	3
New Hampshire								
Belknap	27	0.84	1.23	1	44	1.85	3.19	10
Carroll	40	2.46	9.04	15	80	5.88	13.96	44
Cheshire	38	1.05	1.45	2	58	1.84	3.28	13
Coos	25	2.02	6.07	8	48	4.49	7.08	26
Grafton	74	1.22	1.92	6	151	2.10	3.25	30
Hillsboro	125	1.38	2.52	17	361	2.97	5.20	124
Merrimack	59	1.35	4.74	11	111	2.47	4.40	30
Rockingham	161	1.44	2.43	23	321	2.93	5.10	115
Strafford	29	1.90	3.39	8	63	3.69	7.56	29
Sullivan	35	0.74	0.90	0	80	2.09	3.22	21
New Jersey								
Atlantic	12	0.55	0.72	0	12	1.13	1.41	0
Atlantic +	33	0.60	0.65	0				
Bergen	718	0.72	0.98	16	795	1.34	2.02	53
Burlington	110	0.62	0.74	1	200	1.39	2.05	17
Camden	106	0.83	1.18	7	269	1.87	2.97	50
Cape May	10	0.47	0.49	0				
Cumberland	6	0.93	1.03	0	15	2.36	3.15	3
Essex	412	0.75	1.27	6	589	1.37	2.77	52
Gloucester	38	0.82	1.63	3	81	1.57	2.74	12
Hudson	55	0.81	1.33	5	62	1.08	1.58	5
Hunterdon	1149	2.53	5.56	358	977	5.60	13.56	554
Mercer	413	1.37	2.76	53	877	3.02	8.36	312
Middlesex	301	0.90	1.37	17	523	1.76	2.77	95
Monmouth	160	0.94	1.60	15	316	2.06	3.72	72
Monmouth +	73	0.70	0.94	1				
Morris	3356	1.30	2.39	383	2743	2.94	5.95	940
Morris +	142	1.20	1.80	5				

Ocean	45	0.55	0.60	0	37	0.92	1.12	0
Passaic	602	0.92	1.32	23	477	1.73	2.76	70
Salem	5	0.63	0.74	0	24	1.52	2.05	4
Somerset	686	1.45	2.76	104	1007	3.13	6.60	382
Somerset +	87	1.10	1.54	8				
Sussex	1627	1.61	2.50	227	806	3.52	6.07	357
Union	240	0.82	1.15	7	462	1.44	2.51	49
Warren	365	2.39	5.36	107	344	5.29	9.64	217
New Mexico								
Bernalillo	163	2.02	3.19	40	31	2.90	5.48	12
Dona Ana	15	0.87	1.01	0				
Grant	10	2.79	4.70	3				
Los Alamos	17	1.93	2.58	3	6	3.75	5.08	3
Otero	10	1.00	1.70	1				
Sandoval	12	1.99	5.56	2	1	10.10	10.10	1
Santa Fe	92	2.83	3.56	32	15	5.18	7.62	10
Taos	11	5.66	6.77	8				
New York								
Albany	44	0.86	1.81	4	100	1.72	3.24	19
Allegany	10	2.11	3.64	3	31	3.84	6.37	12
Bronx	19	0.61	0.81	1	21	1.11	1.32	0
Bronx +	86	1.21	1.94	14				
Broome	104	1.13	1.81	12	105	2.26	4.02	29
Cattaraugus	12	1.74	3.08	2	24	2.47	6.06	9
Cayuga	10	1.75	2.91	2	32	2.53	4.90	10
Chautauqua	17	0.85	1.00	0	48	2.59	4.01	18
Chautauqua +	62	1.20	2.00	16				
Chemung	19	2.10	3.49	7	45	6.93	11.58	28
Chenango	15	1.39	1.87	1	13	5.82	12.71	8
Clinton	13	0.63	0.89	1	18	0.91	1.41	2
Columbia	19	1.55	2.30	2	21	3.28	5.16	7
Cortland	8	4.55	7.06	6	26	5.49	10.08	18
Delaware	59	1.65	2.76	15	32	2.88	5.05	9
Dutchess	146	1.56	2.42	27	122	3.36	5.59	50
Erie	75	0.96	1.91	11	318	1.22	3.23	47
Essex					11	0.96	2.24	2
Genesee	16	1.98	4.10	3	38	2.63	5.22	13
Greene	14	1.44	1.96	1	11	2.60	3.87	6

(continued)

TABLE D.1 (continued)

	Living Areas				*Basements*			
Counties	*Number of Measurements*	*Mean*[a] *(pCi/L)*	*Average*[b] *(pCi/L)*	*Number of Measurements Above 4 pCi/L*	*Number of Measurements*	*Mean (pCi/L)*	*Average (pCi/L)*	*Number of Measurements Above 4 pCi/L*
Herkimer	10	1.44	2.27	2	9	3.12	6.07	4
Jefferson	14	0.75	0.89	0	20	1.98	5.66	4
Kings	29	0.67	0.78	0	32	1.49	3.08	4
Livingston	5	0.94	4.12	1	23	3.01	3.97	11
Madison	14	1.65	2.76	4	28	2.65	5.51	11
Monroe	76	0.94	1.27	2	293	1.34	1.97	23
Nassau	76	0.61	0.71	1	117	1.08	1.39	4
New York	30	0.69	1.26	3	24	1.47	2.80	5
Niagra	11	0.77	1.06	0	76	1.02	1.40	2
Oneida	39	1.19	1.62	2	61	2.70	5.49	16
Oneida +	59	1.22	2.50	10				
Onondaga	122	1.83	4.02	31	325	3.21	7.89	131
Ontario	15	0.97	1.20	0	38	2.54	6.26	9
Orange	215	1.29	2.05	18	158	2.44	4.06	46
Oswego	11	0.52	0.55	0	42	0.94	1.19	0
Otsego	12	2.03	2.38	1	17	4.38	8.38	7
Putnam	256	1.40	2.12	28	249	2.38	3.57	73
Queens	47	0.58	1.11	1	47	1.01	1.41	4
Rensselaer	35	1.46	2.33	3	59	2.50	4.25	17
Richmond	44	0.70	0.87	1	38	1.35	2.59	7
Rockland	153	0.92	2.83	11	153	1.38	2.06	14
Saratoga	31	1.06	1.39	1	72	1.82	2.89	15
Schenectady	27	0.96	1.32	1	71	1.60	2.71	9
Steuben	34	1.56	3.70	12	45	4.47	8.00	23
St. Lawrence	17	0.61	0.68	0	14	1.38	2.20	3
Suffolk	60	0.62	0.75	0	85	1.02	1.31	1
Sullivan	17	0.72	0.95	0	15	3.18	4.13	6
Tioga	57	1.92	3.56	14	56	4.14	7.75	27
Tompkins	46	1.36	2.29	8	76	2.03	3.51	19
Ulster	66	1.26	2.07	4	64	2.43	3.33	18

Warren	15	0.70	0.99	1	24	2.14	8.90	6
Washington	5	2.08	2.96	1	17	2.72	4.24	7
Wayne	9	0.87	1.09	0	14	1.62	2.86	2
Westchester	468	0.92	1.21	14	482	1.75	3.97	82
Wyoming					11	4.80	7.36	5
North Carolina								
Alamance	10	0.50	0.50	0	9	0.95	1.12	0
Ashe	12	1.43	1.60	0	17	2.73	3.54	4
Buncombe	40	2.08	3.28	9	34	3.45	4.96	14
Catawba	5	0.82	1.02	0	12	1.83	2.08	0
Cleveland	113	1.81	2.84	19	60	3.92	5.77	31
Cumberland	29	0.77	0.99	1	6	1.55	3.28	1
Durham	45	0.82	1.07	0	12	1.49	1.75	0
Forsyth	35	1.36	2.67	4	40	3.04	4.19	15
Franklin	10	2.10	2.76	2				
Gaston	114	1.69	2.47	24	43	3.37	5.27	20
Guilford	33	0.73	1.07	2	20	0.94	1.26	1
Haywood	17	1.89	3.77	3	23	3.19	4.19	12
Henderson	82	3.00	4.62	33	54	5.78	7.65	38
Jackson	37	1.16	1.70	2	10	2.32	2.88	4
Lincoln	16	2.93	5.18	4	13	2.84	3.95	4
Macon	23	1.26	1.63	1	18	1.96	2.46	3
Mecklenburg	50	0.70	1.06	2	36	6.95	36.15	19
Moore	16	0.93	1.17	0				
New Hanover	10	0.57	0.59	0				
Orange	27	1.01	1.57	1	10	2.76	4.85	3
Transylvania	45	2.55	4.06	13	53	4.88	6.94	34
Wake	155	1.16	2.09	12	55	2.21	3.68	14
Watauga	33	1.99	3.75	9	15	5.02	12.67	8
North Dakota								
North Dakota +	159	2.72	5.20	36				
Benson	30	1.68	2.26	4	9	4.71	6.64	6
Burleigh	10	2.26	2.75	2	44	2.89	4.08	18
Burleigh +	17	1.85	2.30	6				
Cass	23	2.08	3.46	6	55	6.77	10.41	41
Grand Forks	24	6.03	7.46	18	32	8.75	12.31	25
Grand Forks +	77	3.60	7.31	51				

(*continued*)

TABLE D.1 (continued)

	Living Areas				Basements			
Counties	Number of Measurements	Mean[a] (pCi/L)	Average[b] (pCi/L)	Number of Measurements Above 4 pCi/L	Number of Measurements	Mean (pCi/L)	Average (pCi/L)	Number of Measurements Above 4 pCi/L
McLean					15	5.53	6.97	9
Morton					11	4.90	5.56	8
Pembina	4	6.95	7.25	4	11	5.91	8.24	7
Ramsey	5	1.72	2.06	1	11	4.60	5.48	7
Stark	9	3.03	5.62	4	31	4.67	6.55	19
Stutsman	10	1.29	1.39	0	16	4.32	5.66	12
Ward	11	1.73	2.35	2	68	3.49	4.36	44
Williams					14	3.52	7.02	6
Ohio								
Ashtabula	6	1.22	1.32	0	22	1.98	3.53	4
Athens	42	1.31	1.92	2	16	1.42	1.87	1
Belmont	5	1.82	1.86	0	25	3.52	5.22	9
Butler	46	2.74	4.02	16	231	2.82	4.48	76
Clark	27	3.64	4.97	15	30	6.60	9.23	22
Clermont	7	1.36	1.66	0	17	4.29	5.79	9
Columbiana	13	2.66	6.13	3	33	4.77	6.97	18
Cuyahoga	121	1.13	1.64	9	348	1.43	2.09	31
Darke	11	4.98	12.12	6	13	5.58	11.02	8
Delaware	10	2.91	3.49	4	13	4.66	5.35	8
Erie	10	1.44	1.87	1	5	4.70	7.54	2
Fairfield	13	3.43	5.95	5	13	7.25	9.05	10
Franklin	168	3.61	5.64	72	161	6.58	9.85	117
Fulton	13	1.56	2.55	1	10	1.63	2.46	2
Gallia	7	1.41	2.11	1	23	3.53	4.02	10
Geauga	18	0.91	1.13	1	36	1.29	2.01	2
Greene	59	2.58	4.90	18	42	4.95	6.60	24
Hamilton	37	1.65	2.51	5	100	1.66	2.51	12
Jefferson	9	2.05	2.59	1	21	3.04	4.65	7
Lake	25	0.75	0.95	0	47	1.46	2.85	7
Licking	12	4.09	5.91	4	12	9.51	32.39	7
Lorain	6	1.98	2.02	0	37	2.29	3.40	9
Lucas	36	1.29	1.75	1	83	2.96	5.10	22

Mahoning	19	1 22	2.02	1	42	2.29	2.92	7
Marion	7	1.74	2.63	1	12	5.59	7.18	8
Medina	20	2.12	2.53	4	35	2.91	4.58	11
Miami	29	3.90	5.84	15	34	8.14	13.10	26
Montgomery	243	2.60	4.50	72	246	4.36	11.16	127
Portage	13	0.83	1.01	0	16	2.16	2.75	3
Richland	6	2.13	4.33	1	22	5.20	8.18	15
Ross	5	2.47	3.26	1	14	5.78	14.64	7
Sandusky					11	4.51	6.02	56
Seneca					11	4.04	5.38	7
Shelby	5	2.36	3.02	1	14	8.02	12.74	10
Stark	56	2.47	4.13	18	126	4.02	5.62	6
Summit	59	1.55	2.15	6	123	2.50	3.59	33
Trumbull	24	1.41	2.40	3	60	1.94	2.52	8
Warren	10	2.27	3.26	3	28	3.74	6.47	10
Washington	14	1.57	2.38	2	10	2.39	4.00	4
Wayne	5	3.14	5.12	1	10	2.97	3.98	4
Wood	12	1.76	2.37	2	13	3.06	4.55	3
Oklahoma								
Oklahoma	25	1.00	1.44	2	6	1.87	3.22	2
Tulsa	60	0.97	1.21	1	16	0.90	1.19	0
Tulsa +	22	1.01	1.30	0				
Oregon								
Clatsop	12	1.36	1.93	1	4	3.35	6.43	2
Jackson	17	1.13	1.39	0	8	1.76	2.44	2
Klamath	7	0.47	0.47	0	10	1.02	1.47	1
Lane	10	0.54	0.56	0	4	0.83	0.88	0
Marion +	45	1.30	1.60	2				
Multnomah	17	1.85	3.02	2	16	2.42	3.89	6
Washington	14	1.30	2.21	1	4	1.63	1.88	0
Pennsylvania								
Adams	40	1.60	2.80	5	67	2.88	4.66	25
Allegheny	1970	1.95	4.28	490	4627	3.24	5.94	1743
Allegheny +	69	1.22	1.91	9				
Armstrong	55	2.51	3.63	16	97	3.90	7.21	47
Beaver	164	3.04	7.12	61	434	4.99	10.21	230

(continued)

TABLE D.1 (continued)

	Living Areas				Basements			
Counties	Number of Measurements	Mean[a] (pCi/L)	Average[b] (pCi/L)	Number of Measurements Above 4 pCi/L	Number of Measurements	Mean (pCi/L)	Average (pCi/L)	Number of Measurements Above 4 pCi/L
Bedford	33	2.78	5.78	12	78	4.93	11.23	40
Berks	129	3.71	8.46	58	206	7.37	15.63	150
Berks +	46	2.02	2.91	24				
Blair	42	1.70	2.70	9	143	3.50	6.28	62
Bradford	18	3.12	6.46	8	33	3.89	8.06	15
Bucks	502	1.62	3.38	81	1457	4.04	10.20	661
Butler	149	2.39	5.25	43	327	4.55	10.64	156
Cambria	44	1.56	2.71	6	102	3.45	5.21	40
Cambria +	84	1.30	3.00	10				
Carbon	41	2.08	8.15	11	54	5.66	12.57	35
Centre	141	3.87	6.32	77	271	10.57	18.58	221
Chester	303	1.97	4.62	70	539	4.47	10.37	276
Clarion	10	1.38	1.87	1	25	4.18	12.61	13
Clearfield	23	1.87	4.05	6	43	4.11	9.52	18
Columbia	13	1.69	3.62	3	25	4.45	9.96	13
Crawford	10	0.69	0.89	0	44	2.26	3.20	9
Cumberland	337	3.97	7.49	172	507	8.91	16.68	402
Cumberland +	165	6.33	9.24	71				
Cumberland+	38	4.30	6.80	60				
Dauphin	123	5.24	10.37	68	223	8.07	15.63	160
Delaware	134	1.31	3.64	13	328	2.05	3.37	70
Elk	11	1.91	3.56	4	26	4.61	12.57	13
Erie	33	1.10	2.79	4	124	2.01	3.83	32
Fayette	44	1.77	2.84	7	86	2.84	29.32	24
Franklin	162	3.55	6.76	75	383	7.38	11.39	293
Franklin +	56	2.40	3.71	29				
Greene	17	1.30	1.92	3	27	3.82	8.16	13
Huntingdon	17	1.66	3.18	4	0	5.81	9.44	22
Indiana	39	1.37	1.97	5	0	4.10	8.72	34
Jefferson	19	2.74	5.56	5	1	3.55	4.21	8
Lackawanna	26	1.68	2.80	5	57	2.45	4.30	21
Lancaster	537	4.09	8.51	266	469	10.05	19.03	376

Lawrence	40	1.67	2.74	6	130	3.33	7.07	58
Lebanon	120	3.70	6.79	58	126	10.07	17.25	100
Lebanon +	64	3.10	6.80	39				
Lehigh	457	3.41	7.41	204	539	8.51	16.65	417
Lehigh +	76	2.90	5.80	36				
Luzerne	94	1.35	4.29	15	222	2.77	9.72	75
Lycoming	19	2.03	4.32	5	65	5.67	11.80	34
Lycoming +	80	1.73	2.93	17				
McKean	8	0.84	1.36	1	16	2.65	3.89	5
Mercer	52	1.32	1.88	4	213	2.64	15.59	52
Mifflin	8	3.01	4.34	4	13	8.56	15.91	8
Monroe	177	1.64	3.64	35	88	5.27	10.20	49
Montgomery	457	1.63	2.90	63	1194	3.08	6.44	438
Montour	6	1.22	1.92	1	10	3.86	5.17	5
Northampton	241	4.26	8.24	124	282	7.65	15.59	208
Northampton +	91	2.00	4.00	23				
Northumberland	30	1.65	2.27	6	49	5.81	21.05	25
Northumberland +	46	1.40	2.10	9				
Perry	18	6.01	10.67	11	29	7.16	16.39	18
Philadelphia	93	1.14	1.79	12	208	1.74	3.00	41
Pike	95	1.20	1.79	10	53	2.41	4.42	19
Schuylkill	42	5.37	13.22	21	80	7.85	16.08	53
Schuylkill +	150	2.00	3.44	27				
Snyder	7	4.81	6.37	5	18	4.06	8.22	10
Somerset	31	3.00	7.13	11	59	2.75	5.03	17
Susquehanna	13	1.39	1.57	0	16	3.61	11.41	7
Tioga	4	1.20	1.33	0	12	4.52	8.97	6
Union	19	3.03	6.26	9	14	6.75	15.12	8
Venango	28	2.19	6.90	7	74	3.95	9.31	30
Warren	17	1.29	2.50	2	35	4.13	12.97	16
Washington	146	1.68	2.59	23	337	3.19	5.40	129
Wayne	46	1.10	1.76	2	40	2.53	3.63	10
Westmoreland	314	2.19	4.16	89	743	3.45	6.78	289
York	199	3.24	6.87	77	457	7.72	16.37	333
Rhode Island								
Bristol	17	0.97	1.56	3	64	1.75	2.65	13
Kent	42	1.59	2.55	8	173	2.15	3.61	49
Newport	60	1.09	1.96	5	116	2.70	4.25	36

(continued)

TABLE D.1 (continued)

	Living Areas				Basements			
Counties	Number of Measurements	Mean[a] (pCi/L)	Average[b] (pCi/L)	Number of Measurements Above 4 pCi/L	Number of Measurements	Mean (pCi/L)	Average (pCi/L)	Number of Measurements Above 4 pCi/L
Newport +	35	0.75	1.03	3				
Providence	78	1.05	1.49	5	460	1.79	6.75	73
Washington	65	1.81	2.77	10	201	4.08	6.60	102
South Carolina								
Anderson	15	0.77	0.95	0	5	1.57	2.02	1
Berkeley	10	0.54	0.58	0				
Charleston	21	0.62	0.72	0				
Charleston +	31	0.80	1.11	4				
Dorchester	11	0.76	1.13	1				
Greenville	50	1.48	2.63	10	21	2 96	5.93	6
Pickens	14	1.27	2.00	2	4	0.99	1.77	0
Richland	22	0.75	1.13	1	4	1.63	1.80	0
Spartanburg	17	1.02	1.36	1	8	3.38	5.04	3
Spartanburg +	19	1.31	1.70	11				
South Dakota								
South Dakota+	86	2.60	3.91	30				
Brookings	5	3.29	4.02	3	13	4.72	7.11	8
Minnehaha	7	3.68	3.99	3	11	2.81	4.39	4
Minnehaha +	14	2.61	3.05	14				
Pennington	25	3.66	6.52	15	15	6.94	8.86	11
Tennessee								
Anderson	21	2.79	3.83	7	15	4.11	6.39	8
Bedford	16	1.77	2.39	3				
Blount	14	2.26	2.83	2	8	4.04	5.64	5
Bradley	11	1.37	1.63	0	6	3.69	6.08	2
Coffee	31	1.12	1.35	1	16	2.51	6.24	4
Davidson	392	2.18	3.35	97	162	4.79	8.15	90
Dickson	4	1.63	1.70	0	10	2.63	3.36	3
Franklin	28	1.36	1.76	1	7	3.53	5.34	4
Hamblen	11	1.35	1.80	1	8	2.66	3.94	4
Hamilton	101	1.17	1.96	10	51	2.30	3.38	11

Knox	86	1.99	2.98	18	46	4.62	6.70	25
Maury	10	3.82	5.45	4	11	7.93	13.71	9
Montgomery	22	3.07	4.70	7	25	2.39	3.87	9
Moore	11	1.37	5.13	1	4	4.13	5.42	2
Putnam	13	1.86	3.29	3	14	2.63	3.29	3
Roane	12	3.28	7.07	4	9	5.74	7.62	4
Robertson	10	1.32	1.44	0	5	4.17	9.08	2
Rutherford	91	2.46	3.13	18	16	3.36	4.79	10
Shelby	42	0.77	0.87	0	12	1.70	3.28	4
Sullivan	42	2.96	5.79	17	50	5.76	10.43	31
Sumner	45	2.04	2.87	8	16	2.59	4.98	4
Washington	28	2.50	3.41	8	19	5.33	7.31	10
Williamson	361	2.56	3.64	100	41	4.18	7.14	22
Wilson	17	1.61	1.99	0	4	2.56	3.58	1
Texas								
Bexar	38	0.73	1.01	1				
Collin	10	0.76	1.09	0				
Dallas	162	1.23	2.04	16	23	1.98	3.32	7
Ellis	16	1.23	1.88	1				
El Paso	18	0.60	0.72	0	4	1.82	2.18	1
Galveston +	13	0.45	0.43	0				
Hardin	33	0.52	0.65	1	2	0.30	0.00	0
Harris	84	0.64	1.31	3	14	0.66	0.94	1
Jefferson	13	0.41	0.41	0				
Jefferson +	23	0.55	0.60	0				
Mc Lennan	10	1.71	2.35	2				
Montgomery	10	0.34	0.20	0	1	0.80	0.80	0
Nueces +	18	0.43	0.40	0				
Tarrant	111	0.91	1.34	8	18	1.13	1.98	3
Travis	44	0.96	1.23	1	6	0.90	1.32	1
Utah								
Utah +	125	1.42	1.93	10				
Davis +	18	1.40	2.20	34				
Salt Lake	22	1.26	1.66	1	15	2.25	3.45	6
Salt Lake +	87	1.46	1.86	6				
Vermont								
Addison	7	0.57	0.60	0	21	1.16	1.50	2

(continued)

TABLE D.1 (continued)

	Living Areas				Basements			
Counties	Number of Measurements	Mean[a] (pCi/L)	Average[b] (pCi/L)	Number of Measurements Above 4 pCi/L	Number of Measurements	Mean (pCi/L)	Average (pCi/L)	Number of Measurements Above 4 pCi/L
Bennington	11	1.14	1.52	1	27	2.15	3.91	8
Caledonia	7	1.80	2.44	2	22	3.20	3.97	10
Chittenden	57	0.54	1.16	3	79	1.06	1.70	5
Franklin	8	1.17	2.05	2	17	1.24	1.88	3
Orange	10	1.72	2.48	2	39	3.03	4.56	11
Orleans	6	0.62	0.67	0	15	1.07	1.45	1
Rutland	21	1.04	1.72	3	30	1.35	1.79	2
Washington	22	1.32	6.63	3	42	2.14	3.62	10
Windham	39	0.99	1.29	1	73	1.80	2.80	14
Windsor	44	1.20	1.84	2	98	2.11	3.30	21
Virginia								
Albermarle	115	1.47	2.20	13	157	2.55	3.58	41
Alexandria	152	0.92	1.18	3	276	1.20	1.64	19
Amherst	5	1.23	1.58	0	11	2.77	3.46	4
Arlington	265	1.10	1.71	24	716	1.77	2.61	105
Augusta	30	1.47	2.09	4	58	4.52	7.02	33
Bedford	16	0.90	1.14	0	16	2.46	4.36	6
Botetourt	6	2.38	4.07	1	11	3.71	4.85	4
Cambell	15	2.08	3.12	3	21	4.14	5.34	10
Carroll	3	1.94	2.20	0	18	4.25	7.78	10
Chesterfield	123	1.50	2.33	14	32	4.59	6.26	21
Clarke	8	1.59	2.64	2	15	3.68	6.13	8
Culpeper	29	1.39	2.03	4	35	2.46	3.67	12
Fairfax	2222	1.49	2.33	261	4306	2.33	3.63	1141
Fairfax +	89	1.40	2.00	12				
Fauquier	119	1.16	1.83	10	194	2.32	3 76	55
Fluvanna	6	3.07	5.02	3	14	2.80	3.27	4
Franklin	8	0.87	1.07	0	14	4.29	5.45	7
Frederick	53	1.82	3.47	12	95	4.96	8.15	62
Giles					10	2.76	4.38	5
Hanover	13	1.31	1.55	1	14	2.80	3.47	3
Henrico	95	1.42	1.91	7	52	3.28	6.90	23

Herico	10	0.75	0.83	0	8	2.07	2.73	1
King George	2	0.92	0.95	0	15	2.87	5.19	6
Loudoun	214	1.22	1.87	22	306	2.72	4.96	106
Madison	11	0.98	1.12	0	16	3.84	4.85	7
Montgomery	57	2.19	3.31	14	162	5.21	7.84	99
New Kent	10	0.73	0.87	0	5	1.96	2.18	0
Orange	22	1.16	1.30	0	39	2.77	3.14	8
Pittsylvania	13	1.43	2.28	2	16	4.35	5.84	7
Prince William	377	1.34	1.91	38	676	2.42	3.73	202
Pulaski	8	4.28	6.01	4	27	9.64	11.71	23
Rappahannock	12	1.05	1.60	1	15	3.58	6.15	6
Richmond	48	1.46	2.15	11	72	2.96	4.66	27
Roanoke	38	1.69	2.32	6	132	3.50	6.71	58
Rockbridge	9	2.04	2.30	1	12	3.19	3.87	6
Rockingham	19	1.61	1.98	2	34	4.23	12.42	17
Shenandoah	22	1.43	1.82	2	24	4.55	6.48	13
Spotsylvania	45	0.99	1.39	3	65	2.21	3.47	19
Stafford	65	1.04	1.28	0	95	1.88	3.31	14
Warren	17	1.55	2.76	1	27	3.06	3.79	10
Washington	7	2.45	2.59	1	21	4.59	7.64	13
Westmoreland	8	0.91	1.06	0	5	2.27	3.34	1
York	15	0.67	1.16	1				
Washington								
Chelan	10	1.74	2.74	2	5	1.63	1.82	0
Clark	17	1.39	1.65	1	2	11.06	11.65	2
King	53	0.59	0.78	2	27	0.89	1.59	4
Okanogan	38	2.01	6.93	8	21	4.09	15.40	9
Pierce	32	0.62	1.26	1	20	0.45	1.02	1
Snohomish	22	0.53	0.54	0	12	0.67	0.83	0
Spokane	147	3.34	5.58	60	165	5.90	10.84	104
Stevens	14	1.38	1.63	1	10	3.97	6.66	4
Whatcom	25	0.47	0.48	0				
Yakima	14	1.11	1.29	0				
West Virginia								
Berkeley	88	2.01	3.33	23	91	6.03	14.32	52
Cabell	20	1.16	1.51	1	13	2.34	3.65	4
Hampshire	8	1.63	3.79	1	15	3.52	5.52	7
Hancock	12	4.37	4.84	7	17	5.84	13.39	10

(continued)

TABLE D.1 (continued)

	Living Areas				*Basements*			
Counties	*Number of Measurements*	*Mean*[a] *(pCi/L)*	*Average*[b] *(pCi/L)*	*Number of Measurements Above 4 pCi/L*	*Number of Measurements*	*Mean (pCi/L)*	*Average (pCi/L)*	*Number of Measurements Above 4 pCi/L*
Harrison	17	1.84	2.51	4	21	2.83	3.88	5
Jefferson	72	2.04	4.25	19	67	8.68	12.92	51
Kanawha	25	1.33	2.03	3	39	1.50	1.93	5
Marion	16	1.86	3.44	1	14	2.79	3.69	6
Mercer	4	1.53	1.72	0	13	2.05	4.58	3
Mineral	6	1.98	2.33	1	23	4.30	6.85	13
Monongalia	16	1.38	1.63	0	25	2.37	2.78	5
Morgan	18	1.71	3.36	4	23	5.69	10.49	15
Ohio	14	2.42	5.04	6	29	2.77	4.41	9
Wood	11	1.37	1.54	0	18	2.06	2.51	3
Wisconsin								
Brown	17	1.13	1.42	0	59	3.15	4.62	20
Chippewa	4	2.85	3.95	2	12	4.33	5.12	6
Dane	24	1.73	2.86	5	104	3.04	4.62	37
Dodge	6	2.43	2.87	1	48	3.27	6.77	20
Door	10	0.98	1.30	1	26	2.61	6.27	7
Eau Claire	7	1.21	1.96	1	33	2.78	4.58	6
Fond Du Lac					37	2.74	3.69	13
Grant					13	4.66	6.67	9
Jefferson	5	0.80	1.12	0	43	2.84	5.07	14
Kenosha	13	1.81	2.52	2	105	2.69	4.13	39
La Crosse	6	1.90	2.50	1	21	2.79	3.70	7
Lincoln	2	2.66	3.10	1	9	1.85	2.48	1
Manitowoc					16	1.97	2.91	6
Marathon	6	2.86	5.53	2	57	6.38	12.48	35
Milwaukee	83	1.25	2.00	9	961	2.69	4.00	334
Oneida	8	1.09	1.57	1	11	2.31	3.02	3
Outgamie	16	1.53	1.81	0	59	2.53	4.17	19
Ozaukee	19	1.96	2.58	2	111	3.33	4.68	49
Polk					16	2.54	3.41	4
Portage	4	1.25	1.35	0	13	9.44	11.63	12
Racine	26	1.52	1.97	2	219	3.25	5.81	99

Rock	6	1.78	2.80	2	46	4.05	9.05	22
Sheboygan	11	0.65	0.88	1	42	2.36	5.65	8
St.Croix					29	3.65	5.01	11
Walworth	20	1.93	3.08	6	91	4.80	7.53	52
Washington	15	1.47	2.15	2	153	3.04	4.65	64
Waukesha	83	2.47	4.20	27	967	4.67	6.97	568
Waupaca	18	1.73	2.43	2	21	4.16	5.70	13
Winnebago	11	1.40	1.69	0	46	3.75	5.48	21
Wood	8	4.65	11.25	4	16	2.17	2.46	2
Wyoming								
Wyoming +	77	1.90	2.63	15				
Albany	8	1.41	1.74	0	14	1.96	2.12	0
Campbell	23	1.82	2.70	3	49	3.47	4.99	22
Fremont	38	4.82	6.79	21	12	3.31	5.66	5
Laramie	20	2.10	2.81	4	33	2.35	2.82	6
Laramie +	13	1.60	2.12	23				
Natrona	15	2.20	2.63	4	26	3.41	4.52	10
Natrona +	13	2.23	3.50	0				
Park	19	1.45	3.47	2	27	2.70	3.48	8
Sheridan	30	3.04	4.42	15	30	5.87	7.71	24
Sweetwater	15	1.14	1.66	2	18	3.57	4.66	6
Teton	19	4.96	7.12	11	7	11.42	21.50	5

[a]geometric mean of measurements.

[b]arithmetic average of measurements.

+only counties in which at least 10 measurements were made are reported.

Source: Cohen, B.L., and Shah, R.S., *Health Phys.,* **60**(2), 243 (1991).

Appendix E

Dew Point Curves for Air Over Various Liquid Desiccants

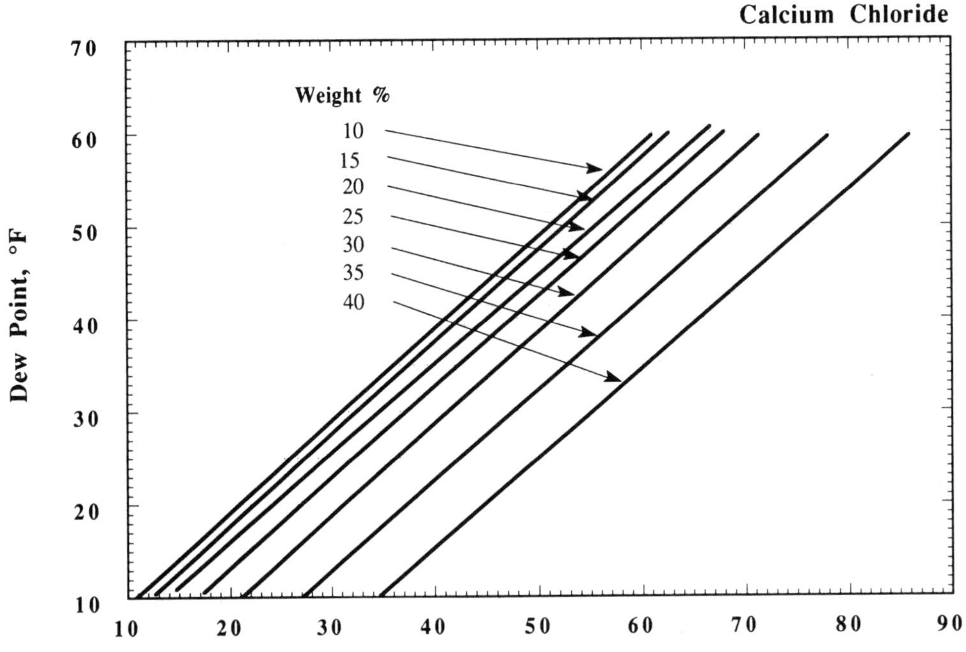

Figure E.1 *Dew point of air in contact with calcium chloride solutions. [Data of Brockschmidt, C. L.,* Gas (Los Angeles), **28,** *28 (1942).]*

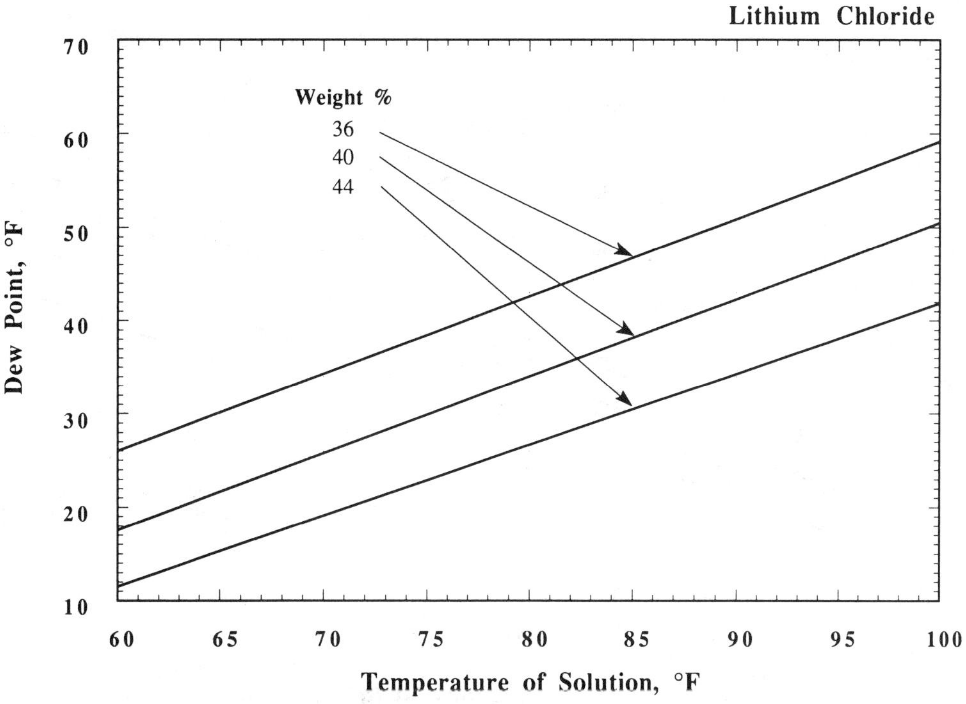

Figure E.2 *Dew point of air in contact with lithium chloride solutions. [Data of Gifford, E. W.,* Heating, Piping Air Conditioning, J. Sect., **29**, *156 (1957).]*

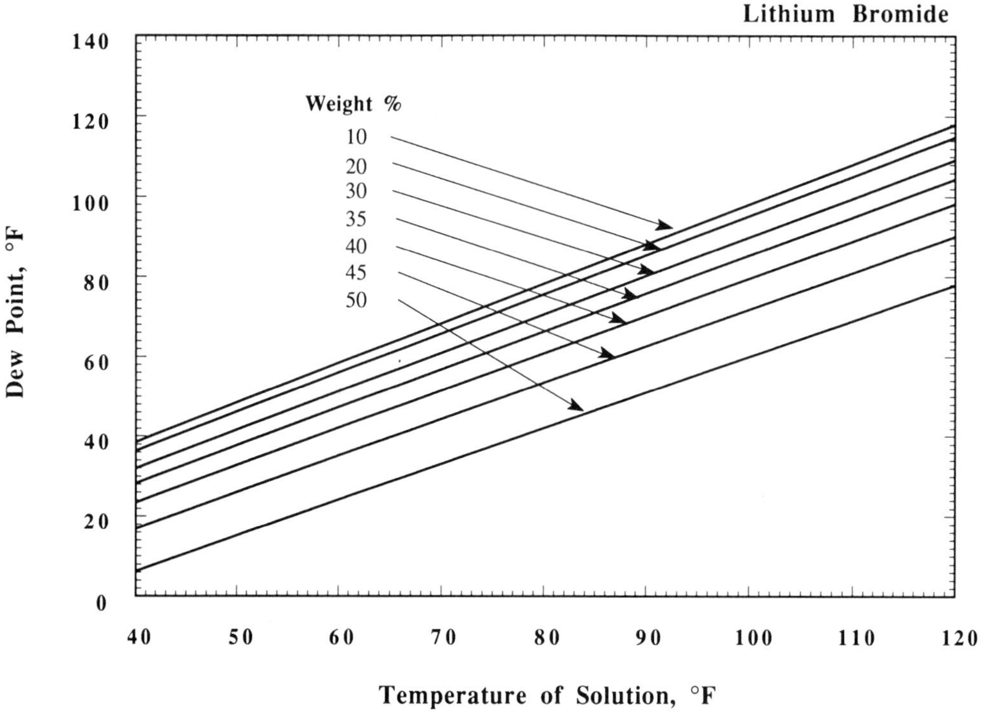

Figure E.3 *Dew point of air in contact with lithium bromide solutions. [Data of McNeely, L. A.,* ASHRAE Trans., **85** *(Part 1), 413 (1979)].*

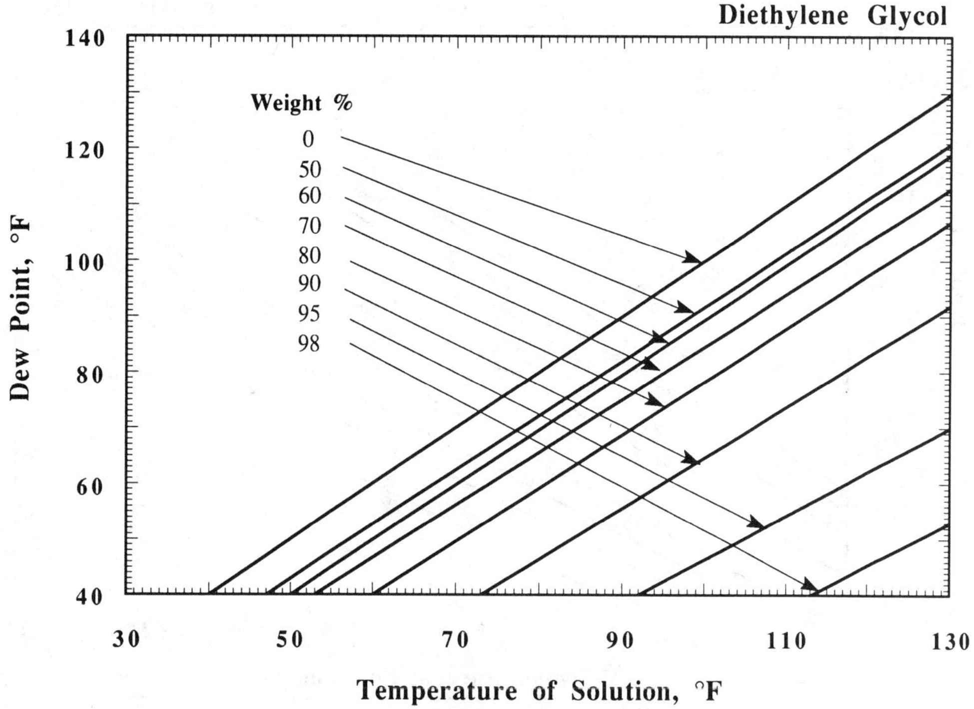

Figure E.4 *Dew point of air in contact with diethylene glycol solutions. (Data of Dow Chemical Company,* Properties and Uses of Glycols, *1956.)*

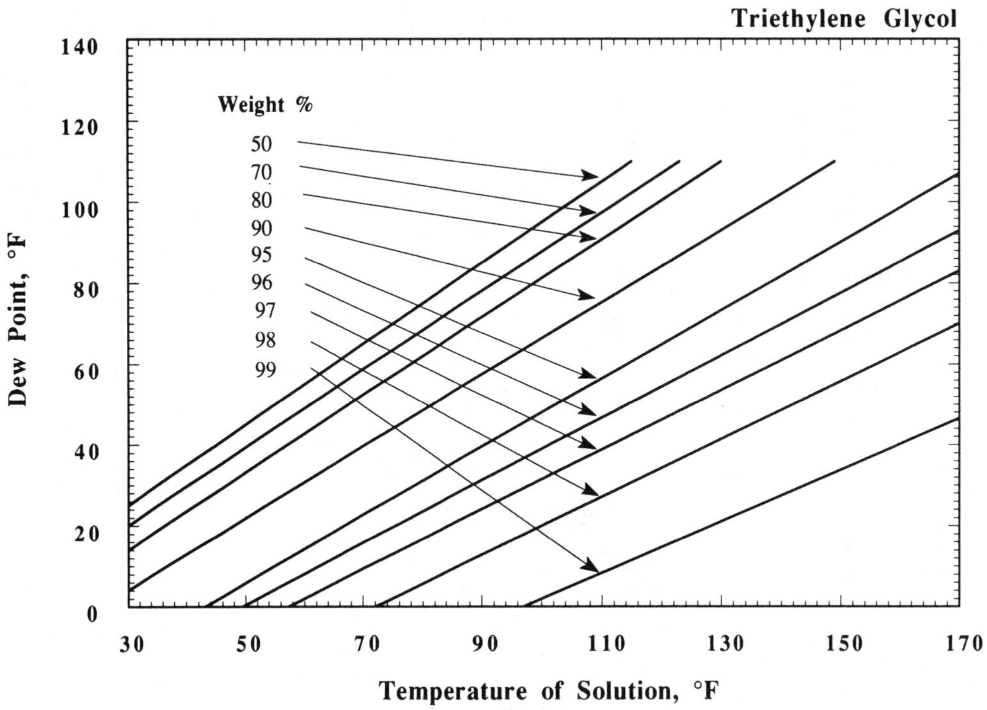

Figure E.5 *Dew point of air in contact with triethylene glycol solutions. (Data of Dow Chemical Company,* Properties and Uses of Glycols, *1956.)*

Author Index

A

B

C

G

H

M

N

O

P

T

U

V

W

Subject Index

A

D

J

K

L

M

N

O

P

Q

R

T

U

V

W

X

Z